NASA SP-447

SPACE PHYSIOLOGY AND MEDICINE

Arnauld E. Nicogossian, M.D.
National Aeronautics and Space Administration

and

James F. Parker, Jr., Ph.D.
BioTechnology, Inc.

National Aeronautics
and Space Administration

Scientific and Technical
Information Branch

1982

Contributing Authors

The following people contributed invaluable assistance through the provision of materials for the preparation of this manuscript:

Bryant Cramer, M.D., Ph.D.
National Aeronautics and Space Administration

Sam L. Pool, M.D.
National Aeronautics and Space Administration

David J. Horrigan, Jr.
National Aeronautics and Space Administration

Paul Rambaut, Sc.D.
National Aeronautics and Space Administration

Carolyn Leach Huntoon, Ph.D.
National Aeronautics and Space Administration

William E. Thornton, M.D.
National Aeronautics and Space Administration

Courtland S. Lewis
BioTechnology, Inc.

Capt. Paul Tyler, Medical Corps
United States Navy

D. Stuart Nachtwey, Ph.D.
National Aeronautics and Space Administration

James M. Waligora
National Aeronautics and Space Administration

Patricia M. Wallace, Ph.D.
BioTechnology, Inc.

FOREWORD

Manned space flight has been one of our national goals since the founding of the National Aeronautics and Space Administration. Indeed, we have discovered that man is not a grounded creature. To wings in the air have been added the rockets and craft to dwell in space.

One challenge of manned space flight is to assure the health, well-being, and performance of those people who will live and work in space. To this end, biomedical personnel within NASA have participated in designing and carrying out all of our manned missions. As a result of this medical attention, there have been no significant health problems on U.S. missions. Our record is unblemished. However, much remains to be done if proper medical support is to be provided during the growing era of the Space Transportation System and for the space station efforts planned for the future.

The practice of medicine in space, and understanding the physiology of life in space, present some of the most interesting medical challenges of our time. Many biomedical problems need to be addressed if space is to become accessible to our general population. This book presents a valuable review of our present state of knowledge in space physiology and medicine. It will aid in developing a more permanent human presence in space.

Gerald A. Soffen, Ph.D.
Director, Life Sciences Division
National Aeronautics and
Space Administration

Library of Congress Cataloging in Publication Data

Nicogossian, Arnauld E.
  Space physiology and medicine.
  (NASA SP ; 447)
  "September 1982."
  Includes bibliographical references and index.
  1. Space flight–Physiological aspects. 2. Space medicine. I.
Parker, James F. (James Fletcher), 1925-   . II. Title. III. Series.
[DNLM: 1. Extraterrestrial environment. 2. Space flight. WD 750
S7325]
RC1150.N52 1983   616.9'80214   82-23047

For sale by the Superintendent of Documents
U.S. Government Printing Office, Washington, D.C. 20402
Stock Number

PREFACE

This book is intended as a general reference manual for persons engaged in or concerned with the practice of space medicine. A large body of ground-based research, including both human and animal experimentation, and inflight life sciences accomplishments support the findings presented in this book.

Space medicine is a relatively new field of environmental medicine dealing with the unique aspects of health maintenance and care of individuals exposed to the rigors of space. The sustenance of life in space and the practice of medicine require a team approach and draw upon the knowledge of many areas of engineering, human factors, and medical specialties. Current activities of physicians and life scientists are devoted mainly to achieving a better understanding of the physiological changes which result from weightlessness and the development of procedures and countermeasures to facilitate the transition to space and to assure a safe return to Earth.

The domain of space is used at this time by a small number of highly select and healthy individuals, with the practice of space medicine focusing on preflight prevention of disease. As space activities expand, a larger and more diversified population will be involved in a variety of activities in the weightlessness of space. At this time, medical professionals and the practice of medicine also will move into space. This next milestone will produce a quantum advance in our understanding of human adaptation to the weightless environment and, at the same time, will require development of new therapeutic means and procedures not dependent on gravity. Space medicine will continue to grow and will become a well-established medical specialty.

CONTENTS

SECTION I
Manned Space Flight

SECTION II
The Space Environment

SECTION III
Spaceflight Systems and Procedures

CONTENTS *(Continued)*

SECTION IV
Physiological Adaptation to Space Flight

CONTENTS *(Continued)*

CONTENTS *(Continued)*

SECTION V
Health Maintenance of Space Crewmembers

CONTENTS *(Continued)*

SECTION VI
Medical Problems of Space Flight

SECTION I
Manned Space Flight

CHAPTER 1
HISTORICAL PERSPECTIVES

In the brief span of eight decades, man has proceeded from the revolutionary invention of a vehicle for powered flight to the point of visiting, by manned and unmanned spacecraft, five other planets in the solar system plus a great number of their moons, including our own. The field of medicine has both contributed to and benefited from man's ventures into space. The challenge of maintaining life and useful human function during space exploration has been met impressively. At the same time, the space program has been the source of many technological advances in the biomedical sciences. The symbiotic relationship between the space sciences and the medical sciences will continue to grow and benefit mankind.

The Support of Man in Space

The foundations for space medicine can be traced back many years into earlier programs in the fields of occupational and aviation medicine. However, it was not until World War II, and the development of the V-2 rocket, that serious consideration was first given to the possibility of manned space flight and, in turn, the need for a special activity to be called space medicine. Major General H.G. Armstrong foresaw the development of space medicine and, in 1948, at the USAF School of Aviation Medicine, organized a panel meeting on the topic of "Aeromedical Problems of Space Travel" (von Beckh, 1979). Presentations were made by the then Colonel Armstrong, Professor Hubertus Strughold, later to be regarded as the "father of space medicine," and the astrophysicist, Dr. Heinz Haber. From an historical perspective, this meeting may be regarded as a beginning point for a new and specialized practice within the field of medicine. Space medicine began at this time, later to emerge as an accepted and growing specialty within the broad domain of occupational medicine.

There followed a rapid growth in interest by biomedical scientists in programs which might lead to manned space flight. By 1950, the United States had launched two primates into space aboard V-2 rockets. While neither animal survived its mission, useful information was obtained concerning the hazards of space flight for mammalian life forms. These early flights demonstrated the need for effective and reliable life support systems. They also began to define the parameters for protection against the rigors and stresses of traveling in and returning from space. The dimensions for the field of space medicine were being established.

Scientists interested in the possibilities of manned space flight soon recognized a need for an organization to coordinate and exchange information concerning space medical research. In 1950, during the 21st Annual Meeting of the Aeromedical Association, the Space Medicine Branch was formed. The committee which petitioned the Association for admission consisted of Drs. A.C. Ivy, J.P. Marbarger, R.J. Benford, P.A. Campbell, and A. Graybiel. In 1951 the petition was accepted. Space medicine now had formal recognition within the larger medical community.

The Schools of Aviation Medicine of both the Navy and the Air Force, as well as a number of laboratories in the services, soon undertook projects dealing with man in space. Topics such as life support, acceleration tolerance, and reactions to confinement—all of which had been studied in the aviation context—now were investigated as a function of the environmental parameters it was believed man would face when he traveled into space. A considerable body of knowledge which would prove quite useful in later space medicine activities was developed through use of these resources.

Many of the early practitioners of space medicine were trained in the aviation medicine programs of the Navy and the Air Force. Beginning in the 1950s, these two organizations expanded their curricula to include more and more topics of specific interest to space medicine. The change in orientation was reflected by new organizational designations. The Air Force facility became the School of Aerospace Medicine, while the Navy school became the Naval Aerospace Medical Institute. The schools of public health at Johns Hopkins, Harvard, and Ohio State Universities, which were cooperating with the military facilities in providing required residency training, also reflected the changing orientation in their curricula.

Interest in the possibility of orbital space flight continued to grow in both the United States and the Soviet Union during the mid-50s. On 4 October 1957, the Soviet Union successfully launched a satellite, Sputnik 1, into orbit around the Earth. This demonstration that the Soviet Union was ahead of the United States in space technology caused considerable surprise and dramatically increased the interest of the American public in our own efforts. It was apparent that there would be no time for leisurely planning and development. The two nations were in a space race. The sense of urgency that moved American space planning after 1957 had considerable implications for space medicine.

Medical scientists became more interested at this time in the biological and medical aspects of space flight. Within this group, however, there was some controversy over the advisability of even attempting such missions. There was concern over man's ability to sustain useful function in space and especially to withstand the stresses of launch and reentry. The National Academy of Sciences—National Research Council Committee on Bioastronautics, meeting in 1958, identified a number of potential problems for astronauts, including such effects on human physiology as anorexia, nausea, inability to swallow food, disorientation, and other issues which would be of real consequence for the conduct of a space mission. A more complete listing of these issues is provided in Table 1 (Dietlein, 1977). As was later found, some of these predictions were confirmed; some were not.

The pace of space activities during the late 1950s left little time for a step-by-step program to develop the basis for manned space flight. Issues of life support, safety, and health had to be dealt with on an *a priori* basis, drawing principally on the well-established science of aviation medicine. The first space suits, for example, were a direct outgrowth of the Navy full-pressure suit program for high altitude aircraft. The pace also meant that new knowledge which was generated came more from mission results than from research conducted in laboratories and in ground-based simulation. Space missions themselves were dictating progress in space medicine.

Table 1

Predicted Weightlessness Effects

Anorexia	Demineralization of bones
Nausea	Renal calculi
Disorientation	Motion sickness
Sleepiness	Pulmonary atelectasis
Sleeplessness	Tachycardia
Fatigue	Hypertension
Restlessness	Hypotension
Euphoria	Cardiac arrhythmias
Hallucinations	Postflight syncope
Decreased G tolerance	Decreased exercise capacity
Gastrointestinal disturbance	Reduced blood volume
Urinary retention	Reduced plasma volume
Diuresis	Dehydration
Muscular incoordination	Weight loss
Muscle atrophy	Infectious illnesses

Dietlein, 1977

Manned Space Flight: The American Program

Since space medicine was tied so directly to flight activities for many years, it is appropriate to consider the manned space programs conducted by the United States and the Soviet Union and to review key biomedical problems and findings. A summary presented at the end of this chapter lists all manned space missions conducted by these two nations to date.

Project Mercury

The National Aeronautics and Space Administration, formed in 1958, was charged by the President of the United States with carrying out a two-fold mission in manned space flight. The mission was given high national priority, ranking second only to national defense. NASA was, at the earliest feasible time, to launch a man into space, provide him with an environment in which he could perform effectively, and recover him safely. This was Project Mercury. At the same time, NASA, using the support of leading life scientists, was to develop a capability for extended manned space flight (Lovelace, 1965).

The mission given to NASA was accomplished with remarkable success (Figure 1). The original goal, that of orbiting a man in space and returning him safely to Earth, was accomplished in just three years of program effort (Kleinknecht, 1963).

Figure 1. Launch of the Mercury-Redstone 3 carrying
Alan Shepard, America's first man in space.

During the period of preparation for the first manned Mercury launch, many problems were faced by biomedical scientists. One of the first was to establish criteria by which a corps of astronauts could be formed. As this problem was being addressed, President Eisenhower made a decision that all astronaut candidates should come from the ranks of military test pilots. An important factor was the demonstrated capability of this group to meet threatening situations in the air with accurate judgment, quick decisions, and practiced motor skills. There were 110 test pilots among the many applicants who fully met all requirements (Link, 1965). These individuals were given interviews, psychiatric examinations, and a complete medical evaluation, including medical stress tests. The purpose of the protracted and extensive medical evaluation, sometimes using research-type testing, was to discover any hidden medical problems, to establish the physical fitness level, and, of considerable importance, to start a medical data base for each individual against which any changes brought about by later space missions might be examined. Right or wrong, this meant that the medical personnel who selected the variables to be included within the selection tests were, in effect, establishing the parameters of interest for the science of space medicine. Thus, while the selection criteria were taken almost directly from those used in military aviation, there remained a real challenge for physicians and biomedical personnel to identify those medical parameters that would be most useful in assessing man's adjustment to space.

A key problem in Project Mercury, possibly one of the 'most demanding and challenging, was the development of a life support system which would operate without fail under the conditions of orbital space flight. The technology to develop such a system was at hand, as demonstrated in the successful balloon flight of an Air Force flight surgeon, David G. Simons, in 1957, which attained an altitude record of 30,942 meters and lasted for 32 hours. Translating this technology into a spacecraft system, however, was a difficult undertaking for engineers and biomedical scientists. The human requirements for protection, proper breathing atmosphere, maintenance of pressure, provision of food and water, removal of metabolic by-products, and thermal control had to be matched with severe constraints concerning reliability, size, weight, power demands, and operation under conditions of thermal extremes, acceleration, and weightlessness. The system as finally designed functioned perfectly.

Project Mercury, designed specifically to prove man's ability to survive in the space environment, lasted from May 1961 to May 1963. There were two sub-orbital flights and four orbital missions, including one which lasted for 34 hours and made 22 orbits of the Earth.

The six astronauts who flew in Project Mercury returned in quite satisfactory health condition. These flights were as valuable for the many medical concerns which were dispelled as they were for those which were verified. The principal findings showed a weight loss due primarily to dehydration and some impairment of cardiovascular function. Cardiovascular data from the last and longest Mercury flight showed orthostatic intolerance and dizziness on standing as well as hemoconcentration (Dietlein, 1977). From the behavioral point of view, it was found that astronauts could perform well under conditions of weightlessness.

Project Gemini

Planning for Project Gemini began in May 1961 just as the first suborbital mission of Project Mercury was successfully completed. The Gemini program represented a logical outgrowth of Mercury, with increased capability and extended objectives. The Gemini program, providing a two-man space capsule, was designed to build upon the experience gained from Project Mercury, to demonstrate new capabilities such as that of extravehicular activity, and to study the limits of astronaut endurance in order to support the manned lunar landing and other future activities. The specific Gemini objectives were to conduct the development and test program necessary to: (1) demonstrate the feasibility of long-duration space flight for at least that period required to complete a lunar landing mission; (2) perfect the techniques and procedures for achieving rendezvous and docking of two spacecraft in orbit; (3) achieve precisely controlled reentry and landing capability; (4) establish capability in extravehicular activity; and (5) achieve the less obvious, but no less significant, flight and ground crew proficiency in manned space flight (Mueller, 1967).

Project Gemini successfully completed ten manned space missions. There were many notable accomplishments. The first U.S. extravehicular activity was performed in the Gemini 4 mission (Figure 2). The first rendezvous and docking maneuver was completed by Gemini 8 crewmembers.

Of particular interest to medical scientists are the Gemini 4, 5, and 7 missions, lasting for four, eight, and 14 days, respectively. A number of inflight experiments were conducted on these missions, as well as preflight and postflight studies.

Figure 2. Astronaut Ed White (Gemini 4) carries out the first American extravehicular activity.

A major point of concern in the Gemini medical investigations was the evaluation of the changes in cardiovascular function first noted in Mercury. The cardiovascular changes seen in Gemini crewmembers were regarded as an adaptive response due to intravascular fluid losses resulting from exposure to the weightlessness of space flight. The key question was: "Is the cardiovascular deconditioning found in space missions a self-limiting adjustment?"

The medical conclusion following the completion of the Gemini program was that man can live and work in the space environment, certainly for periods as long as would be required for a lunar mission. A number of changes were noted, as shown in Table 2, but none was considered of real consequence for mission durations on the order of two weeks. No medical basis was found which might preclude the forthcoming Apollo missions.

Table 2
Significant Biomedical Findings in the Gemini Program

Loss of red cell mass (ranging 5 to 20% from baseline)
Postflight orthostatic intolerance in 100% of crews
Loss of exercise capacity compared with preflight baseline
Loss of *os calcis* bone density (7% from baseline)
Sustained loss of bone calcium and muscle nitrogen
Higher than predicted metabolic cost of extravehicular activity

The importance of the Gemini results lay with the manner in which these findings served to structure experiments designed for later missions of longer duration. Such experiments would be needed to determine the basis and time course of the observed changes. Gemini, though providing answers to some medical questions, left other issues unresolved.

Project Apollo

The goal of the Apollo program was singular and straightforward—to land a man on the Moon and return him safely to Earth. Further, President John F. Kennedy directed, in 1961, that this goal be achieved "before this decade is out." The goal was met with complete success. Twenty-nine astronauts flew in the program, with 12 of them spending time on the lunar surface (Figure 3). The Apollo program represents one of man's greatest achievements in science, engineering, and exploration.

Figure 3. Astronaut Irwin explores the lunar surface at the base of Mount Hadley.

9

The Apollo program was supported by a broad biomedical effort. The biomedical activities had three distinct and rather separate goals (Johnston, 1975). These were:

1. **Insure the Safety and Health of Crewmembers.** The Apollo flights brought forth health issues which had not been addressed earlier. Principal among these was the potential for inflight illness. During orbital flight it is always possible to recover an astronaut within a reasonable time should there be inflight distress. During a lunar mission, this was not feasible because the trajectory of the spacecraft required circumnavigation of the Moon. Therefore, it was necessary to develop a program which would minimize the likelihood of any illness occurring during flight and which would also allow some measure of emergency treatment to be given should an illness occur.

2. **Prevent Contamination of Earth by Extraterrestrial Organisms.** Project Apollo plans called for a spacecraft to land on another celestial body, stay for some time, and then return to Earth. This raised for the first time the possibility of contamination of the Moon by microorganisms carried from Earth or, even more intriguing, the possibility of strange microorganisms from the lunar surface being introduced to Earth either with lunar samples or from exposed crewmembers. In order to insure that unwanted microorganisms were not transported in either direction, strict quarantine and decontamination procedures were followed both before and after each mission. A special Lunar Receiving Laboratory was constructed at the Johnson Space Center to house both astronauts and lunar samples for appropriate periods postflight.

3. **Study Specific Effects of Exposure to Space.** A number of medical issues, such as cardiovascular deconditioning and bone demineralization, were identified during the Gemini program. The Apollo flights provided an opportunity to study these problems more closely and to develop improved measurement procedures for assessing change. Although the operational complexity and demands of the Apollo program limited the available time for biomedical experiments, some studies were conducted which provided a considerable amount of information. Predominant were cardiovascular, metabolic balance, and microbial load studies. In addition, limited opportunities were available to conduct biological experiments such as one dealing with radiation effects on the pocket mouse and another which studied the effect of individual heavy nuclei of galactic cosmic radiation on a number of biological systems.

A key point in the Apollo observations was the addition of vestibular disturbances to the litany of significant biomedical findings incident to space flight (Dietlein, 1977). No U.S. astronauts had experienced or reported any motion sickness symptoms prior to the Apollo missions, although Soviet cosmonauts had reported the occurrence of vestibular dysfunction inflight as early as 1961 (Titov on Vostok 2). In Apollo, this problem quickly drew attention and was labeled "space motion sickness." In the Apollo 8 and 9 flights, five of the six crewmen suffered some degree of motion sickness, ranging from stomach awareness in three to actual sickness in two others. In one case, the severity of the vestibular disturbance required a postponement of inflight completion of some parts of the flight plan.

10

There were other significant biomedical findings from the Apollo program, as shown in Table 3, which confirmed the Gemini results and were useful in characterizing these responses in further detail. Of additional interest was the fact that no microorganisms could be found in the materials returned from the lunar surface.

Table 3

Significant Biomedical Findings in the Apollo Program
(Pre- versus Postflight)

Vestibular disturbances

Less than optimal food consumption (1260 to 2903 kcal/day)

Postflight dehydration and weight loss (recovery within one week)

Decreased postflight orthostatic tolerance (tilt/LBNP tests)

Reduced postflight exercise tolerance (first 3 days)

Apollo 15 cardiac arrhythmias (frequent bigemini)

Decreased red cell mass (2-10%) and plasma volume (4-9%)

Dietlein, 1977.

Project Skylab

The Skylab program offered the first opportunity to study problems of habitability and physiological adaptation in space over an extended period of time. Skylab was more than simply a space vehicle; it was a habitat and a laboratory in space. It was comprised of a number of components. The Orbital Workshop provided the primary on-orbit living and working quarters for Skylab crews. It was made from the structure of an S-IVB stage, the third stage of the Saturn V booster rocket, and was equipped to house three astronauts for an uninterrupted period lasting for at least three months. The workshop, cylindrical in shape, was 6.7 meters in diameter and 14.6 meters long. As living and working quarters, the Orbital Workshop was huge (approximately 294 m^3) by comparison with the habitable spaces available in earlier Mercury, Gemini, and Apollo spacecraft (approximately 1-8 m^3). The spaciousness of the Orbital Workshop meant that astronauts could attempt a lifestyle more comparable to Earth standards, with a radical improvement in freedom of movement.

A second feature of importance in Skylab was the duration of the missions. Skylabs 2, 3, and 4 lasted for 28, 59, and 84 days, respectively. This meant that the physiological changes seen in astronauts during earlier space programs now could be studied in more detail. It now became possible to establish the time course of physiological adaptation to the weightless environment of space.

A major contribution of the Skylab program was its demonstration once again of the need for man in space systems. Without direct human intervention, the Skylab vehicle would have been uninhabitable because of the thermal problem caused by the loss of the micrometeoroid shield and failure of the solar array wing to deploy properly (Belew, 1977). After rendezvous and survey of the damage, the Skylab Commander and Scientist Pilot spent nearly four hours outside the spacecraft attempting one of the most difficult and daring of all orbital repair jobs (Figure 4). The extent of the damage was unknown; the outcome was uncertain; and there were no EVA aids where the work had to be done. Drawing on considerable guidance from ground personnel, the Skylab team was successful in releasing the solar wing and rectifying the problem to the extent feasible. The Scientist Pilot, incidentally, was Dr. Joseph P. Kerwin, the first physician to fly as part of an American space crew (Figure 5).

The large volume of the Skylab Orbital Workshop afforded an opportunity to investigate a number of spacecraft habitability issues. In general, the living arrangements proved quite satisfactory. By previous standards, only minor problems were experienced. For example, the sleeping compartments were not sufficiently isolated from each other and from the waste management compartment for optimum noise control. It also was learned that, in the zero-gravity environment, mobility and restraint systems are major contributors to perceived habitability. In all, Skylab showed that, with sufficient attention to such issues as food service, waste management, and sleep arrangements, a spacecraft can be perfectly habitable for long periods of time.

Figure 4. Skylab 2 crewmen Charles Conrad and Dr. Joseph Kerwin repairing the Orbital Workshop's damaged Solar Array System.

Figure 5. Dr. Joseph Kerwin. Scientist Pilot on Skylab 2, checks out the bicycle ergometer.

Skylab crewmen were monitored closely for signs of space motion sickness (Graybiel, 1981). In the first mission, none of the astronauts was motion sick. Although one crew member did take medication immediately after entry into orbit, the real reason for the lack of symptoms may lie with the extraordinary effort required of this crew to repair Skylab damage before they could enter the Orbital Workshop.

The second Skylab crew experienced severe motion sickness symptoms. These crew members did not take medication prophylactically and entered the large volume workshop on the first day of orbit, immediately starting a full work schedule. In one case, the motion sickness appeared within an hour after insertion into orbit while the crew member was in the act of removing his space suit. This is the earliest appearance of motion sickness in orbital flight on record (Graybiel, 1981).

The third Skylab crew took a number of special precautions, including flying aerobatic maneuvers on the day prior to the mission and following a planned schedule for anti-motion sickness medication during the early days of the mission. Two of the crew members experienced motion sickness, with symptoms for one astronaut persisting well into the fourth day of the mission.

Based on the subjective reports of the three Skylab crews, and vestibular experiments conducted during flight, it was concluded that space motion sickness remains a problem, is not predictable by the usual Earth-bound tests, and can be alleviated somewhat by the prophylactic administration of medications. The search for the optimum medication and schedule of administration continues.

Particular attention was given in Skylab to cardiovascular deconditioning since earlier space missions showed a consistent change in cardiovascular function under spaceflight conditions. The parameters most closely observed were orthostatic tolerance, electrical activity, and changes in heart size. During Skylab, the response of astronauts to orthostatic stress was examined for the first time inflight. Crewmen were tested using a lower body negative pressure device before, during, and after all Skylab missions. This device provides an orthostatic stress for a period of five minutes through the application of a negative 50 mmHg pressure. Again, the customary indices of a reduction in cardiovascular efficiency were obtained. However, the observed cardiovascular deconditioning was found to stabilize after a period of four to six weeks with no apparent impairment of crew health or performance (Dietlein, 1977). The change apparently is a self-limiting adaptation to the reduced cardiovascular load imposed by residence in weightlessness.

Another topic of concern during Skylab was that of bone mineral loss and mineral balance. Here again, measurements of bone mineral content, using a photon absorptiometric technique, taken preflight were compared with similar measurements taken at varying intervals postflight. No mineral losses were observed in the upper extremities, but some bone loss was noted in the lower extremities, specifically the *os calcis* (Smith et al., 1977). Data from the 84-day mission led to the conclusion that the mineral losses are comparable to those observed in bedrest studies (Dietlein, 1977). No evidence was found during these missions that the loss of bone mineral is self-limiting.

Skylab metabolic studies showed a significant increase in the excretion of urinary calcium during flight for all crewmen who were measured. The loss continued throughout the period of flight with no evidence of decrease during later stages. Significant losses of nitrogen and phosphorus also occurred, presumably associated mainly with muscle tissue loss (Whedon et al., 1977). Other evidence of muscle loss was obtained from stereometric analyses of body form in the Skylab crews. These analyses showed a marked loss in leg volume, much of it restored within four days following mission completion. About one-third of the loss was attributed to partial atrophy of the leg muscles due to disuse in zero gravity, with the remainder due to a body fluid deficit (Whittle et al., 1977).

The Skylab missions provided a wealth of biomedical data which helped in evaluating the health and physiological responses of man as he performs normal work activities during long-term space missions. Skylab was particularly useful in delineating those physiological changes which appear to be self-limiting from those which seem to continue throughout the longer missions. This information provides guidance both for ground-based research and for inflight studies to characterize and understand human response to the stresses of space.

The Apollo-Soyuz Test Project (ASTP)

The Apollo-Soyuz Test Project shared political objectives with those of science and technology. This mission was conducted jointly by the United States and the Soviet Union as a means of promoting a spirit of international cooperation in space ventures. The primary mission objective was to test systems for rendezvous and docking of manned spacecraft as might be needed during international space rescue missions. This required an ability to transfer crews between two spacecraft with dissimilar atmospheres. A second mission purpose was to conduct a program of science experiments and technology applications. Both the Apollo and Soyuz spacecraft used in ASTP were identical with those flown previously (Figure 6). A docking module for use in crew transfer was constructed especially for the mission.

Figure 6. The joint US/USSR Apollo-Soyuz Test Project. The Apollo capsule (left) is shown docked with the Soyuz vehicle (right).

The ASTP mission lasted for nine days and was successful in accomplishing the rendezvous and docking maneuver. The two spacecraft were docked for two days, during which time each crew visited the other spacecraft. During the recovery phase, the United States crew was exposed to toxic gases, mostly nitrogen tetroxide, from inadvertent firing of the reaction control system during descent. The toxic gases entered the command module through a cabin pressure relief valve, which was opened during the landing sequence. All crew members developed chemical pneumonitis as a result of exposure to nitrogen tetroxide and required intensive therapy and hospitalization at the Tripler Army Medical Center, Honolulu, Hawaii (Nicogossian et al., 1977). Because of this exposure and the new health condition, it was necessary to drop most of the planned postflight medical experiments and focus on the clinical examinations and treatment of the astronauts. Even so, considerable information was obtained concerning the reaction of the crew to the spaceflight conditions.

Investigation of skeletal muscle function in both leg extensor and arm flexor muscles, using electromyographic analyses, showed that the muscle dysfunction characteristics found after 59 days of exposure to weightlessness in the Skylab 3 mission were also evident after only nine days of exposure (LaFevers et al., 1977). It was also shown that this short-term exposure results in greater fatigue response in muscle tissue, particularly in anti-gravity muscles.

15

Generalized hyperreflexia was reported following Skylab flights. Measurements were made of the Achilles tendon reflex duration during ASTP (Burchard & Nicogossian, 1977). The reflex was measured within two hours after recovery and compared with preflight measurements. As in Skylab, two crewmen showed a shortening in the reflex duration time. In addition, all three crew members showed significant fine tremor, which it was felt might reflect the effects of the inhaled vapor of nitrogen tetroxide.

From a purely medical point of view, the most important contribution of the ASTP program may be in its illustration of the manner in which the medical support complex is prepared to deal efficiently with the results of the unforeseen hazards and occurrences inherent in manned space programs.

The Space Transportation System

The successful first orbital flight of the Space Shuttle on 12 April 1981 marked the beginning of a new era in manned space activities for the United States, an era which will last for a decade at least. The Space Shuttle, the principal component of the Space Transportation System and the world's first reusable spacecraft, will be the focal point for all manned missions within this time frame (Figure 7). A key capability of the Shuttle is the deployment of the pressurized Spacelab module, a manned space laboratory in which, for the first time, scientists, engineers, and technicians, in addition to astronauts, will be able to conduct experiments in Earth orbit. Since the Space Shuttle and the Spacelab are of such paramount importance, both for the practice of space medicine and for the conduct of life sciences research, later sections describe both in detail.

Figure 7. The Space Shuttle Columbia glides toward a landing.

Biomedical information obtained from the early test flights of the Space Shuttle (STS-1, 2, 3, and 4) substantiated earlier findings of postflight orthostatic intolerance, intravascular fluid losses, and incidence of space motion sickness.

Soviet Manned Space Program

Preliminary Activities: Sputnik

As was mentioned earlier, the orbiting of Sputnik 1 in October 1957 startled the world—in particular the American aerospace community. The launching of a dog into orbit a month later did nothing to dispel this surprise. The cabin carrying the experimental animal contained air regeneration and thermal regulation systems, and permitted the monitoring of pulse, respiration, blood pressure, and electrocardiograms (ECGs). Cabin environmental parameters were also tele-metered back to Earth.

These striking technological feats were subsequently repeated and expanded upon in a series of five *Korabl Sputnik* flights which served as precursors to the first manned mission. Korabl Sputnik-2, launched on 10 August 1960, was a veritable Noah's Ark bearing two dogs, 40 mice, two rats, a number of flies, and a variety of plant life into orbit and back to Earth—the first time that living organisms had flown in space and returned. Instrumented mannikins and live television transmissions supplemented the substantial amount of biomedical data that the Soviets were able to obtain regarding the effects of space flight and the space environment on living things (DeHart, 1974).

Vostok Program

Despite the rather obvious preparations represented by the Korabl Sputnik series, the launching of a man into orbit less than four years after the first Sputnik once again caused consternation in the West. On 12 August 1961, 27-year-old Air Force Senior Lieutenant Yuri Gagarin completed a single orbit in a flight lasting 108 minutes. His craft, Vostok 1, consisted of a near-spherical capsule containing the life support system, instrumentation, and an ejection seat, plus a conical service module containing gas bottles, batteries, rockets, and support equipment. After reentry, Gagarin ejected from the cabin and carried out the final portion of his descent by parachute.

The presence of a human in the spacecraft focused concern of Soviet scientists on four problems (Prishchepa, 1981):

- Assuring the safety of the cosmonaut in case of launch vehicle malfunction
- Protecting the cosmonaut against the space environment and ensuring normal vital functions during orbital flight
- Ensuring reliable operation of equipment in space
- Ensuring safe and accurate reentry and descent.

Most of these were technical and engineering issues. However, a critical question concerned the reaction of the cosmonaut to weightlessness. The previous animal experiments had indicated that

17

normal mental and physiological function could be maintained, but confidence on this subject was by no means total. Consequently, television coverage was directed at observing Gagarin for signs of distress or disorientation. The flight program included (in addition to the first human observations of the Earth and stars from space) the intake of food and water. Medical parameters monitored during the brief flight included continuous heart rate, pneumography, ECG, seismocardiography, electroencephalography, electrooculography, electromyography, thermography, and galvanic skin response. The monitoring of these parameters has remained constant in all Soviet missions since the very first one, although the emphasis varies in each flight (Dodge, 1976).

Gagarin showed no deviations from normal psychological and physiological function during his short ride in space. Nearly four months later, Air Force Major Gherman Titov completed 17 orbits in Vostok-2. During this one-day flight, the cosmonaut experienced spatial disorientation and motion sickness. This was the first indication of what would become the most persistent problem in the initial period of space flight, and it prompted a strong Soviet emphasis on the vestibular system in cosmonaut selection and training. Titov was also the first human to sleep in space, and was the first to exercise autonomous control (periodically) of his craft. His pulse rates, respiration rates, and ECGs were all within normal preflight ranges.

Subsequent Vostok missions were paired flights of two vehicles launched a day or so apart. The first pair was co-orbited so that the two ships approached to less than four miles of each other. This was a test of both the feasibility of rendezvous in space (as a prelude to later dockings) and of the ground control coordination necessary for such maneuvers. The influence of identical conditions of space flight could also be observed; in the three- and four-day flights no significant biomedical problems or changes emerged. The second pair of flights, Vostok 5 and 6, were notable for the setting of a manned flight duration record of five days and for the fact that the pilot of the latter mission was the first woman in space, a 26-year-old textile factory worker who did not have the usual military flying background (Smith, 1976). The female cosmonaut exhibited no major biomedical differences from her male counterparts in response to the space environment.

Voskhod Program

The second-generation spacecraft, Voskhod, differed from the Vostok craft primarily in the removal of the ejection seat to afford room for a three-man crew. Due to the success of the previous series, the crew now wore coveralls rather than the cumbersome pressurized suits. With the crew of Voskhod 1 including a military physiologist and a civilian technical scientist in addition to the commander, more comprehensive medical data were obtained on that flight. The detailed onboard physiological measurements were combined with extensive biotelemetry to evaluate future telemetry needs. In addition to the now-standard biomedical parameters monitored on Vostok, Voskhod also included onboard studies of hearing, lung function, the state of the vestibular apparatus, and muscle strength during weightlessness.

The second (and last) flight of Voskhod was remarkable for a 20-minute EVA conducted by one crewman, wearing a self-contained life support system. Due to a failure of the automatic reentry system, the crew of Voskhod 2 also accomplished the first manually controlled reentry. They landed in a dense, snowy forest hundreds of miles north of the target and spent the night in their capsule awaiting recovery.

Soyuz Program

After a pause of nearly two years, during which the American Gemini program had accomplished a great many successful flight operations, the Soviets embarked in 1967 on what was to be their most ambitious and long-running program of manned missions. The initial version of the Soyuz vehicle contained two passenger compartments: a work compartment (or orbital module) and a command (or reentry) module in which piloting was done. The two compartments, connected by an airlock, had a total volume of about 9 m^3. A docking unit was optional, and the Soyuz craft could be outfitted with one, two, or three seats. Unlike the Vostok/Voskhod ships, which had a ballistic trajectory on reentry, the Soyuz reentry module employed aerodynamic lift to permit a controlled reentry and more precise landing. The load on reentry was thus reduced from 8-10 G's (in the previous series) to between 3 and 4 G's.

The objective of the Soyuz program was to provide a multi-purpose spacecraft which could be employed in connection with an orbital space station: as a base for assembling it, as a supply and transport vehicle to it, and as a vehicle for conducting additional independent studies of space. The ship had to have broad maneuverability, a docking capability, and the capability to support long-duration flight—in essence, it had to be a combination transport ship and orbital station (Malyshev, 1981).

The first trial of this new and more sophisticated vehicle, in April 1967, ended in tragedy. A loss of attitudinal stability seems to have resulted in a spin about the longitudinal axis upon reentry; the parachute system became tangled and failed to deploy fully, so that the Soyuz and its pilot were destroyed upon impact.

Soyuz 3, the next manned flight, was delayed for 18 months. During this four-day mission the pilot twice approached to within a few feet of the unmanned Soyuz 2, in what were apparently unsuccessful attempts at docking (Smith, 1976). In the following missions (Soyuz 4 and 5), rendezvous and docking of two ships was successfully accomplished, forming an experimental orbital space station. Soyuz 6, 7, and 8 were a test of the simultaneous command and control of three spacecraft for rendezvousing in space, and also for performing tasks (welding) in the space vacuum.

None of the cosmonauts participating in these Soyuz missions demonstrated unusual or unexpected physiological changes. The typical cardiovascular response was an increase in pulse rate during launch and orbital insertion, and an increase in pulse rate variability over much or all of the flight. ECGs and seismocardiograms were normal except for an occasional depression of the S-T segment. Respiration rate ordinarily increased during launch and insertion, and both respiration and cardiac contraction rates were increased during the performance of complex work tasks and EVAs.

Medical monitoring of cosmonauts during the Soyuz program was specifically formulated to assess the effects of weightlessness. In addition to the cardiorespiratory measurements, extensive pre- and postflight examinations were made of the central nervous system, metabolism, blood chemistry, and fluid-electrolyte balance. As in the American program, with increasing mission length the cosmonauts have assumed greater responsibility for carrying out inflight medical tests. Soyuz 9 had as its primary objective the examination of physiological stresses associated with space flight. Pre- and post-exercise studies of cardiorespiratory response were carried out. The exercise program in this mission was extensive, with each cosmonaut exercising for two one-hour periods per day. Chest expanders, isometric exercises, elastic tension straps, and an early version of the "Penguin" constant-loading suit all were employed. Other tests assessed pain sensitivity, hand strength, mental capability, and vestibular sensitivity.

Ironically, after this 18-day flight the two cosmonauts demonstrated a pronounced orthostatic intolerance and had to be carried from the spacecraft. Postflight readaptation to gravity took at least 11 days. This anomalous experience (in view of an absence of similar responses in much longer subsequent missions) gave rise to renewed concern about the outlook for long-term manned missions, prompting a vigorous search for countermeasures to physiological deconditioning in the next phase of the Soviet manned space program.

The Salyut Era

Work began on the first Salyut orbital station in 1969, after technical problems associated with rendezvous and docking had been solved. The vehicle was conceived as a long-term habitat for rotating crews, with a power plant and extensive onboard systems and scientific equipment. Provision for maintenance and housekeeping by the crew had to be made, as well as facilities for eating, sleeping, hygiene, and physical exercise.

As developed, the basic Salyut station is cylindrical, with an overall length of about 21 m, a maximum diameter of 4.2 m, and a total internal volume of 100 m^3 (Smith, 1976). The Soyuz craft is used as a transfer vehicle; its use for independent missions was virtually eliminated with the advent of Salyut.

Salyut 1 was launched into orbit unmanned on 19 April 1971. Its first use was by the three-man crew of Soyuz 11, which remained on board for 23 days conducting a comprehensive program of geographical, space, and biomedical research. However, as the crew was returning to Earth, a valve malfunctioned upon separation of the reentry module from the work module; rapid depressurization occurred, and the three crewmen died of dysbarism.

This accident resulted in a lengthy delay while measures to increase crew safety were worked out. After two years, Salyut 2 was launched, but it was evidently damaged by explosion of the carrier rocket's upper stage upon entry into orbit, and was abandoned. Not until September 1973 was another manned mission, Soyuz 12, flown. The Soyuz vehicle had now been modified to hold two crewmen wearing pressure suits, rather than three in coveralls.

The second successful Soviet space station, Salyut 3, was launched in June 1974 and remained in orbit for seven months. Improvements on this station included movable solar panels, more efficient life support systems, and a more "homelike" interior design and decoration. On the 15-day Soyuz 14 mission to this station, extensive physical exercises were again employed; these were designed to assess the effect of physical conditioning on eventual readaptation to gravity. A treadmill and improved Penguin suit were combined to provide a universal trainer that could simulate walking, running, jumping, and weight-lifting. Some evidence of readaptation to simulated gravity was seen (Dodge, 1976). Medical experiments included studies of blood circulation to the brain and blood velocity in the arteries. A regular daily schedule of eight hours sleep, eight hours work, and eight hours for exercise and administrative/housekeeping duties was employed.

The next significant manned missions were associated with Salyut 4, which was launched in December 1974 and functioned in orbit for more than two years. This model featured larger solar panels that were individually and automatically rotatable to produce 4 kilowatts of power. The first mission to the new space station was the 30-day flight of Soyuz 17. For the first time, the "Chibis" vacuum suit was used to reduce the volume of headward fluid shifts. This variant of a lower body negative pressure device is worn during exercise and for extended periods during normal activity in an effort to reduce deconditioning. A swivel chair similar to that used on Skylab was installed for checking vestibular reactions. With this flight, the exercise program began to assume the appearance of a standard regimen: three days of regulated exercise (three times per day, 2.5 hours total) followed by a fourth day on which the selection of exercises was optional. A bicycle was now included, in addition to the treadmill. Another new approach to physical conditioning was electrical stimulation of various muscle groups by means of the "Tonus" apparatus.

The follow-on Soyuz 18 mission continued these programs, but was longer (63 days—a new record). An innovation was to use the final ten days of the flight period to prepare the cosmonauts for the return to gravity. To physical exercise was added a high-salt diet and forced intake of water to increase body fluids. This combination was found to be successful (Smith, 1976). Similar measures were taken on 49- and 18-day missions to Salyut 5, which flew for over a year during 1976 and 1977.

With increasing mission duration, the logistical problem of resupply of consumables became more significant. Limitations on the weight and volume of an orbital station meant that it could carry life support system reserves which could support a two-man crew for no more than 120 days. Each crew member requires over 10 kg of consumables per day. Fuel reserves needed to be replenished periodically. Fresh storage batteries had to be supplied. The Soyuz transport craft alone was unable to ferry enough additional cargo to make up for these deficiencies. In addition, in case of malfunction or damage aboard the docked Soyuz ship, there would be no place for a reserve ship to dock with the station (Feoktistov & Markov, 1982).

These and other concerns led to the development of a second-generation orbital station, Salyut 6 (Figure 8). Its main structural difference was the presence of an additional docking unit at the opposite end of the station. To provide the expanded resupply capability, an unmanned "Progress" cargo ship was developed (actually a modified Soyuz). This vehicle is automatic and non-returnable, delivering about 2,300 kg of cargo to the Salyut in each trip. It made possible not only longer missions, but more extensive and versatile scientific and technical programs—since additional scientific equipment could be brought to and from the station to accommodate different requirements. The station itself had provisions for expanded maintenance of systems by the crew, and additional amenities, such as a shower. These capabilities meant that the Salyut station was now indeed a full-scale orbiting research laboratory, and the Soviets made full use of its capabilities.

Figure 8. A mockup of the Salyut station, used for cosmonaut training in Star City.

Over its four and one-half years of active manned operation, between September 1977 and June 1981, Salyut 6 was visited by 27 cosmonauts participating in 15 separate missions. Six of these individuals made two flights to the station, and seven spent more than 100 days in space. The "Interkosmos" program carried out during this period involved the inclusion of cosmonauts and technical equipment from Soviet-bloc countries such as Czechoslovakia, Hungary, Mongolia, Rumania, Vietnam, and Cuba in the crews of shorter-duration flights. More than 1,600 experiments were carried out in areas of space technology, Earth resources research, space research, and bio-medical research.

With the expanded habitability of Salyut 6, emphasis was on further extending the duration of manned flight and determining whether firm barriers exist to the endurance of weightlessness. Accordingly, several long-term "prime" crews were installed aboard the station in succession. Both for psychological relief and for practice in multimanned station operation, these crews were periodically visited by other crews which typically stayed for a week. There were a total of five prime crews, carrying out missions lasting (in order) 96, 140, 175, 185, and 75 days. Due to an accident to a trainee, one member of the 175-day crew returned to the station just six months after his flight to participate in the 185-day mission, thus becoming the world's record-holder for time in space, with nearly a year.

The experience of these crews, particularly on the six-month missions, has given medical planners and researchers confidence that even longer exposures to weightlessness and other space-flight factors is possible. Most physiological alterations are seen to be self-limiting. Particularly with adequate exercise and diet, cardiovascular deconditioning and muscle mass losses can be controlled. Changes in water-salt metabolism and red blood cell numbers and shapes also appear to be self-limiting and reversible. However, some processes do not appear to level off. Osteoporosis, the loss of calcium from bone, is one such process, although it is probably not a problem in missions of less than a year.

An important offsetting advantage of long-term missions is their efficiency. The uninterrupted performance of a lengthy research program in space permits longer and more complex experiments to be conducted. In addition, the more complete adaptation to residence in space that occurs during longer missions produces a bonus in efficiency and performance capability. The commander of the 175-day Salyut 6 mission observed that "during space flights we get accustomed to weightlessness to such an extent that it does not hinder our work. After several months of stay in outer space a space pilot experiences a wonderful lightness of the body (and) the pleasure that all objects surrounding him are weightless" (Lyakhov, 1981). In order to maximize the psychological adaptation to space flight, Soviet medical specialists instituted aboard Salyut 6 a sleep-wake-work cycle keyed to normal Moscow time, with a five-day work week and two-day weekend. A comprehensive psychological support program was also implemented, including frequent two-way communications with families, access to radio and television, and delivery of gifts, letters, and news. These measures appeared to make the stress of long-term isolation and heavy workload tolerable.

The final significant change during the Salyut 6 period was the development of a new model of the Soyuz ferry craft, designated Soyuz T. This is essentially a three-seater version of the old Soyuz. The first manned flight of the new ferry craft, T-2, took place in June 1980. Subsequently it was used to transport the crew of the final long-term mission to Salyut 6. It has now completely replaced the old Soyuz, which was last flown in Soyuz 40.

In April 1982, a new Salyut station (Salyut 7) was launched. It was occupied the following month by its first prime crew, arriving aboard Soyuz T-5. In June 1982 a three-man visiting crew (Soyuz T-6), including a French cosmonaut, was sent to the station for a one-week stay. This new Salyut is not significantly different from its predecessor. It is expected to be used as a prototype of future multimanned, long-duration stations with multiple docking ports to which can be attached modules dedicated to specific research and other scientific activities. Such a large-scale complex would constitute a "space operations center" supporting around-the-clock, year-round work by teams of cosmonauts (Oberg, 1982). A major objective would be the refinement of materials processing techniques for eventual use in the large-scale manufacture of superconductors, optical lenses, superpure vaccines, and many other products which would benefit from production under weightless conditions. Earth studies from space have already produced substantial benefit to the Soviet economy in such areas as agriculture, environmental studies, and the location of natural resources. Such studies will be expanded. The creation of a large-scale solar power generating station in orbit is possible. And biomedical studies such as those conducted throughout the Soviet space program will continue with a view to increasing man's adaptability to and long-term endurance of the conditions encountered in space.

Space Medicine Today

A review of the history of the manned space flight programs of the United States and the Soviet Union serves two purposes for space medicine. First, the review illustrates and clarifies the dimensions of space medicine as it is practiced today. The present professional priorities in this field are in large measure a result of earlier problems identified and solved or not solved. Second, the review shows the direction in which flight programs are moving and, by implication, the nature of the problems to be faced by space medicine in the future. This information is most useful in guiding space medicine in the development of a technology base today.

Emphasis in space medicine at this time is on understanding and controlling the physiological changes produced by exposure to weightlessness. Through 20 years of manned space flight, a rather definitive list of problem areas has been developed. In a recent biomedical status report reflecting the efforts of space medicine within the U.S. program, Dietlein and Johnston (1981) list the following biomedical problems as being of principal concern:

- Space motion sickness

- Cardiovascular deconditioning

- Hematology (red cell mass loss)

- Bone mineral loss.

Other occupational issues in space medicine are directly related to the type of missions planned for the future. The history of preceding flight programs illustrates the medical support areas of importance and shows how support requirements change as mission objectives grow. Tracing the history of man's missions through the development of a Space Transportation System and on into plans for space platforms and perhaps a Space Operations Center indicates the following issues will be important in space medicine:

24

- Selection of space personnel
- Medical training of flight crews
- Medical care during flight
- Extravehicular activity support for space station operations
- Radiation protection
- Use of artificial gravity
- Human factors considerations in man/machine interface design.

An ongoing program in space medicine for the development and improvement of a capability in each of these areas is necessary if the health, safety, and well-being of crew members in future missions is to be assured.

Summary of U.S. and USSR Manned Space Flight Experience

Launch Date	Astronaut(s)/ Cosmonaut(s)	Mission	Length of Flight	Points of Biomedical Interest
12 April 1961	Gagarin	Vostok 1	1 hr. 48 min.	First manned orbital flight
5 May 1961	Shepard	Mercury MR-3	15 min. ⎫	Suborbital flights. No significant physiological problems noted.
21 July 1961	Grissom	Mercury MR-4	16 min. ⎬	
6 August 1961	Titov	Vostok 2	25 hr. 18 min.	First reports of space motion sickness, relieved by restriction of head and body movement.
20 February 1962	Glenn	Mercury MA-6	4 hr. 55 min.	First U.S. manned orbital flights. No significant physiological effects noted.
24 May 1962	Carpenter	Mercury MA-7	4 hr. 56 min.	
11 August 1962	Nikolayev	Vostok 3	94 hr. 22 min. ⎫	First dual mission. Objective to study human capabilities to function as an operator in space.
12 August 1962	Popovich	Vostok 4	70 hr. 57 min. ⎬	
3 October 1962	Schirra	Mercury MA-8	9 hr. 13 min.	First episode of orthostatic intolerance noted postflight.
15 May 1963	Cooper	Mercury MA-9	34 hr. 20 min.	
14 June 1963	Bykovskiy	Vostok 5	119 hr. 6 min.	
16 June 1963	Tereshkova	Vostok 6	70 hr. 50 min.	First flight of a woman in space. Labile cardiovascular responses inflight.
12 October 1964	Komarov Feoktistov Yegorov	Voskhod 1	24 hr. 17 min.	Crew operated in "shirtsleeve" environment. First medical examination in space by a physician. Better characterization of vestibular dysfunction.
18 March 1965	Belyayev Leonov	Voskhod 2	26 hr. 2 min.	First USSR extravehicular activity (EVA). Neurovestibular function studies performed.
23 March 1965	Grissom Young	Gemini GT-3	4 hr. 53 min.	Cardiopulmonary monitoring inflight utilizing ECG, blood pressure, respiration rate measurements.
3 June 1965	McDivitt White	Gemini GT-4	97 hr. 48 min.	First U.S. extravehicular activity (EVA). Overheating inside the EVA suit.
21 August 1965	Cooper Conrad	Gemini GT-5	190 hr. 59 min.	
4 December 1965	Borman Lovell	Gemini GT-7	13 days, 18 hr. 35 min.	Comprehensive medical evaluations.

Launch Date	Astronaut(s)/ Cosmonaut(s)	Mission	Length of Flight	Points of Biomedical Interest
15 December 1965	Schirra Stafford	Gemini GT-6	25 hr. 51 min.	First U.S. rendezvous flight.
16 March 1966	Armstrong Scott	Gemini GT-8	10 hr. 42 min.	First docking in space with a target satellite.
3 June 1966	Stafford Cernan	Gemini GT-9	72 hr. 22 min.	Two hour, seven minute walk in space performed. EVA suit visor fogging.
18 July 1966	Young Collins	Gemini GT-10	70 hr. 46 min.	EVA performed without heat or work rate problems.
12 September 1966	Conrad Gordon	Gemini GT-11	71 hr. 17 min.	High and exhausting work loads during EVA.
11 November 1966	Lovell Aldrin	Gemini GT-12	94 hr. 34½ min.	Extravehicular activity included astronauts working with tools. No biomedical problems encountered during EVA.
23 April 1967	Komarov	Soyuz 1	27 hr.	Reentry braking system failed. Vehicle destroyed, resulting in death of cosmonaut.
11 October 1968	Cunningham Schirra Eisele	Apollo 7	260 hr. 9 min.	Crew experienced symptoms of upper respiratory viral infection inflight.
26 October 1968	Beregovoy	Soyuz 3	94 hr. 51 min.	First Soviet attempt to manually dock in space with an unmanned target, Soyuz 2. Consistently high heart rates during flight.
21 December 1968	Borman Lovell Anders	Apollo 8	147 hr.	First circumnavigation of the moon. First U.S. report of symptoms of motion sickness. Two-week preflight crew health stabilization program (HSP) instituted.
14 January 1969 15 January 1969	Shatalov Volynov Khrunov Yeliseyev	Soyuz 4 Soyuz 5	71 hr. 14 min. 72 hr. 46 min.	Spacecraft docked for approximately four hours. Two cosmonauts transferred to Soyuz 4 utilizing EVA procedures.
3 March 1969	McDivitt Scott Schweickart	Apollo 9	241 hr. 1 min.	Launch postponed for three days because of viral infection. Flight plans for EVA revised because of symptoms of motion sickness.

Launch Date	Astronaut(s)/ Cosmonaut(s)	Mission	Length of Flight	Points of Biomedical Interest
18 May 1969	Stafford Cernan Young	Apollo 10	192 hr. 3 min.	Fiberglass insulation produced skin, eyes, and upper respiratory passages irritation. No impact on mission functions.
16 July 1969	Armstrong Aldrin Collins	Apollo 11	195 hr. 18 min.	The first walk on the moon. Occurrence of bends reported by one crewman. Lunar quarantine instituted postflight. Three weeks of HSP initiated for this and subsequent missions.
11 October 1969	Shonin Kubasov	Soyuz 6	118 hr. 42 min.	First mission with multiple crews. The intent was to study human performance capabilities inflight, welding of metals in space, feasibility of building a manned space station.
12 October 1969	Filipichenko Volkov Gorbatko	Soyuz 7	118 hr. 41 min.	
13 October 1969	Shatalov Yeliseyev	Soyuz 8	118 hr. 41 min.	
14 November 1969	Conrad Bean Gordon	Apollo 12	244 hr. 36 min.	Contact dermatitis from biosensor electrolyte paste. Lunar quarantine postflight.
11 April 1970	Lovell Haise Swigert	Apollo 13	142 hr. 52 min.	Forced to return to Earth because of explosion in service module. Circumnavigation of moon completed. Urinary tract infection occurred due to the combined effects of cold, dehydration, and prolonged wearing of the urine collecting device.
1 June 1970	Nikolayev Sevastyanov	Soyuz 9	424 hr. 59 min.	Extensive biomedical investigations to determine cardiovascular and musculoskeletal responses. Test of different exercise regimen. Protracted recovery period postflight.
2 February 1971	Shepard Roosa Mitchell	Apollo 14	216 hr. 1 min.	
22 April 1971	Shatalov Yeliseyev Rukavishnikov	Soyuz 10	48 hr.	Docked with Salyut 1 for 5½ hours.
6 June 1971	Dobrovolsky Volkov Patsayev	Soyuz 11	552 hr.	Docked with Salyut 1. During return to Earth, pressure leak in Soyuz vehicle hatch resulted in decompression and death of all three crewmen. Focal areas of atrophy in antigravity muscles noted.

Launch Date	Astronaut(s)/ Cosmonaut(s)	Mission	Length of Flight	Points of Biomedical Interest
26 July 1971	Scott Worden Irwin	Apollo 15	295 hr. 12 min.	Cardiac arrhythmias and extra-systoles observed during flight. Postflight lunar quarantine discontinued. First use of Lunar Rover vehicle.
16 April 1972	Mattingly Duke Young	Apollo 16	254 hr. 51 min.	Study of light flash phenomenon.
7 December 1972	Cernan Evans Schmitt	Apollo 17	301 hr. 51 min.	Record stay on the lunar surface (75 hours) and 34 km travel utilizing the Lunar Rover vehicle.
25 May 1973	Conrad Kerwin Weitz	Skylab 2	672 hr. 49 min.	First detailed metabolic studies. First U.S. physician as a crewmember.
28 July 1973	Bean Garriott Lousma	Skylab 3	59 days	Reversal of red cell mass loss noted.
27 September 1973	Lazarev Makarov	Soyuz 12	2 days	Flight of the second Soviet physician/cosmonaut. Spacecraft modified to hold two crewmen in space suits rather than three in coveralls.
16 November 1973	Carr Gibson Pogue	Skylab 4	84 days	Apparent beneficial effects of vigorous exercise in minimizing cardiovascular deconditioning postflight. First inflight determination of lung vital capacity changes.
19 December 1973	Klimuk Lebedev	Soyuz 13	8 days	Test of countermeasures for cardiovascular deconditioning: anti-g suits, loading suits, bungee exercises. Studies of cerebral circulation utilizing rheography.
4 July 1974	Popovich Artyukhin	Soyuz 14	16 days	Docked with Salyut 3 space station.
26 August 1974	Sarafanov Demin	Soyuz 15	2 days	Failed to dock with Salyut 3.
2 December 1974	Filipichenko Rukavishnikov	Soyuz 16	7 days	Test flight to verify Soyuz systems prior to the joint US/USSR ASTP mission.
9 January 1975	Gubarev Grechko	Soyuz 17	30 days	Docked with Salyut 4 space station.

Launch Date	Astronaut(s)/ Cosmonaut(s)	Mission	Length of Flight	Points of Biomedical Interest
5 April 1975	Lazarev Makarov	Soyuz X	Aborted	Cosmonauts suffered injuries and exposure. No fatalities.
25 May 1975	Klimok Sevastyanov	Soyuz 18	63 days	Docked with Salyut 4. Continuation of studies to test human endurance to weightlessness.
15 July 1975	Kubasov Leonov	Soyuz 19 (ASTP)	6 days	First international space mission. US/USSR joint spaceflight experiment. Cosmonauts and astronauts transferred from respective spacecrafts. U.S. crewmen exposed to nitrogen tetroxide accidentally.
15 July 1975	Stafford Brand Slayton	Apollo-Soyuz Test Project	9 days	
6 July 1976	Volynov Zholobov	Soyuz 21	48 days	Docked with Salyut 5 space station.
15 September 1976	Bykovskiy Aksenov	Soyuz 22	8 days	USSR/GDR joint experiments.
14 October 1976	Zudov Rozhdestvenskiy	Soyuz 23	2 days	Failed to dock with Salyut 5.
7 February 1977	Gorbatko Glazkov	Soyuz 24	18 days	Docked with Salyut 5. Evaluation of medical countermeasures to prevent deconditioning in long duration missions. Space processing.
9 October 1977	Kovalenok Ryumin	Soyuz 25	2 days	Failed to dock with Salyut 6 space station.
10 December 1977	Romanenko Grechko	Soyuz 26	96 days	First prime crew of Salyut 6. Significant cardiovascular deconditioning postflight.
10 January 1978	Dzhanibekow Makarov	Soyuz 27	6 days	First visiting crew to Salyut 6.
2 March 1978	Gubarev Remek	Soyuz 28	8 days	Second visiting and first international crew to Salyut 6.
15 June 1978	Kovalenok Ivanchenko	Soyuz 29	140 days	Second prime crew of Salyut 6. New EVA suit with minimum prebreathing requirements introduced.
27 June 1978	Klimuk Hermaszewski	Soyuz 30	8 days	
26 August 1978	Bykovskiy Jaehn	Soyuz 31	8 days	

Launch Date	Astronaut(s)/ Cosmonaut(s)	Mission	Length of Flight	Points of Biomedical Interest
25 February 1979	Lyakhov Ryumin	Soyuz 32	175 days	Third prime crew of Salyut 6. First reports of recurrence of vestibular symptoms inflight and postflight.
9 April 1979	Rakavishnikov Ivanov	Soyuz 33	2 days	Fourth international flight. Failed to dock with Salyut 6.
9 April 1980	Popov Ryumin	Soyuz 35	185 days	Longest period in space to date.
26 May 1980	Kubasov Farkash	Soyuz 36	8 days	
5 June 1980	Malyshev Aksenov	Soyuz T-2	5 days	Docked with Salyut 6. Manned test of a new orbital transfer vehicle.
22 July 1980	Gorbatko Tuan	Soyuz 37	8 days	
17 September 1980	Romanenko Mendez	Soyuz 38	8 days	
27 November 1980	Kizim Makarov Strekalov	Soyuz T-3	13 days	
12 March 1981	Kovalenok Savinykh	Soyuz T-4	75 days	Last prime crew onboard Salyut 6 space station. Extensive use of mechanical countermeasures for motion sickness and cardiovascular deconditioning.
22 March 1981	Dzhanibekov Gurragcha	Soyuz 39	9 days	
12 April 1981	Crippen Young	STS-1	54 hr. 21 min.	First demonstration of successful man-piloted hypersonic flight. First experience with G_z forces on reentry. First runway landing.
14 May 1981	Popov Prunariu	Soyuz 40	8 days	Last use of original Soyuz vehicle. Last visiting crew to Salyut 6.
12 November 1981	Engle Truly	STS-2	54 hr. 13 min.	
22 March 1982	Fullerton Lousma	STS-3	8 days	Expanded manned test of the STS vehicle.
14 May 1982	Berezovoy Lebedev	Soyuz T-5	211 days	Enhanced Salyut 7 space station. Detailed health maintenance studies.

Launch Date	Astronaut(s)/ Cosmonaut(s)	Mission	Length of Flight	Points of Biomedical Interest
24 June 1982	Djanibekov Iventchenov Chretien	Soyuz T-6	8 days	Visiting international crew (French) to Salyut 7. First inflight cardio-vascular measurements utilizing echocardiography.
27 June 1982	Mattingly Hartsfield	STS-4	7 days	
19 August 1982	Savitskaya Popov Serebrov	Soyuz T-7	8 days	Second woman in space.
10 November 1982	Brand Overmyer Lenoir Allen	STS-5	5 days	First launch of 4-man crew. Entirely manual reentry and landing.

References

Belew, L.F. (Ed.). *Skylab, our first space station* (NASA SP-400). Washington, D.C.: U.S. Government Printing Office, 1977.

Burchard, E.C., & Nicogossian, A.E. Achilles tendon reflex. In A.E. Nicogossian (Ed.), *The Apollo-Soyuz Test Project medical report* (NASA SP-411). Springfield, VA: National Technical Information Service, 1977.

DeHart, R. Biomedical aspects of Soviet manned space flight (ST-CS-13-373-75). Washington, D.C.: Defense Intelligence Agency, 1974.

Dietlein, L.F. Skylab: A beginning. In R.S. Johnston & L.F. Dietlein (Eds.), *Biomedical results from Skylab* (NASA SP-377). Washington, D.C.: U.S. Government Printing Office, 1977.

Dietlein, L.F., & Johnston, R.S. U.S. manned space flight: The first twenty years. A biomedical status report. *Acta Astronautica*, 1981, *8*(9-10), 893-906.

Dodge, C.H. The Soviet space life sciences (Ch. 4). In *Soviet space programs, 1971-1975: Overview, facilities, and hardware, manned and unmanned flight programs, bioastronautics, civil and military applications, projections of future plans* (Vol. 1). Washington, D.C.: Library of Congress, Science Policy Research Division, Congressional Research Service, 1976.

Feoktistov, K.P., & Markov, M.M. Evolution of Salyut orbital stations. *JPRS, USSR Report: Space*, 29 March 1982, *15*, 1-11.

Graybiel, A. Coping with space motion sickness in Spacelab missions. *Acta Astronautica*, 1981, *8*(9-10), 1015-1018.

Johnston, R.S. Introduction. In R.S. Johnston, L.F. Dietlein, & C.A. Berry (Managing Eds.), *Biomedical results of Apollo* (NASA SP-368). Washington, D.C.: U.S. Government Printing Office, 1975.

Kleinknecht, K.S. Preface. In J.M. Grimwood (Ed.), *Project Mercury, a chronology* (NASA SP-4001). Washington, D.C.: U.S. Government Printing Office, 1963.

LaFevers, E.V., Nicogossian, A.E., Hursta, W.N., & Baker, J.T. Electromyographic analysis of skeletal muscle. In A.E. Nicogossian (Ed.), *The Apollo-Soyuz Test Project medical report* (NASA SP-411). Springfield, VA: National Technical Information Service, 1977.

Link, M.D. *Space medicine in Project Mercury* (NASA SP-4003). Washington, D.C.: U.S. Government Printing Office, 1965.

Lovelace, W.R., II. Introduction. In M.M. Link, *Space medicine in Project Mercury* (NASA SP-4003). Washington, D.C.: U.S. Government Printing Office, 1965.

Lyakhov, V. Manned space explorations: The way to the future. *Spaceflight*, 1981, *23*(1), 22-23.

Malyshev, Yu. Evolution of the Soyuz spacecraft. *JPRS, USSR Report: Space*, 2 March 1981, 9, 8-13.

Mueller, G.E. Introduction. In *Gemini summary conference* (NASA SP-138). Washington, D.C.: U.S. Government Printing Office, 1967.

Nicogossian, A.E., LaPinta, C.K., Burchard, E.C., Hoffler, G.W., & Bartelloni, P.J. Crew health. In A.E. Nicogossian (Ed.), *The Apollo-Soyuz Test Project medical report* (NASA SP-411). Springfield, VA: National Technical Information Service, 1977.

Oberg, J. Soviet development points for space operations center. *Astronautics & Aeronautics*, 1982, *20*(4), 74-77.

Prishchepa, V.I. Twentieth anniversary of Gagarin's flight: A collection of articles. *JPRS, USSR Report: Space* (FOUO 4/81), 13 October 1981, 7-10.

Smith, M.S. Program details of man-related flights (Ch. 3). In *Soviet space programs, 1971-1975: Overview, facilities, and hardware, manned and unmanned flight programs, bioastronautics, civil and military applications, projections of future plans* (Vol. 1). Washington, D.C.: Library of Congress, Science Policy Research Division, Congressional Research Service, 1976.

Smith, M.C., Jr., Rambaut, P.C., Vogel, J.M., & Whittle, M.W. Bone mineral measurement—experiment M078. In R.S. Johnston & L.F. Dietlein (Eds.), *Biomedical results from Skylab* (NASA SP-377). Washington, D.C.: U.S. Government Printing Office, 1977.

von Beckh, H.F. The space medicine branch of the aerospace medical association. *Aviation, Space, and Environmental Medicine*, 1979, *50*(5), 513-516.

Whedon, G.D., Lutwak, L., Rambaut, P.C., Whittle, M.W., Smith, M.C., Reid, J., Leach, C.S., Stadler, C.R., & Sanford, D.D. Mineral and nitrogen metabolic studies—experiment M071. In R.S. Johnston & L.F. Dietlein (Eds.), *Biomedical results from Skylab* (NASA SP-377). Washington, D.C.: U.S. Government Printing Office, 1977.

Whittle, M.W., Herron, R., & Cuzzi, J. Biostereometric analysis of body form. In R.S. Johnston & L.F. Dietlein (Eds.), *Biomedical results from Skylab* (NASA SP-377). Washington, D.C.: U.S. Government Printing Office, 1977.

SECTION II

The Space Environment

CHAPTER 2
ORBITAL FLIGHT

It is important that life scientists, just as astronomers and physical scientists, understand the nature of space. The planning and development of manned space missions requires this understanding in order to achieve proper design of crew living quarters, personal protective systems, work procedures, and, indeed, all aspects of space systems which support human survival and performance.

The space program of the United States for the foreseeable future calls for increasing manned missions. During the decade of the 1980s, Space Shuttle flights will become more frequent, with increasing numbers of astronauts participating. By the beginning of the 1990s, crewmen will remain in space for considerably longer periods as space platforms are developed from which a larger force of space workers can follow a variety of pursuits. During this period of growth and change in manned space missions, there is one feature, however, which will remain constant. All missions and activities will occur in Earth orbit. While it is inevitable that man ultimately will return to the moon and then proceed on into other parts of the solar system, the U.S. space program does not call for such missions in this century. Therefore, it is important for the practitioners of space medicine to understand the environment of orbital space. The more exotic interplanetary environmental issues, which certainly must be considered as planning begins for missions of the 21st century, are not relevant to the biomedical support of space activities during the coming two decades.

The Transition to Near-Earth Orbital Space

The gaseous envelope that forms our atmosphere is acted upon principally by two forces: the terrestrial gravitational force that binds it to Earth, and solar thermal radiation, which causes its gases to tend to expand into the surrounding space. Because these two forces are in relatively constant balance, the atmosphere exhibits a distinct vertical profile of density and pressure. As the distance from Earth increases, the density of the gaseous medium decreases (Figure 1). The border of the atmosphere is defined as the point where collisions between air molecules become immeasurably infrequent. This "collision limit" occurs at about 700 km above the Earth's surface. Above this level is the exosphere, a zone of free-moving air particles that gradually thins out into true space. Even in space, however, the density of gas particles is about 1-10 particles per cubic centimeter.

The transition zone from the atmosphere of the Earth into space contains two points of particular interest for spacecraft design. At an altitude of approximately 80 km, the so-called von Karman line is encountered. This is the minimum altitude at which aircraft control surfaces are effective aerodynamically. Above this line, the orientation of space vehicles is controlled through use of reaction jets. At about 180 to 200 km, the resistance of the air becomes insignificant. This is considered to be the mechanical border between the atmosphere and space. These and other significant transition altitudes are shown in Table 1.

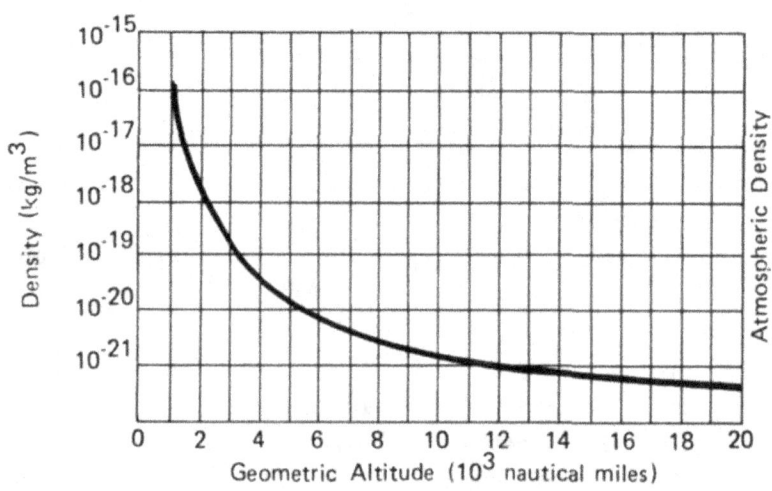

Figure 1. Atmospheric density as a function of altitude. The rapid decrease in density is correlated with a decrease in both atmospheric pressure and partial pressure of oxygen. (Adapted from *Air Force Surveys in Geophysics*).

Table 1

Significant Altitudes in the Transition from
the Earth's Atmosphere to Space

Boundaries	Approximate Altitude	
	km	mi
Oxygen limit—maximum altitude for sustained O_2 breathing through a pressure mask	13	8
Tropopause—upper limit for most atmospheric weather effects	15	9
Physiological limit—space suit or cabin a requirement for protection of human body	20	12
von Karman line—aerodynamic control no longer effective	80	50
Mechanical boundary—air resistance becomes negligible	200	125
Collision limit—"space vacuum" achieved with essentially no collisions between air molecules	700	435

Manned flight in near-Earth orbit takes place at altitudes in the order of 240 km, below the region defined as "true space." However, even at these altitudes a space vehicle is well beyond the functional limits of the Earth's atmosphere. Designers of manned space vehicles must account for the new operating environment (weightlessness), the lack of a life-supporting atmosphere, and specific problems such as radiation and the danger of collision with small objects in space (micrometeoroids).

Force Fields

The inertial or rotational forces acting on an astronaut in space are a key feature of this environment since these forces can affect work efficiency, health, and even survival. There are two kinds of issues to be considered with respect to the force environment encountered in an orbital mission. The first is the acceleration or deceleration experienced as a space vehicle launches toward orbit or reenters the Earth's atmosphere at the completion of a mission. The second is the loss of the normal gravitational force which acts on us at all times on Earth. This loss of gravity, or "weightlessness," occurs when the gravitational force vector is exactly counterbalanced by the centrifugal force imparted to a spacecraft as it travels tangentially to the Earth's surface.

The launch and reentry acceleration forces of spacecraft all have been well within the tolerance limits established for healthy, well-conditioned subjects. Indeed, as new craft are developed, the imposed acceleration loads have been decreasing. Figure 2 shows the acceleration profile during the launch phase of a manned flight during the Mercury program. This shows that for a brief period the acceleration reached an 8 G_x value. Reentry forces also were high, reaching a value greater than

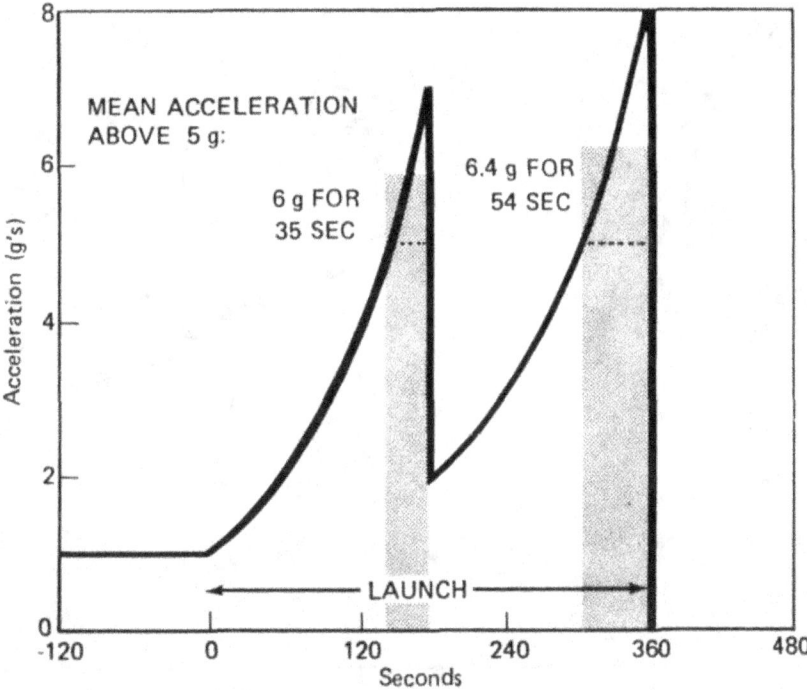

Figure 2. Acceleration profile of launch phase of the manned Mercury-Atlas 6 orbital flight. (From Waligora, 1979)

39

6 G_x during the peak period of deceleration during atmospheric reentry. Contrast these values with those experienced during Space Shuttle missions. Measures taken during the launch of the first Space Shuttle orbital mission (STS-1) showed an acceleration profile typical of previous spacecraft launches, but with a maximum of only +3.4 G_x (chest to back). The programmed reentry profile, with the acceleration forces acting on a different line due to the changed orientation during reentry (Figures 3 and 4), show only nominal forces of +1.2 G_z (head to foot). While these forces are considerably less than those for earlier missions, they do act over a much longer period of time (17-20 minutes) and this must be considered medically.

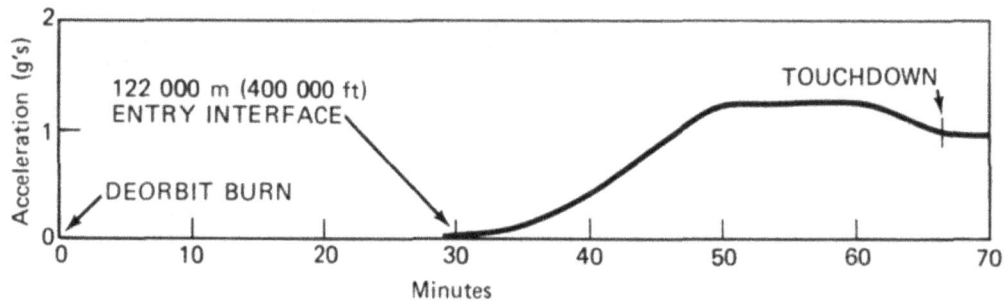

Figure 3. Acceleration profile of the Space Shuttle vehicle as a function of time. (From Waligora, 1979)

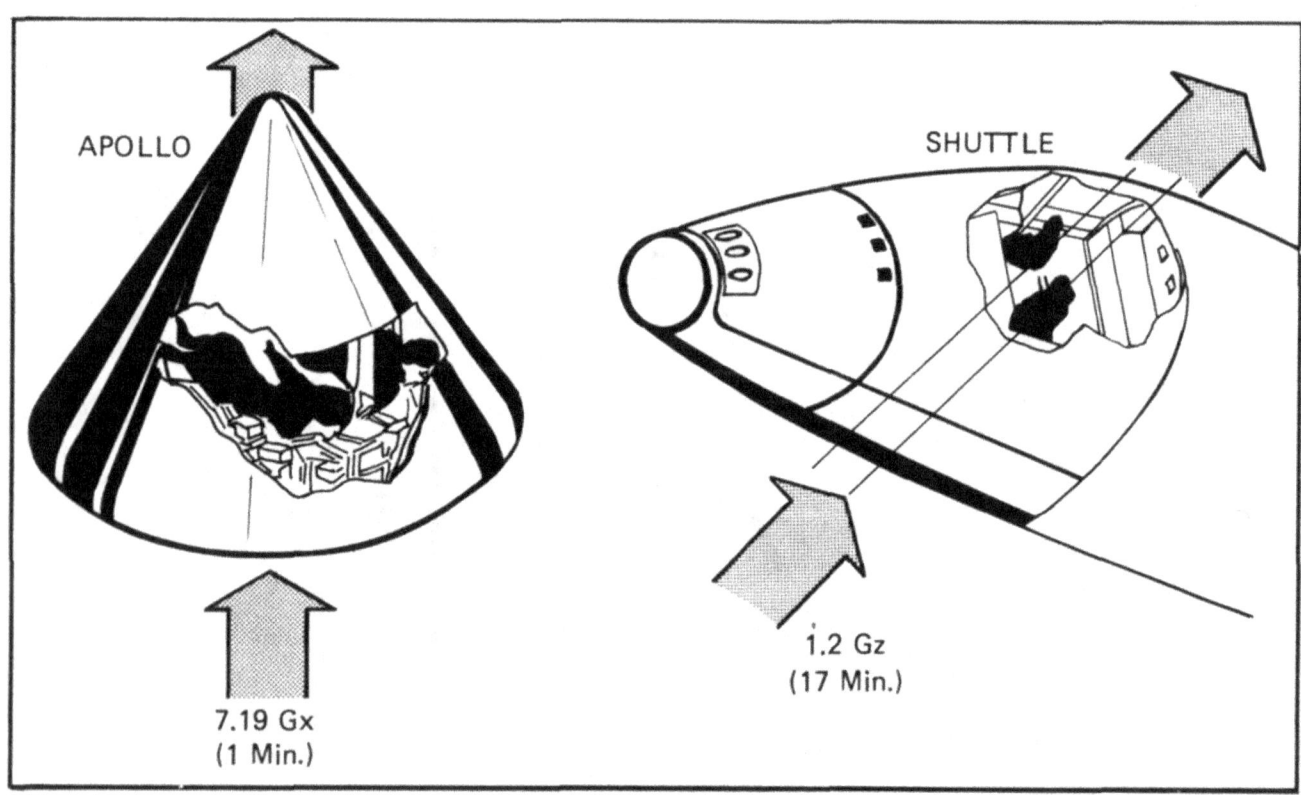

Figure 4. Comparison of reentry force direction in Apollo versus Space Shuttle.

40

Zero Gravity

The most dramatic environmental feature of orbital space flight is weightlessness. To live and work in a world in which there is no gravity is a totally new experience for space travelers. It is an experience characterized mostly for its novelty, but one with a broad range of important medical and behavioral consequences. The novelty aspects are apparent in Figure 5, in which two Skylab astronauts illustrate the ease with which an individual can deal with a heavy object in weightlessness.

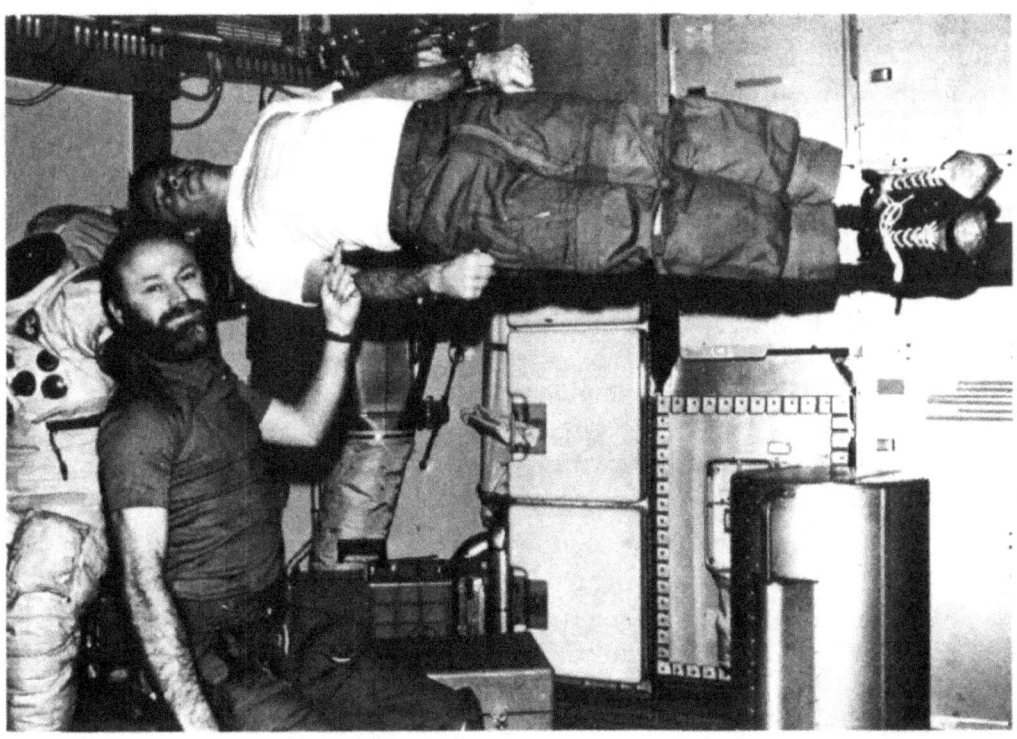

Figure 5. Skylab 4 astronauts demonstrate the effect of zero gravity on weight.

The dynamics of human experience are continuously shaped by the one-gravity force which is all-pervasive. Every conscious movement is made in a manner which accounts for gravity. Such a simple motor act as leaning forward brings into play a number of muscle systems which control the movement of the body as the center of the gravitational effect also moves. These are highly skilled acts and require no conscious thought. On initial entry into a gravity-free environment, however, each movement and act must be done differently and a period of relearning is required. Fortunately, the problems of relearning have not proven as difficult as was predicted by some. Dr. Joseph Kerwin, in recounting his experience as a crewmember in Skylab 2, noted that "The primary theme was one of pleasant surprise at all the things that didn't change, at all the things that were pleasant and easy to do." (Kerwin, 1977).

Following whatever relearning is necessary, the zero-gravity environment can be employed in a beneficial manner. Berry (1971), summarizing the early Apollo experiences, noted that the absence of gravity could represent a bonus for locomotion since moving in zero gravity requires much less work than on Earth. Movement is accomplished with minimal effort, frequently in a swimming manner. Acrobatic maneuvers, such as rolling, tumbling, and spinning, are done with ease. Additionally, inflight activities frequently are aided by the easy capability to impart minimal velocities to objects to be moved.

The zero-gravity environment affects the physiological functioning of major body systems in a manner less obvious than for locomotion but of greater medical consequence. On entry into weight-lessness there is an immediate redistribution of body fluids. The function of the vestibular system, which is uniquely sensitive to gravity, is disturbed. Other systems begin a slow process of adaptation to the altered environment. For example, the cardiovascular system adjusts to a new set point in which the demands placed on it are greatly decreased. The result is a system, "deconditioned" by Earth standards, which functions appropriately for life in weightlessness but which may have real difficulty when called on to readjust suddenly to a one-gravity environment. Later chapters in this book address in detail the physiological changes brought about by long-term exposure to zero gravity.

Micrometeoroids

A number of solid objects regularly pass through the orbital environment of the Earth. The largest of these are the meteors, or "shooting stars," appearing on occasion as long streaks in the sky. These meteors, which are solid matter heated to incandescence by friction with the atmosphere, are of two types, with about 61 percent being stone, 35 percent iron, and 4 percent a mixture of the two (Lundquist, 1979). Remnants of meteors are called meteorites or micrometeorites. Most of the extraterrestrial matter reaching the Earth's surface, estimated at 10,000 metric tons per day, is in the form of micrometeorites.

The terms "meteoroid" and "micrometeoroid" describe the individual solid objects found in interplanetary space. Micrometeoroids often are referred to as interplanetary dust.

The presence of micrometeoroids in space has long been known. The extent to which such objects might represent a hazard for manned space flight, however, has not been known. Therefore, all missions have included some means of protection for crewmembers. In some instances, the protection has been provided through the shielding afforded by the spacecraft itself. During the Apollo program, astronauts who were scheduled for periods on the lunar surface were provided with an integrated thermal micrometeoroid garment (Carson et al., 1975). This was a lightweight multilaminant assembly which covered the torso-limb suit assembly to provide protection against micrometeoroids. Materials such as rubber-coated nylon, aluminized mylar, dacron, and teflon-coated filament Beta cloth provided the required protection. No instances of significant damage by micrometeoroids were recorded during the many hours spent in lunar exploration or in extra-vehicular activities.

The Skylab program presented an opportunity to obtain measures over an extended period of the extensiveness of micrometeoroids in orbital space. Since a number of these objects were expected to strike Skylab, an experiment was developed in which thin foils and polished metal plates were exposed on the exterior of the Skylab vehicle to record penetrations by micrometeoroids (Lundquist, 1979). On return to Earth, the exposed materials were studied with optical microscopes and scanning electron microscopes at magnifications of 200X and at 500X. Table 2 shows the results of these analyses. Since the exposed plates were located at different positions on Skylab and at different orientations, these figures represent an approximation of the micrometeoroid flux in orbit.

Table 2
Micrometeroid Impacts Recorded
During Skylab Experiments
(Exposed area $= 1200$ cm^2)

Sample Period	No. of Impacts
1–34 days	23
2–46 days	17
3–34 days	21

Lundquist, 1979

The size of the craters measured in the Skylab experiment showed all of the impacting particles to be quite small, with one believed to have been between 0.1 and 0.2 m in diameter. The particles did have considerable power, as witnessed by craters measured in a stainless steel surface. However, even though Skylab's micrometeoroid shield was lost during launch, the Orbital Workshop's wall, 3.18 mm thick, was not penetrated. There appears to be little if any meteoroid hazard to spacecraft in orbit with an exterior of such thickness.

Radiation

The vacuum of space, a more perfect vacuum than can be achieved in any facility on Earth, nonetheless contains a great deal of matter which is of interest to planners of space missions. Biomedical scientists are particularly concerned with ionizing radiation which, when absorbed in a living cell, ionizes the atoms or molecules in the cell and may result in death of the cell. Table 3 lists the ionizing radiation of concern during space flight.

The radiation encountered during orbital flight can be classed as primary cosmic radiation, geomagnetically trapped radiation (Van Allen belts), and that due to solar flares (Warren & Grahn, 1973). The latter two classes determine the unique quality of radiation found in near-Earth space. Indeed, these two classes are important factors in determining the scheduling and trajectories of orbital flights. Anticipated radiation levels always are taken into account prior to a mission.

Table 3
Ionizing Radiation in Space

Name		Charge (Z)	RBE	Location
X-rays		0	1	Radiation belts, solar radiation
Gamma rays		0	1	and in the secondaries made by
				nuclear reactions, and by stopping
				electrons.
Electrons				
1.0 MeV		1	1	Radiation belts
0.1 MeV		1	1.08	
Protons				
100	MeV	1	1-2	Cosmic rays, inner radiation belts,
1.5	MeV	1	8.5	solar cosmic rays
0.1	MeV	1	10	
Neutrons				
0.05	eV (thermal)	0	2.8	Produced by nuclear interactions;
.0001	MeV	0	2.2	found near the planets and the sun
.005	MeV	0	2.4	and other matter
.02	MeV	0	5	
.5	MeV	0	10.2	
1.0	MeV	0	10.5	
10.0	MeV	0	6.4	
Alpha particles				
5.0	MeV	2	15	Cosmic rays
1.0	MeV	2	20	
Heavy primaries		$\geq$3	(see text)	Cosmic rays

Johnson & Holbrow, 1977.

The atmosphere of Earth, coupled with its magnetic field, serves as a protective blanket to shield man from virtually all of the types of radiation which could be damaging during space flight. In the electromagnetic spectrum, there are only two "windows" which allow radiation from the sun and from deep space to pass through to the Earth. One covers the visible light and part of the ultraviolet and infrared frequencies. Another covers radio frequencies of approximately 10^9 Herz (Olson & McCarson, 1974). All other radiation is effectively blocked. The protection afforded Earth's inhabitants means that most of what has been learned concerning space radiation has come from space missions and probes flown during the past 30 years.

Galactic Cosmic Radiation

This class of radiation consists of particles which originate outside the solar system, probably resulting from earlier cataclysmic solar events such as the supernova explosion witnessed by Chinese astronomers in the year 1054 A.D. Measures from space probes show that these particles consist of 87 percent protons (hydrogen nuclei), 12 percent alpha particles (helium nuclei), and 1 percent heavier nuclei ranging from lithium to tin. The individual particle energies are extremely high, in some instances up to 10^{20} electron volts. The high particle energies mean that galactic cosmic particles, though of very low flux density, are virtually unshieldable by passive shields. At present, principal interest is in determining the extent to which periodic or continuous exposure to this radiation might affect career limits for space workers. Active shielding, such as might be offered by induced magnetic fields, is a consideration for future space habitats.

Trapped Radiation

A project team led by Dr. James Van Allen of the University of Iowa conducted experiments in 1958 in the U.S. Explorer satellite series in which they discovered the existence of bands of geomagnetically trapped particles encircling the Earth. The electrons and protons of the solar wind encounter the Earth's magnetic field and become trapped, leading to a period of oscillation back and forth along the lines of magnetic force. The trapped particles follow the magnetic field completely around the Earth (Figure 6).

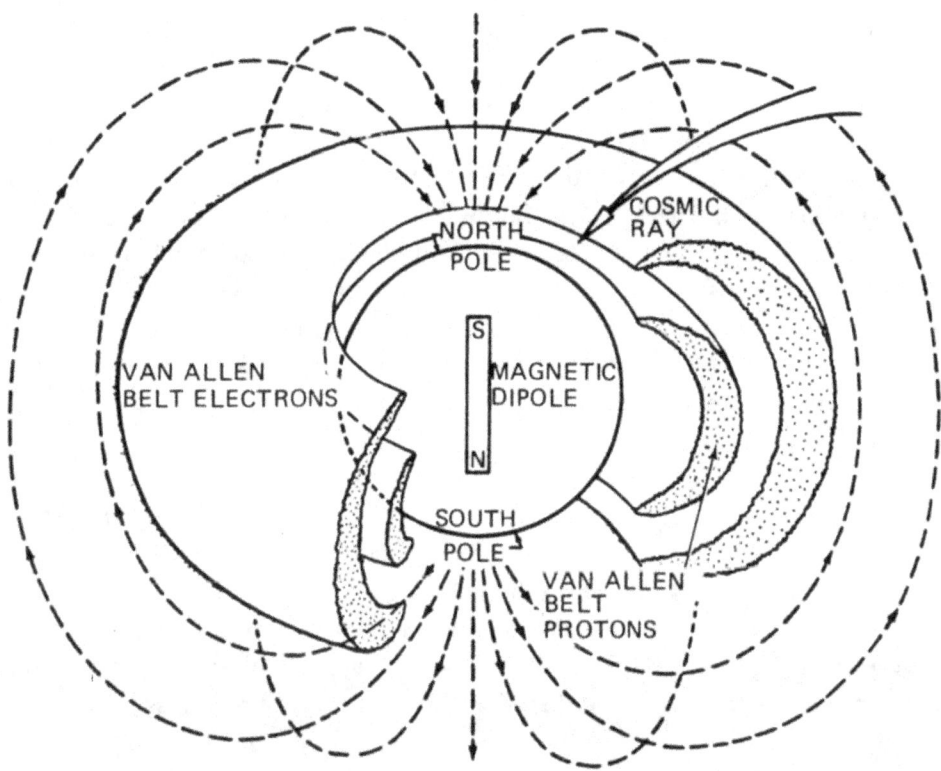

Figure 6. Solar wind electrons and protons are trapped by the Earth's magnetic field, forming the Van Allen belts.

There are two portions to the Van Allen belts, with effects being evident at altitudes as high as 55,000 kilometers. The inner Van Allen belt begins at an altitude of roughly 300 to 1,200 kilometers, depending on latitude. The outer belt begins at about 10,000 kilometers, with its upper boundary dependent upon the activity of the sun.

In the low Earth orbit followed in Space Shuttle missions, radiation from the Van Allen belts is slight. However, there is a discontinuity in the Earth's geomagnetic field in the southern hemisphere, known as the South Atlantic Anomaly, which is to be avoided to the extent possible in planning orbital missions. At this location, which extends from about zero to 60 degrees west longitude and 20 to 50 degrees south latitude, the trapped proton intensity for energies more than 30 MeV is the equivalent at 160–320 kilometers altitude to that at 1,300 kilometers altitude elsewhere (Warren & Grahn, 1973).

Almost all of the radiation a Space Shuttle crew in a low-Earth, low-inclination orbit might receive would result from passes through the South Atlantic Anomaly. With a 38° orbit, as used for the initial flights of the Space Shuttle, there were about six orbital rotations which passed through the South Atlantic Anomaly and about eleven which did not. On these eleven, there was essentially no radiation exposure. Extravehicular activities thus can be scheduled at a time in the orbital trajectory when radiation will not be a problem.

Solar Flares

Solar flares are a major source of radiation concern, possibly the most potent of the radiation hazards. The sun follows approximately a ten to eleven year cycle of activity. When activity peaks, there can be spectacular disturbances on the surface of the sun, as seen in Figure 7. A solar flare is in fact a solar magnetic storm. These storms build up over several hours and last for several days. While their occurrence cannot be forecast, the onset of build-up can be detected. As the flare builds, there is first an increase in visible light, accompanied by disturbances in the Earth's atmosphere, probably due to solar X-rays. The principal problem, though, is with the high energy protons produced during the storm. The energy of these protons ranges from about ten million electron volts to about five hundred million electron volts (Olson & McCarson, 1974). The flux also may be quite high.

The largest solar flares appear to occur either just before the peak of a sun spot cycle or on the downward limb of the measured solar activity. Figure 8 shows the close correspondence between high energy protons, measured at Earth, and solar activity, as reflected in number of sun spots. Note that in August of 1972, during a period of decline of solar activity, there occurred the largest proton flux yet recorded. This would have represented a particularly dangerous time for the conduct of extravehicular activity in orbital space. There would be a real possibility of an astronaut receiving a lethal dose.

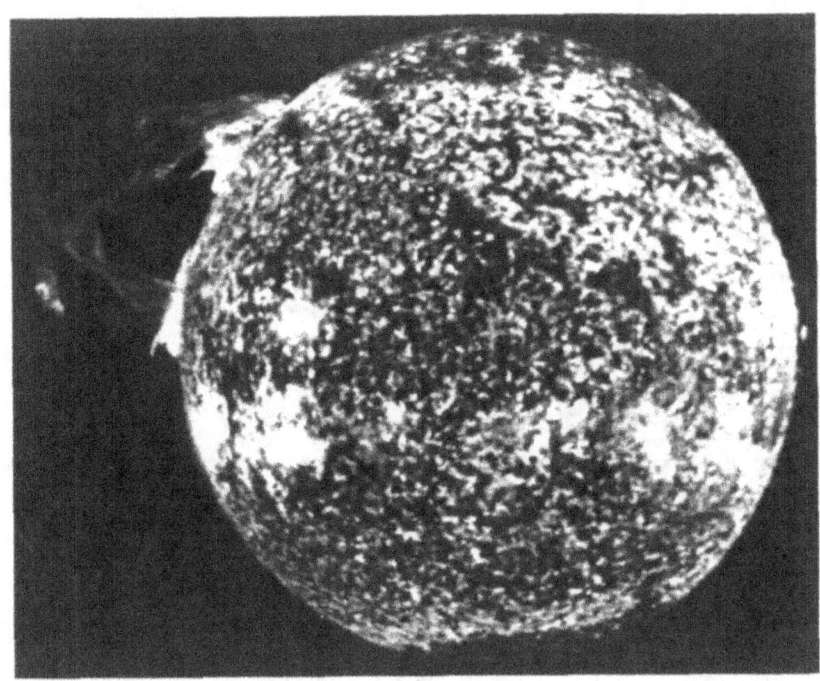

Figure 7. This Skylab 4 photograph of the sun shows one of the most spectacular solar flares ever recorded (upper left).

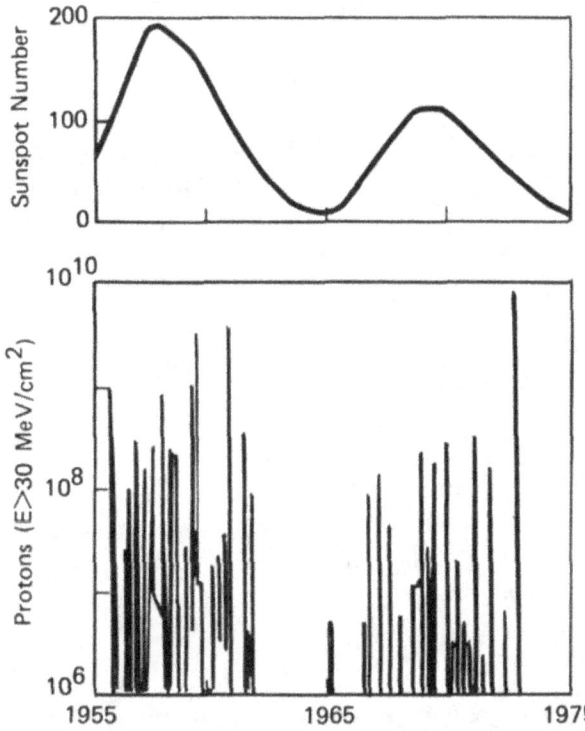

Figure 8. Increases in sunspot activity (top) produce corresponding increases in solar flare proton flux (bottom). (From Nachtwey, 1982).

Neutron Flux

The Skylab missions provided an opportunity to evaluate yet another radiation hazard in space, the energetic neutron. Neutrons were known as one component of space radiation before Skylab, but the magnitude and, in particular, the source were not well understood. Neutrons are of biomedical importance because, upon colliding with a hydrogen nucleus, a proton, there is a high probability of an energy exchange (Lundquist, 1979). Since humans contain an abundance of hydrogen-rich compounds such as proteins, fat, and especially water, neutron exposure could cause considerable damage. It is therefore necessary to understand ambient neutron flux within a space vehicle.

It is known that free neutrons are not stable. With a half-life of eleven minutes, neutrons decay into a proton and an electron. This means that neutrons detected in a space vehicle must be generated either within the spacecraft or within the Earth's atmosphere and represent products of the nuclear reactions caused by strikes of primary radiation. Skylab measurements showed that neutron flux within a spacecraft is higher than had been predicted. This flux was too high to be attributed to solar neutrons, Earth albedo neutrons, or even neutrons induced by cosmic rays in space station materials (Lundquist, 1979). It was concluded that the neutrons were produced through bombardment of spacecraft material by trapped protons in the Van Allen belt. However, the flux level was not high enough to be considered a biological hazard for crew members.

References

Air Force Surveys in Geophysics, AFCRL, No. 115, August 1959.

Berry, C.A. Biomedical findings on American astronauts participating in space missions: Man's adaptation to weightlessness. Paper presented at The Fourth International Symposium on Basic Environmental Problems of Man in Space, Yerevan, Armenia, U.S.S.R., 1-5 October 1971.

Carson, M.A., Rouen, M.N., Lutz, C.C., & McBarron, J.W., II. Extravehicular mobility unit. In R.S. Johnston, L.F. Dietlein, & C.A. Berry (Eds.), *Biomedical results of Apollo* (NASA SP-368). Washington, D.C.: U.S. Government Printing Office, 1975.

Johnson, R.D., & Holbrow, C. (Eds.). *Space settlements—a design study* (NASA SP-413). Washington, D.C.: U.S. Government Printing Office, 1977.

Kerwin, J.P. Skylab 2 crew observations and summary. In R.S. Johnston & L.F. Dietlein (Eds.), *Biomedical results from Skylab* (NASA SP-377). Washington, D.C.: U.S. Government Printing Office, 1977.

Lundquist, C.A. (Ed.). *Skylab's astronomy and space sciences* (NASA SP-404). Washington, D.C.: U.S. Government Printing Office, 1979.

Nachtwey, S. Radiation exposure, detection, and protection. Paper presented at the 53rd Annual Scientific Meeting of the Aerospace Medical Association, Bal Harbour, FL, May 10-13, 1982.

Olson, R.E., & McCarson, R.D., Jr. (Eds.). *Space handbook* (Tenth Revision) (AU-18). Maxwell Air Force Base, AL: Air University Institute for Professional Development, July 1974.

Page, L.W., & Page, T. *Apollo-Soyuz pamphlet no. 6: Cosmic ray dosage* (9-part set). Washington, D.C.: U.S. Government Printing Office, 1977.

Waligora, J.M. Physical forces generating acceleration, vibration, and impact. In J.M. Waligora (Coordinator), *The physiological basis for spacecraft environmental limits* (NASA RP-1045). Washington, D.C.: National Aeronautics and Space Administration, Scientific and Technical Information Branch, November 1979.

Warren, S., & Grahn, D. Ionizing radiation. In J.F. Parker, Jr., & V.R. West (Eds.), *Bioastronautics data book* (2nd ed.) (NASA SP-3006). Washington, D.C.: U.S. Government Printing Office, 1973.

CHAPTER 3
PLANETARY ENVIRONMENTS

The spectacular successes of American and Soviet manned and unmanned missions have provided a wealth of scientific knowledge concerning the nature of our nearest celestial neighbors. In recent years, the rate of acquisition of data concerning the solar system has been such that it may be a decade or more before all of it can be analyzed. Yet, even as we develop detailed descriptions of the solar planets, new questions are raised. For example, extensive analyses of geological specimens have not as yet solved the problem of the moon's origin. Changes seen in the Martian polar regions have still not been explained; nor have the Viking missions provided insight as to the evolutionary processes which produced puzzling differences between the Martian surface and soil and that of Earth. Voyager spacecraft sent back photographs of the moons of Jupiter and the rings of Saturn that are striking beyond belief. However, the nature of processes occurring on certain of Jupiter's moons remains unclear, as does the physics involved in the configuration of some of Saturn's rings. An understanding of our solar system and, indeed, of our universe requires that questions such as these be resolved.

Because it is the essence of the human spirit that partial descriptions and unresolved scientific issues provoke action, it is inevitable that at some point man will venture past the moon and on into interplanetary space. Interplanetary missions will require a new era of research and planning if man is to be sustained during these long ventures and if he is to carry out all assigned activities. The information we now have concerning interplanetary space as well as the orbital and surface environments of the planets tells us that such missions will not be simple. As an indication of the kinds of issues to be confronted by biomedical scientists participating in the planning of interplanetary missions, the following brief descriptions of the moon and the four planets which have been probed most extensively to date are offered.

The Moon

The moon is the only extraterrestrial body to have been explored directly by man. Twenty-four astronauts have observed the moon at close distance and 12 have walked on its surface. One-hundred and sixty man hours were spent in exploration, with 100 km (60 mi) of the lunar surface covered by foot and through use of the electric-powered Rover vehicle. During the Apollo missions, 30,000 photographs were taken of the moon and its various features. The Apollo missions returned a total of 380 kg (841 lbs.) of lunar rock and soil to Earth for detailed analyses. This has been supplemented by other soil returned by Russia's unmanned Luna 16 and Luna 20 spacecraft, soil from a part of the moon not sampled by Apollo.

Prior to the detailed exploration of the moon, there were boundless questions concerning its origin, history, composition, and surface characteristics. Today, many of the questions have been answered, but, in turn, new ones have arisen. Certainly, the issue of life on the moon has been resolved. No major life forms have been discovered and there has been no hint even of micro-organisms. Tests conducted at the Lunar Laboratory at the Johnson Space Center found no traces of organic matter. These tests failed to detect any fossil organisms or residual molecular building blocks of living matter, such as amino acids and nucleotides, except in amounts so small that it was uncertain whether contamination by a technician's fingerprint might have been responsible for them (Jastrow & Thompson, 1975).

The conclusion that the lunar surface contains no signs of life was, in one sense, incorrect. Since the arrival of Apollo 11, signs of life abound. There are plaques, figurines, flags, spacecraft sections, four-wheeled vehicles, and, of course, thousands of footprints. Well before these relics disappear under the relentless onslaught of micrometeorites, new visitors will appear—to stay longer and perhaps permanently. That lunar stations and lunar colonies will exist someday is inevitable, for the pressures to understand and to use the richness in the rest of our solar system are growing. A NASA Study Group (Johnson & Holbrow, 1977) concluded "Space colonization appears to offer the promise of near limitless opportunities for human expansion, yielding new resources and enhancing human wealth. The opening of new frontiers, as it was done in the past, brings a rise in optimism to society."

The comprehensive studies of the physical characteristics of the moon, which accelerated with the Apollo data, are a necessary step toward increased use of our closest neighbor. Detailed know-ledge of its resources and its environment will be invaluable when planning begins for a manned lunar work station.

The moon, like the Earth, has a crust, a mantle, and very likely a molten core. The moon's outer layer, or crust, is almost four times thicker than that of Earth (Weaver & Bond, 1973). Beneath the crust is a mantle extending apparently to a depth of almost 965 km (600 mi), with temperatures as high as 1650° C (3000° F). The center of the moon contains an internal heat source, a molten core, which extends inward the final 772 km (480 mi). This core differs from the Earth's iron-nickel core and probably derives its heat from decaying radioactive elements.

An important finding in Apollo was that the entire surface of the moon, including the maria and highlands, is covered with basalt. Basalts are created on the Earth only by the cooling of molten lava. Thus, this single finding from Apollo indicates that the surface of the moon was entirely molten at some point in its past (Jastrow & Thompson, 1975).

The moon's surface today is far from molten, as thousands of Apollo photographs have shown (Figure 1). The surface is covered by a thin layer of dust, which extends well up the sides of the many peaks. Astronaut David R. Scott (1973) wrote that "A dark-grey moon dust—its consistency seems to be somewhere between coal dust and talcum powder—mantles virtually every physical feature of the lunar surface. Our boots sink gently into it as we walk; we leave sharply chiseled footprints" (Figure 2).

Figure 1. An Apollo 17 astronaut stands next to a huge boulder on the lunar surface. The Lunar Rover is at the left.

Figure 2. Astronaut on the surface of the moon, with the Lunar Module at the left.

Numerous craters dot the surface of the moon. The largest of these is Mare Imbrium, a crater about 1050 km (650 mi) across, formed through a cataclysmic impact with a huge meteor some four billion years ago. Other craters, resulting from the continuous shower of meteorites, are so small as to be barely noticeable. All of these testaments to celestial violence remain virtually unchanged over millions of years, as the earthly erosive forces of wind and water are not found on the moon.

It was long believed that the moon contained no atmosphere at all. This was in large measure verified by the Apollo findings, although tiny amounts of the gases argon, neon, and helium were detected. Much of this is believed to come from the solar wind. This lack of atmosphere has a number of implications for the earthly visitor. First, there is no absorption of solar radiation. Temperatures near the equator range from about 110° C (230° F) at lunar noon to as low as -178° C (-290° F) just before dawn. These extreme changes place great demands on life support systems. The lack of atmosphere also affects the visibility of objects, since there are no atmospheric particles to scatter light on objects which are shadowed. Such objects are not entirely dark, however, since the lunar surface itself serves to reflect a certain amount of light.

The moon also differs from Earth in that it has virtually no magnetic field, although existence of a weak magnetic field may be detected in lunar rocks. This is interpreted as showing the existence of a lunar magnetic field at some time in the remote past. It might be that the liquid core of the moon, if mostly iron, could have acted as a dynamo early in the history of the moon to create a magnetic field.

The moon by most objective standards is hostile to human residence. Certainly a lunar visitor requires total protection from its environment. Yet this celestial neighbor seems uniquely appealing to man. Astronaut Scott, in preparing for his departure from the moon in Apollo 15, felt that "In a peculiar way, I have come to feel a strange affection for this peaceful, changeless companion of the Earth."

Venus

Venus, together with Earth, Mercury, and Mars, is known as a terrestrial planet, being composed of rocky materials and iron and having a density quite similar to that of Earth. Passing 39 million km (24 million mi) from Earth at its point of closest approach, Venus is Earth's nearest planetary neighbor. It represents a logical candidate for exploration missions within the solar system.

The first successful mission to Venus took place in 1962, when the U.S. Mariner 2 spacecraft flew to within 35,000 km (22,000 mi) of the planet and radioed back information concerning the near-environment of Venus. One of the more interesting of these discoveries was that Venus, for all practical purposes, has no surrounding magnetic field. Since 1962, the Soviet Union, with its Venera series, and the United States, with its Mariner and Pioneer programs, have sent 17 spacecraft to Venus.

Most of the information obtained in the American exploration of Venus has been through the Pioneer program. In December 1978, the Pioneer-Venus spacecraft arrived in orbital flight around Venus. While the bus spacecraft remained in orbit, four entry probes were launched toward the Venusian surface. The gross topology of Venus was defined by a radar mapping system in the orbiting craft (Figure 3). The probes provided much information on the structure of the atmosphere and more detailed surface effects. One probe transmitted for 67 minutes from the surface of Venus before heat rendered it ineffective.

Figure 3. This global view of Venus was taken by the radar altimeter of the Pioneer spacecraft, so that the planet's cloud cover is eliminated. The large feature at the top of the picture is Venus' northern "continent," Ishtar.

The Soviet Venera spacecraft were the first to accomplish a landing on the planet's surface. The Venera 9 vehicle transmitted black-and-white photographs to Earth on 22 October 1975, showing a Venusian landscape with considerably more light than had been expected under the heavy cloud cover and with sharp-sided rocks indicating little of the expected erosion effects. More recently, in March of 1982, Venera 13 and 14 landed and transmitted photographs taken by imaging systems that permitted color reconstruction (Figure 4). Both landers also succeeded in analyzing samples of soil scooped up from their landing sites.

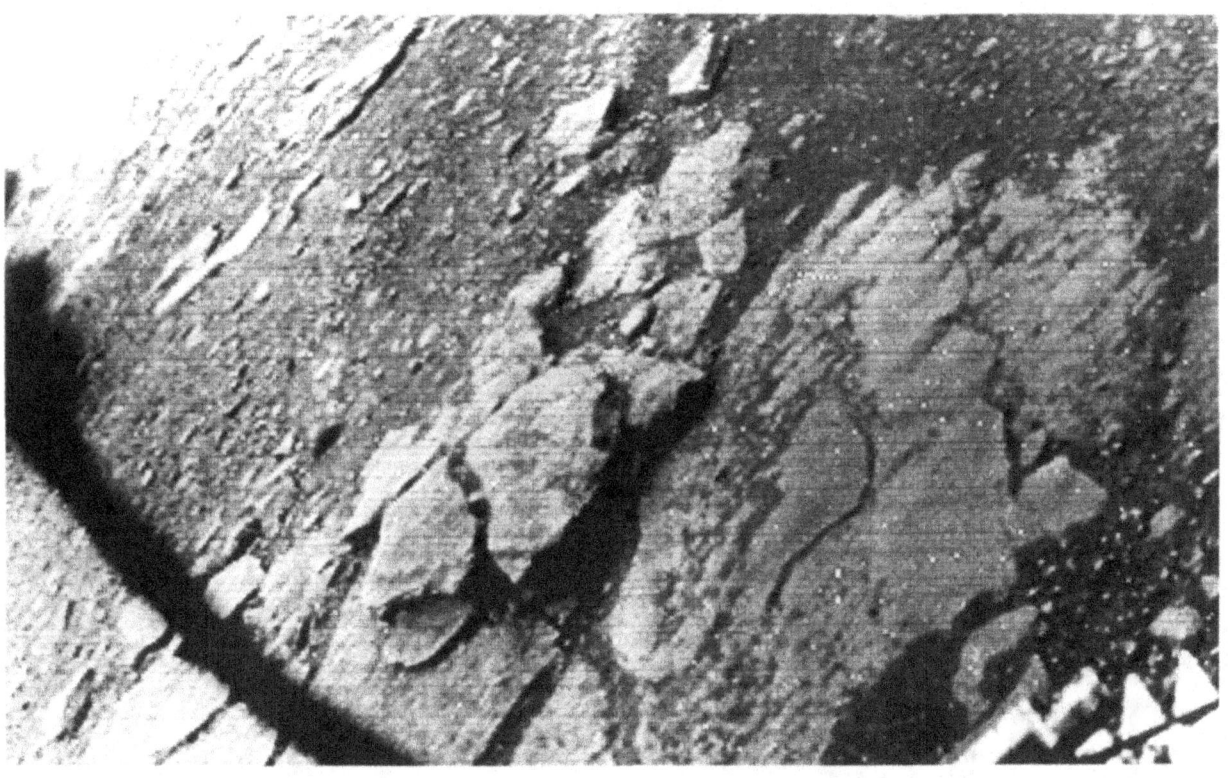

Figure 4. Photograph of the surface of Venus taken by Venera 13.

Although Venus has a number of features similar to Earth, such as its size and density, there are some striking differences. The Venus atmosphere is nearly 90 times more massive than that of Earth, and its predominant gaseous component is carbon dioxide. Sulphuric acid clouds encompass the entire planet, whereas the water clouds on Earth only partially cover the planet (Hoffman et al., 1980). Water vapor and sulphur dioxide have been detected in the lower atmosphere. Nitrogen also is present, on the order of three percent by mass. However, the large volume and density of the atmosphere means that the amount of atmospheric nitrogen there is about three times greater than on Earth. The planet's surface temperature is extremely high—more than 450° C (842° F). Material analyzed by Venera 13 and 14 was shown to be basalt, an igneous rock similar to material found on volcanically active mid-ocean ridges on Earth. This finding suggests recent or ongoing volcanic activity on Venus.

Three-fourths of the solar energy that impinges on Venus is reflected back into space by the planet's atmosphere and clouds, with 60 percent of the remainder absorbed in the heavy cloud layer. This layer consists chiefly of one- to three-micrometer-sized particles that are believed to be sulphuric acid.

One of the more interesting findings during the earlier probes of Venus was evidence of lightning activity in the lower atmosphere. Both the U.S. and Soviet data indicate that lightning discharges may occur as often as 25 times per second in relatively small areas above the Venusian surface. Terrestrial lightning, on the other hand, occurs only on the order of 100 times per second over the entire Earth. The electrical characteristics of Venusian storms, therefore, may be both different and more severe than those to which we are accustomed.

It was also learned that Venus has no planetary magnetic field to shield it from the solar wind. As a result, the Venusian ionosphere reacts strongly with the stream of particles from the sun. Both ion density and the height of the top of the ionosphere are affected by solar wind speed and pressure.

Another strange feature of Venus is that its rotation is much different than that to which we are accustomed, resulting in a day of 243 Earth days and a year of 224 Earth years. Thus a day on Venus is longer than a Venusian year.

The picture of Venus pieced together from the various probes demonstrates that, while orbital observations certainly would be possible, surface characteristics are not conducive to manned exploration of this planet. The great heat and barometric pressure at the surface and the composition and turbulence of the atmosphere make the planet quite inhospitable to human life. A very sophisticated life support system would be required to allow surface expeditions. Venus is not a planet that beckons visitors.

Mars

Mars is the closest neighbor of Earth as one proceeds away from the sun. It also is a planet about which we have an unparalleled fund of scientific information, due mainly to data provided through Project Viking. By the beginning of the 1980s, the United States and the Soviet Union had sent 16 probes to Mars. The three most successful as planetary exploration missions were Mariner 9 and Viking 1 and 2.

The Viking program was a masterful blend of science and technology, for the first time carrying an automated scientific laboratory to a soft landing on the surface of another planet. The Viking 1 spacecraft landed on Mars on 20 July 1976, and was followed by Viking 2 on the 3rd of September. These spacecraft were each comprised of an orbiter-lander combination which traveled through space as a single unit, separating only after orbit had been achieved around Mars and the appropriate landing site selected. The orbiter served as a relay station to transmit to Earth information received from the lander. In addition, each orbiter took many thousands of high-resolution photographs of Mars as its orbit carried it over different areas of the planet. In combination, these four spacecraft have provided information concerning the Martian surface and its atmospheric environment which may be analyzed and studied fruitfully for many years.

The Martian atmosphere is totally unlike that found on Venus. For one thing, it is extremely thin. Pressures measured at various times by the Viking spacecraft were in the range from 6.5 to 7.7 millibars, a value less than one percent of the Earth's atmospheric pressure at sea level. Nevertheless, it is an active atmosphere. The Mariner 9 spacecraft, which orbited Mars in late 1971, photographed huge dust storms which obscured the surface for a two-month period until they quieted enough for photographs to be made of the surface. The primary constituent of the Martian atmosphere is carbon dioxide, just as on Venus. There also are nominal amounts of neon, argon, and oxygen.

The remarkable photographs transmitted by Viking have made the surface features of Mars familiar to us all (Figure 5). Mars has a very heterogeneous surface, with extensive cratering and

Figure 5. This view of the Martian surface and sky was taken by Viking 1. Color reconstruction showed that orange-red surface materials (possibly limonite) overlie darker bedrock, and that the sky is pinkish-red.

evidence of volcanos over vast areas of the planet. Much of the northern hemisphere is covered by volcanic fields. There is widespread evidence of catastrophic flooding, but no collection basins such as lakes or oceans have been found, and the source and sink of the water are still conjectural (Soffen, 1977). The only water found on Mars to date occurs in the polar ice caps. These caps, which contain substantial amounts of water ice, are seasonally covered by carbon dioxide frost and vary in size seasonally.

A major objective of the Viking project was to search for any indication of life on Mars. A number of photographs of the surface were taken, some at close range, to determine if any organism—large or small—could be seen. Results were negative.

A more scientific approach to the detection of life was made through use of an automated laboratory carried in the Viking landers. Samples of Martian soil were used for three experiments in the Biology Instrument Package. The first tested whether the presence of moisture and appropriate environmental conditions would produce metabolic activity in simple, pre-biological organic complexes that might be present in the soil in a dormant state. The second experiment assumed that, since both carbon dioxide and carbon monoxide are found in the atmosphere of Mars, organisms might have developed a capacity to assimilate one or both of these gases and convert them to organic matter. The third experiment tested the assumption that Martian organisms would be capable of decomposing the simple organic compounds reported to be produced in the so-called "primitive reducing atmospheres" in laboratory simulations (Klein, 1977).

The results of the experiments were of principal value in suggesting the presence of highly reactive oxidizing agents in the Martian soil. The general conclusion is that the Martian surface is very different from Earth in terms of its chemistry, and that this difference seriously affects the interpretation of biology experiments. The bulk of the evidence from these experiments favors a non-biological explanation of the observed findings.

Mars remains an attractive candidate for further exploration. It has a thin atmosphere mostly comprised of carbon dioxide, and a surface which could be traversed readily by exploration vehicles. The soil resembles that on Earth but apparently has an intriguing increase in oxidative qualities. Of particular importance is the water supply contained in the large polar ice caps. This resource would be invaluable to an exploration team.

Jupiter

As one travels somewhat more than 386 million km (240 million mi) past Mars and on toward the outer solar system, the next planet to be encountered is Jupiter, the first of the giant planets. Jupiter is the largest planet, with a mass 318 times larger than the mass of Earth. However, its density (1.33 g/cc) is only one-fourth that of Earth, and just slightly greater than the density of water.

The United States has launched four successful fly-by missions to Jupiter. The first of these, Pioneer 11, reached its closest point to the planet (130,000 km) on 3 December 1973. The most recent, Voyager 2, passed within 650,000 km of the planet on 9 July 1979. The two Pioneer and the two Voyager spacecraft returned a wealth of scientific information plus thousands of photographs of Jupiter. Although much of the interior of the planet remains a mystery, the outer surface of Jupiter, its satellites, and conditions in the space surrounding it are now well-documented. This information is contributing much to an understanding of the processes whereby the solar system was formed.

Jupiter is a gas planet, more massive than all the other planets combined. It has no surface, in the usual sense, and many scientists believe that Jupiter is an entirely fluid planet, with no solid core whatever (Morrison and Samz, 1980). However, recent studies postulate that the planet may have a small core of rocks and ice that constitutes about four percent of its mass (Ingersoll, 1981).

In many respects, Jupiter resembles a star; in fact, had it been 70 times more massive, it would have contracted during the time of its early formation into a star. Had this occurred, the sun would have been a double star and the solar system would be much different. Since 1969, it has been known that Jupiter radiates more heat than it receives from the sun. Thus, Jupiter must have an internal heat source. The internal heat is believed to represent the conversion of gravitational potential energy from the contraction of a giant cloud of gas beginning some 4.6 billion years ago (Ingersoll, 1981). The surface heat, about 10^{17} watts of power, flows from the dynamic processes within its luminous interior, believed to be about $30,000^{O}$ K. Jupiter is composed of the same elements as the sun and stars, primarily hydrogen and helium. Most of its interior is metallic hydrogen held under enormous pressure, with normal molecular hydrogen appearing nearer the surface. In the upper regions the hydrogen is a gas.

The great mass of Jupiter and its powerful gravity mean that all gases and solids available during its early stages of condensation should remain today. Jupiter therefore has the same basic composition as the sun, with hydrogen and helium being the two principal constituents. However, there is an abundance of additional elements and compounds.

The atmosphere of Jupiter is a mass of complex motion. Some of the atmospheric changes are quite fleeting; others, such as the Great Red Spot, last relatively unchanged for centuries. The dominant features in the atmosphere of Jupiter are banded belts and zones, the Great Red Spot, and three white ovals (Morrison and Samz, 1980) (Figure 6). The Great Red Spot is larger than the Earth, while the ovals are about the size of our moon. All show anti-clonic, counterclockwise motion. Materials within the Great Red Spot can be seen to rotate about once every six days.

Within the Jovian atmosphere, clusters of lightning bolts have been seen on the night side of the planet. One Voyager photograph showed the electrical discharges of 19 "superbolts" of lightning. This evidence of extensive electrical activity confirms the disturbed condition of the atmosphere of Jupiter.

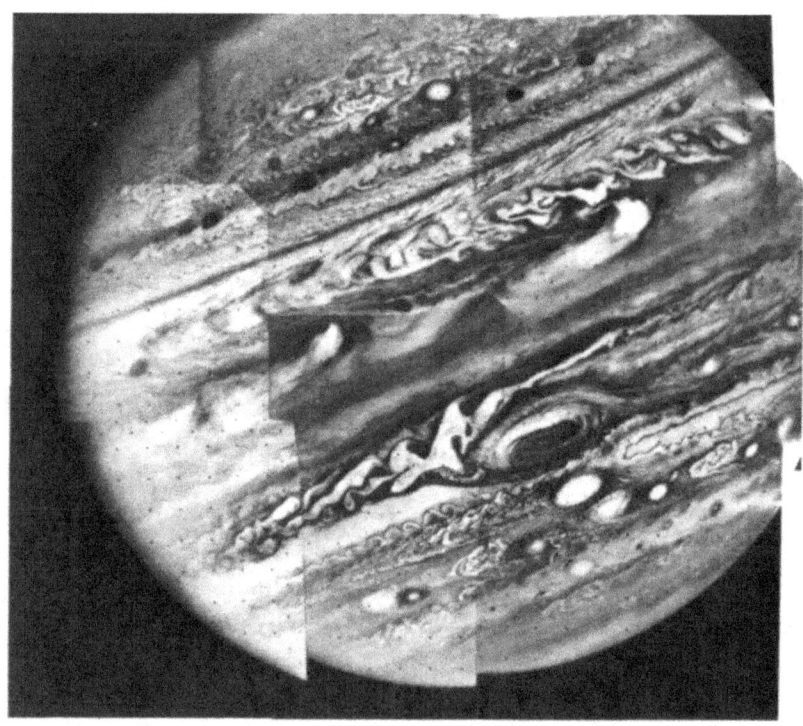

Figure 6. This mosaic of Jupiter was assembled from nine individual photographs taken by Voyager 1. Shown are the Great Red Spot, banded cloud features, and several white ovals.

The similarities between the weather systems of Jupiter and Earth are matched by likenesses in the surrounding magnetic fields. It is presumed that the interior pressures of Jupiter are so great that hydrogen becomes an electrical conductor. The rotation of the planet thus causes a current to flow through the metallic core and to produce a surrounding magnetic field. However, the strength of the Jupiter field is about 4,000 times that found around Earth. A key effect of this field is to trap atomic particles arriving as part of the solar wind. The boundaries of the magnetosphere, in the direction toward the sun, lie between four and eight million kilometers from the planet. Within the magnetosphere, charged particles can be accelerated to high energies, with some subsequently escaping and being encountered far from Jupiter. On Voyager 1, one stream of hot plasma was encountered almost 50 million kilometers from the planet. Inside the magnetosphere, there are certain "hot spots." Voyager 1 also detected such a plasma about five million kilometers from Jupiter and measured its temperature at 300-400 million degrees (Morrison & Samz, 1980). This is the highest temperature encountered anywhere within the solar system. (It should be noted that the low particle densities found in such a stream make the concept of temperature almost meaningless from a physiological standpoint.)

One of the most spectacular features of the Voyager missions was the photographs taken of the principal satellites of Jupiter. Jupiter now is known to have 14 satellites, with the four largest being termed the Galilean moons, in consideration of their discovery by Galileo in 1610. These moons (Io, Europa, Ganymede, and Callisto) are considered "terrestrial" space bodies. They are similar to the planets of the inner solar system, including Earth, in both size and composition. Io, the innermost Galilean satellite, gained a measure of fame when it was found, by chance observation, to have an active volcano. A more careful search then revealed that there were no fewer than eight active volcanos on Io throwing up plumes from 70-300 km (Soderblom, 1980). The next satellite, Europa, was found to be crisscrossed by stripes and bands that may represent filled fractures in the satellite's icy crust. It is believed that water, in solid and liquid form, constitutes about 20 percent of Europa's mass. The next satellite, Ganymede, is the largest of Jupiter's moons. A most noticeable feature of this moon is an immense dark area, the remnant of an ancient crust showing the impact of many meteorites. Finally, there is Callisto, at a distance of 1.8 million kilometers from the planet. Callisto is about the size of the planet Mercury and shows the effects of billions of years of cratering. It also shows a number of concentric rings which encircle the satellite. These rings are believed to have been formed dynamically by the impact of a very large body early in the development cycle of the satellite.

Future explorations of Jupiter, particularly if there is any requirement for close observation, will almost certainly be through unmanned space probes. The planet itself is most interesting, since it retains many features from the early development period of the solar system. It also possesses a system of satellites which are quite different from one another and which should offer unique insights into the dynamics of early planet development. However, the region itself is not hospitable. There is an immense band of trapped radiation surrounding Jupiter. The planet itself is extremely hot toward the interior and offers no reasonable landing site. Any conceivable manned mission would have to be to one of the satellites.

Saturn

The rings of Saturn are without doubt one of the most spectacular features in the solar system. These rings, which were first observed by Galileo in 1610, are the distinguishing characteristics of Saturn. Observations from Earth could easily distinguish three of the rings but provided little insight into the ultimate number and complexity of these rings until close observations were made by American spacecraft.

The Pioneer 11 spacecraft accomplished a successful fly-by mission to Saturn on 1 September 1979. Considerable scientific information was returned, including a number of images from a photopolarimeter, which achieved up to 20 times the resolution provided in Earth-based photographs. A major accomplishment of Pioneer was its demonstration that a spacecraft can safely cross the ring plane of Saturn. While Pioneer was struck at least five times during the encounter by particles at least ten microns in diameter, there was no real threat to its survival. Pioneer paved the way for the much more ambitious Voyager flights which followed immediately.

61

The missions of the Voyager 1 and 2 spacecraft provided thousands of startling photographs of the features of Saturn (Figure 7). Voyager 1 made its closest approach to the cloud tops of Saturn on 12 November 1980. Thirty-eight months after launch and nearly a billion miles from Earth, Voyager 1 passed the giant planet at a distance of 200,000 km. The Voyager 2 encounter with Saturn occurred on 25 August 1981, at a distance of less than 100,000 km. Each spacecraft photographed the planet, its rings, and its satellites. Passage of each spacecraft carried it through an area of the ring structure.

Figure 7. This is a montage of Saturn and its principal moons. In the forefront is Dione. Tethys and Mimas are at the right of Saturn. Enceladus and Rhea are off the rings to the left, and at the upper right is Titan.

The surface of Saturn is covered by dense clouds, preventing any direct observation. However, much is known of the structure of the planet. Saturn is a gas planet, like Jupiter. Its mass is 95 times that of Earth. Recent estimates indicate that Saturn may have at the bottom of its gaseous and molten levels a solid core which could constitute about 25 percent of its mass (Ingersoll, 1981). The outermost layer of the planet has a liquid mixture of hydrogen and helium, with the hydrogen molecular in form. Beneath this is a metallic liquid layer of hydrogen. The core is presumed to be made of rock and ice. It also is known that Saturn has a source of internal heat, presumably from the conversion of gravitational potential energy.

The atmosphere of Saturn is similar to that of Jupiter, although muted by a thick haze layer. It contains dark belts, white-banded zones, and circulating storm regions. Maximum wind speeds are at the Equator and can reach speeds of about 1,600 km/h. Temperatures near the cloud tops range from -193° C to -182° C. While auroral emissions have been seen near the poles, lightning has not yet been observed. Saturn also has been found to have a red spot, approximately 11,263 km (7,000 mi) in length, which resembles Jupiter's. This relatively stable spot is believed to be the upper surface of a convective cell (Smith, 1980).

The rings of Saturn are most fascinating. The particles which make up these rings are believed to be primarily rock and ice. They have been described as having the size and appearance of "dirty snowballs" (Smith, 1980). Voyager measurements indicate that particle size in the ring structure ranges from microns to meters, with all sizes in between. The well-known A, B, and C rings were found to consist of hundreds of rings or ringlets, a few of which are elliptical in shape. The F ring is more complex, and may consist of three interwoven rings which seem to be bounded by two "shepherding" satellites. It is believed that the "spokes" observed in the B ring may be due to fine electrically charged particles above the ring, perhaps resulting from lightning occurring within the ring.

The flight of Voyager 1 resulted in the discovery of three new satellites of Saturn, bringing the total to 15 at that time. Analyses of data from the Voyager 2 encounter, combined with information from the Voyager 1 flight, now brings the number of known Saturnian satellites to between 21 and 23. The two "possible" satellites were seen in only one observation each, so their orbits could not be confirmed. Scientists at the Jet Propulsion Laboratory are continuing their extensive review of Voyager 2 data and it is possible that additional satellites may yet be confirmed.

With the exception of one—the satellite Titan—all of the moons of Saturn are covered with water or ice and, in some instances, are composed mainly of water ice (Stone & Miner, 1981). For the most part, these satellites show evidence of heavy cratering through the years.

The most interesting of the moons is Titan. Voyager 1 passed within 7,000 km of Titan, the closest encounter between either of the Voyager spacecraft and any planet or moon. This was done in order to obtain as much information as possible. Titan is the second largest moon observed in the solar system, with a radius of 2,575 km. It is the only moon in the solar system with a measurable atmosphere, found to be largely nitrogen with lesser amounts of methane, ethane, acetylene, ethylene, and hydrogen cyanide. The atmosphere of Titan is three times as dense and ten times as deep as that on Earth. It also is quite cold, with a temperature near the surface of -182° C. The surface pressure is 1.6 bars, or 60 percent more than found at the Earth's surface. Titan's surface, which is not visible through the atmosphere, may be liquid methane or liquid nitrogen.

The real value of observations of Titan is that this moon, which in many ways resembles a primitive Earth maintained in a deep-freeze condition, may provide considerable information concerning the manner in which Earth has developed. Titan might serve as a natural laboratory within which to realistically study the interaction of environmental forces and chemical factors.

The Voyager missions have shown Saturn and its environment to be a complex and dynamic scene. There may be as many as 1,000 identifiable bands within its complicated ring structure. The known moons of Saturn now number at least 21. The planet has a strong magnetic field and an extensive magnetosphere containing energetic charged particles. It also has the only satellite in the solar system, Titan, known to have a measurable atmosphere.

In all, there is much to recommend future missions to Saturn as we attempt to understand better the nature and origin of our solar system. Indeed, planning is being done now for possible insertion of an instrumented probe into the atmosphere of Titan at some future date. It is entirely feasible for such a probe to be launched from a manned laboratory orbiting within the near-space of Saturn. A manned mission, allowing direct control over the data collection plan and the operation of onboard instrumentation, would provide a wealth of scientific information and would represent a tremendous stride in the exploration of the solar system.

Interstellar Space

Today, manned space flight beyond the solar system remains a topic for science fiction. There can be little doubt, however, that there will be a time—probably within the next 100 years—when serious planning will begin for manned missions beyond the farthest planets. Definitive data showing that the nearest stars have planetary systems similar to that of our sun certainly would serve to spur such planning.

Missions into interstellar space will require new technologies. The challenges in engine development will be tremendous. Lightweight engines capable of imparting a continuous accelerative force over a period of days or months will be necessary. For the life sciences, a completely regenerative life support system will be needed. Obviously, opportunity for resupply of any kind will be quite limited.

The interstellar environment within which a manned spacecraft will operate is sparse but by no means empty. There is considerable gaseous matter in the interstellar medium, although at densities substantially lower than the best vacuum achievable on Earth. Approximately 90 percent of this gas is neutral hydrogen, radiating at a characteristic wavelength of 21 centimeters and thus readily detectable. Nearly all of the remaining ten percent is made up of atoms of helium. Most of the hydrogen is believed to have been formed during the explosive events that occurred in the creation of the universe approximately 13 to 15 billion years ago. Some of the helium perhaps was formed from primordial hydrogen at the same time, with the rest being manufactured in stars and distributed through supernova explosions. The remaining one percent or less of interstellar gases consists principally of carbon, nitrogen, oxygen, aluminum, and iron. All of these, including still scarcer elements ranging to aluminum, were formed in nuclear reactions with stars.

In recent years, complex molecules such as cyanogen and formaldehyde also have been detected in interstellar space. These molecules are of considerable interest as possible precursors of living matter. They also demonstrate that, while much is being learned about the interstellar environment, a number of mysteries remain. For example, certain small, dense clouds of gas within the Milky Way appear to have substantial concentrations of formaldehyde molecules (Webster, 1977). Radio observations of these clouds show that the formaldehyde absorbs energy from its microwave background, rather than radiating energy. The formaldehyde appears to be about 1.7° K colder than the background. Since one would expect that over time the formaldehyde normally would be warmed to match the temperature of the microwave background radiation, some unspecified action is working with the gas clouds to keep the formaldehyde chilled. Whether this interstellar refrigerator, as well as other features yet to be observed, is of any consequence for manned missions remains to be determined. The finding of such effects does underscore the work remaining before such missions can be attempted.

References

Hoffman, J.H., Hodges, R.R., Donahue, T.M., & McElroy, M.B. Composition of the Venus lower atmosphere from the Pioneer Venus mass spectrometer. *Journal of Geophysical Research*, 1980, *85*(A13), 7882-7890. (Reprinted by American Geophysical Union, December 30, 1980).

Ingersoll, A.P. Jupiter and Saturn. *Scientific American*, 1981, *245*(6), 90-108.

Jastrow, R., & Thompson, M.H. *Astronomy: Fundamentals and frontiers* (2nd ed.). New York: John Wiley & Sons, Inc., 1975.

Johnson, R.D., & Holbrow, C. *Space settlements—a design study* (NASA SP-413). Washington, D.C.: U.S. Government Printing Office, 1977.

Klein, H.P. The Viking biological investigation: General aspects. *Journal of Geographical Research*, 1977, *82*(28), 4677-4680. (Reprinted by American Geophysical Union as *Scientific results of the Viking project*, 1977).

Morrison, D., & Samz, J. *Voyager to Jupiter* (NASA SP-439). Washington, D.C.: U.S. Government Printing Office, 1980.

Scott, D.R. What is it like to walk on the Moon? *National Geographic*, 1973, *144*(3), 326-329.

Smith, B.A. Voyager 1 finds answers, new riddles. *Aviation Week & Space Technology*, 1980, *133*(2), 16-20.

Soderblom, L.A. The Galilean moons of Jupiter. *Scientific American*, 1980, *242*(1), 88-100.

Soffen, G.A. The Viking project. *Journal of Geophysical Research*, 1977, *82*(28), 3959-3970. (Reprinted by American Geophysical Union as *Scientific results of the Viking project*, 1977).

Stone, E.C., & Miner, E.D. Voyager 1 encounter with the Saturnian system. *Science*, 1981, *212* (4491), 159-162.

Weaver, K.F., & Bond, W.H. Have we solved the mysteries of the Moon? *National Geographic*, 1973, *144*(3), 309-325.

Webster, A. The cosmic background radiation. In *Readings from Scientific American: Cosmology +1*. San Francisco, CA: W.H. Freeman & Co., 1977.

SECTION III

Spaceflight Systems and Procedures

CHAPTER 4

SPACE VEHICLES

The United States

During the first 20 years of manned space flight, the vehicles designed to house and protect man during his ventures into space evolved along a straight line. From the conical "tin can" of Project Mercury, through the Gemini program, and over the course of the Apollo and Skylab missions, the external configuration of the manned vehicle changed little. With increasing mission complexity, objectives, and duration, and with increases in crew size, the internal configuration and systems design of these capsules changed considerably—although even these changes were essentially elaborations upon the basic requirements for life support and instrumentation necessary for spacecraft control and task performance. Apart from the Skylab Orbital Workshop, which was designed for habitation rather than transportation, it was not until the advent of the Space Shuttle that American space vehicles underwent a fundamental change in design and appearance. The following discussion reviews the design of these space systems and the philosophy underlying their development in each of the programs.

Project Mercury

In the late 1950's, considerable importance was attached by U.S. leaders to being the first to place a man into orbit. Accordingly, the approach taken in Mercury was to use existing technology and off-the-shelf equipment in conjunction with the simplest design that would be reliable. All systems would be automated, with the astronaut functioning primarily as an observer and as a backup system in case manual control became necessary. The design requirements for the Mercury craft (Smith, 1981) were:

- To include an escape system which would remove the spacecraft from the launch vehicle in the event of a pre-launch emergency.

- Drag braking for reentry.

- A retrorocket system for de-orbiting.

- A water landing and recovery capability.

- Provisions for manual control of spacecraft attitude by the pilot.

Figure 1 (Grimwood, 1963) shows the configuration of the spacecraft and the escape system. The escape system was positioned over the manned capsule at launch, as shown in the left-hand portion of the figure. In the event of a prelaunch emergency, its escape rocket would propel the capsule away from the launch site; shortly after a successful launch, a tower jettison rocket would detach it from the capsule. At right in the figure is the actual spacecraft, including an antenna can, a recovery compartment containing the descent parachute system, and the habitable capsule. At the base of the capsule was a retropack containing rockets for separating the spacecraft from the launch vehicle and retrorockets for de-orbiting. This retropack was jettisoned just prior to reentry.

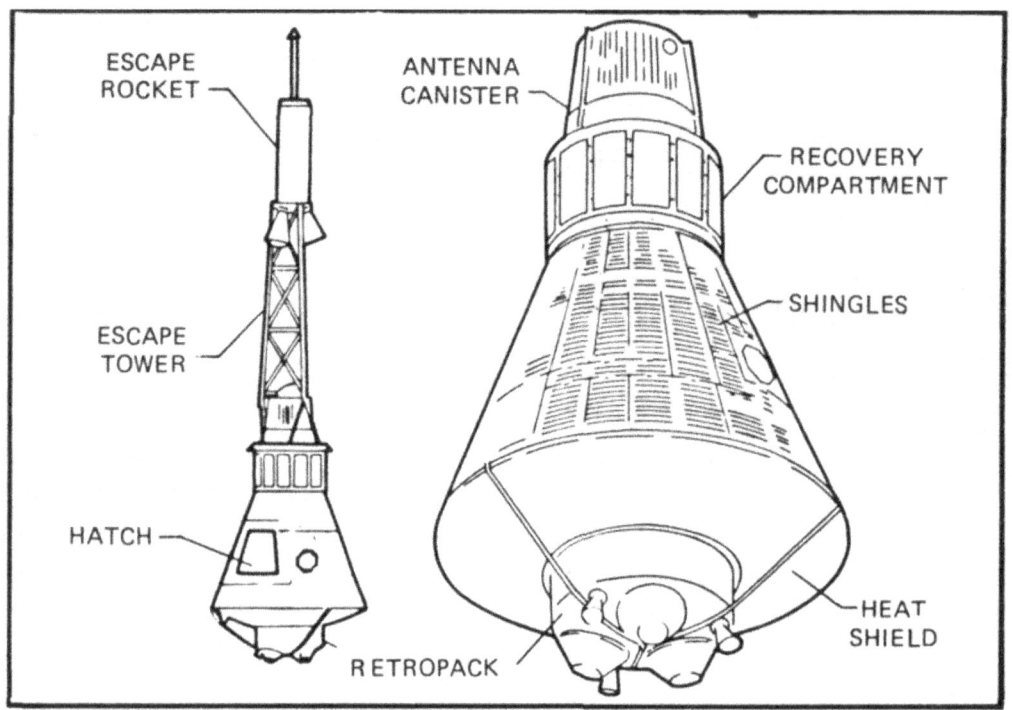

Figure 1. Mercury spacecraft and escape system configuration.

The bottom of the cone was covered with a heat shield composed of beryllium for the suborbital Mercury flights and ablating fiberglass for the orbital flights. Inside the capsule, the single astronaut sat upright in a couch (Figure 2). Ahead of him was the central section of a three-piece instrument panel and a trapezoidal window. The bottom portion of the central panel contained a periscope for direct observation of the Earth. Power was supplied by six batteries with 24 V DC output.

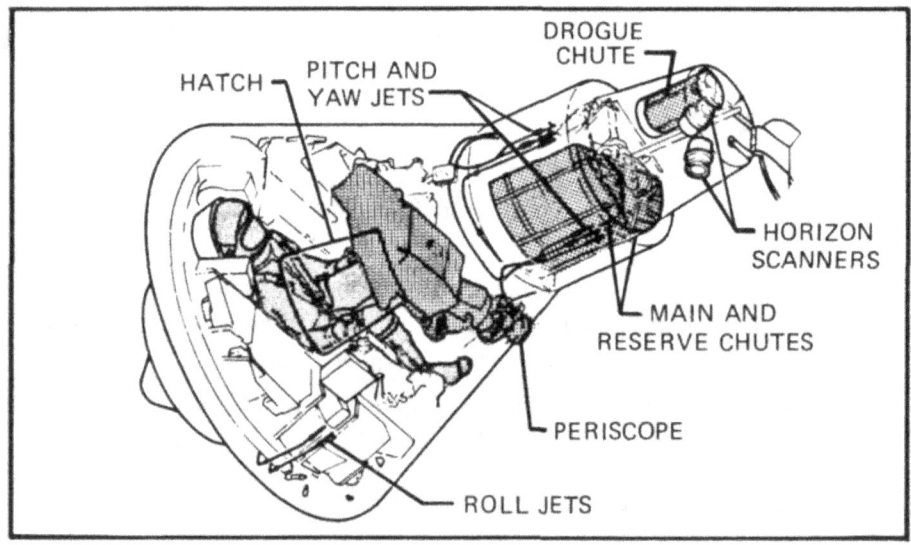

Figure 2. Mercury spacecraft interior layout.

The Mercury spacecraft could not change its orbit. However, provision was made for attitude control. Of the four modes of attitude control tested by Mercury astronauts, the most successful and most frequently used was manual fly-by-wire (FBW). A three-axis control stick was located at the astronaut's right hand.

During descent, a drogue parachute was deployed at 6,700 meters to slow the ship, followed by deployment of the main chute at 3,300 meters. The G-forces of splashdown were greatly reduced by a "landing-shock attenuation system" consisting of a fiberglass cushion which would fill with air during descent.

The first two manned launches in the Mercury program were suborbital flights. (The Soviets had already carried out an orbital mission before even the first of these flights.) The only significant mishap with either of these vehicles was the loss of Liberty Bell 7 after splashdown; due to premature blowing of a hatch, it filled with water and sank. Subsequently, four orbital Mercury missions were successfully flown, culminating in a maximum mission duration of over 34 hours and 23 orbits. During three of these missions, manual control of spacecraft attitude in space and during reentry was critical to mission success.

Gemini Program

The Gemini Program was undertaken in order to develop American manned spaceflight capabilities to the point at which a lunar landing (in the follow-on Apollo Program) would be possible. Therefore, its two most important objectives were: (1) to extend mission duration and (2) to develop and practice techniques and procedures necessary for rendezvous and docking. Secondary objectives were to achieve a capability for precisely controlled reentry and landing, to gain experience in extravehicular activity (EVA), and to improve the proficiency of flight crew and ground control operations (Smith, 1981).

The Gemini vehicle itself was an extension of the Mercury vehicle. There were two primary functional differences: the Gemini capsule was designed to carry two astronauts, rather than one; and Gemini greatly emphasized command by the crew over control by automated systems. Another major area of change was to make the spacecraft easier to maintain, in anticipation of multiple launches and an accelerated program timetable as the "race for the moon" developed. To this end, easily replaceable modular subsystem assemblies were employed which could be accessed from outside the spacecraft by several technicians at a time.

The Gemini craft was comprised of a reentry module, which included the crew cabin, and an adapter module, which was jettisoned prior to reentry (Figure 3). The reentry module, which was the forward or top section, was itself comprised of three separate sections. The first section (A) housed rendezvous and radar systems, parachutes, a UHF antenna, and docking mechanisms. The second section (B) contained the attitude control system used during reentry (not during orbital flight), and was one of the few fully automated systems aboard Gemini.

71

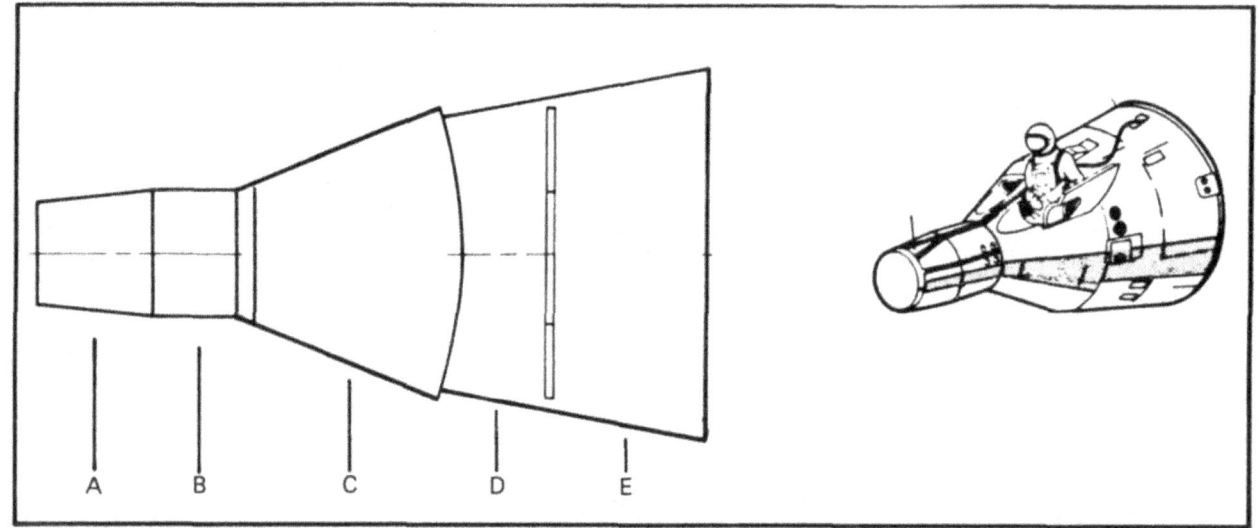

Figure 3. Gemini spacecraft configuration. (Functions of sections A-E are described in text.)

The third section (C) of the reentry module was the manned cabin, which was larger than the Mercury cabin in order to accommodate two passengers. It was a conical section with bases of 229 and 97 centimeters and a height of 190 centimeters. Above each crewmember's head was an EVA hatch which could be opened manually, with a window in each hatch (see Figure 3). An aft hatch beneath the crew couches opened into a compartment containing part of the environmental control system. An ablative heat shield covered the bottom of the reentry module.

The Gemini spacecraft's Adapter Module contained two sections as well. The first, or retro-grade section (D), housed four retrorockets for reentry or emergency abort during launch. It also contained six thrusters for the Orbital Attitude and Maneuver System (OAMS), which permitted orbital parameter adjustment (unlike Mercury). The second section, the equipment section (E), contained ten additional OAMS thrusters, fuel cell modules providing electricity and drinking water, part of the environmental control system (ECS), and supporting electronic equipment. The overall module was also conical in shape, with a 229-centimeter top, a 305-centimeter base, and a height of 229 centimeters. The entire Adapter Module was jettisoned just prior to reentry.

The eight translational thrusters of the OAMS were controlled by the commander using a maneuverable joy-stick. Attitudinal thrusters were controlled in manual modes by a side-arm stick mounted between the pilots. Overall, there were five separate modes for attitude, three manual and two automatic. Both control and guidance systems were considerably more sophisticated in the Gemini spacecraft than in the Mercury vehicle, since the spacecraft was designed for rendezvous and docking in orbit.

Like Mercury, the Gemini capsule used a parachute-assisted ocean landing. Gemini's parachute was somewhat different, however. Rather than impacting heat-shield down, the Gemini capsule hung by a two-point suspension from the parachute, with attachments at either end of the craft, so that it landed on its side in the water. A drogue parachute was deployed at 15,000 meters, with the main parachute deploying from the nose at 3,300 meters. At 2,250 meters the landing attitude switch was activated, shifting the ship to two-point suspension.

The extremely successful Gemini manned flight program gave the U.S. more flexibility in space operations and resulted in many important "firsts" in space. It also paved the way to the successful achievement of the objectives of the Apollo program. Between March 1965 and November 1966, there were ten manned Gemini flights. The successive buildup to a 14-day Gemini VII mission removed all doubts regarding the ability of spacecraft and crews to function in space long enough to carry out a lunar landing and return. Ten rendezvous were completed, and nine different dockings of spacecraft in orbit were achieved. EVAs were performed in five different Gemini missions, with total success in EVA achieved by Gemini XII. In addition, a wide variety of inflight experiments were carried out, laying the groundwork for extensive onboard research conducted in future programs (Mueller, 1967).

Apollo

Of all the scientific explorations ever undertaken by mankind, the Apollo Lunar Landing Program was the largest and most complex. It was mounted primarily with the goal of landing the first man on the moon, and this was accomplished in less than a year from the first manned flight test of the Apollo vehicle. The accelerated program was not, however, carried out without mishap. A flash fire aboard Apollo 204 (scheduled as the first manned Apollo flight) during prelaunch testing caused the deaths of three astronauts. An investigation was conducted which eventually resulted in a number of hardware and procedural changes to further ensure crew safety. To regain lost time, the planned inflight biomedical experiments were also cancelled.

A total of 11 manned Apollo flights were ultimately launched between October 1968 and December 1972, 6 of which put a total of 12 men on the moon. A program of Earth-orbital flights (the Apollo Applications Program) which were originally planned to fly concurrently with the lunar program was reduced in scope during the early 1970's, and eventually became the Skylab program. The Apollo manned spacecraft was used to transport crews in this later program, as well as in the even later Apollo-Soyuz Test Project flight (Smith, 1981).

The three-man Apollo spacecraft used in the lunar program was comprised of three modules: the Command Module and the Service Module (referred to in combined form as the Command and Service Module, or CSM), and the Lunar Module (LM). A Launch Escape System, similar to that used in the Mercury program, was also present. This overall configuration (Figure 4) was determined by the decision to adopt lunar orbit rendezvous (LOR) as the method for accomplishing the mission (Smith, 1981). The LOR concept involved launch to lunar orbit, separation of the LM from the CSM for descent to the lunar surface, and subsequent return of the LM to orbit for docking with the CSM.

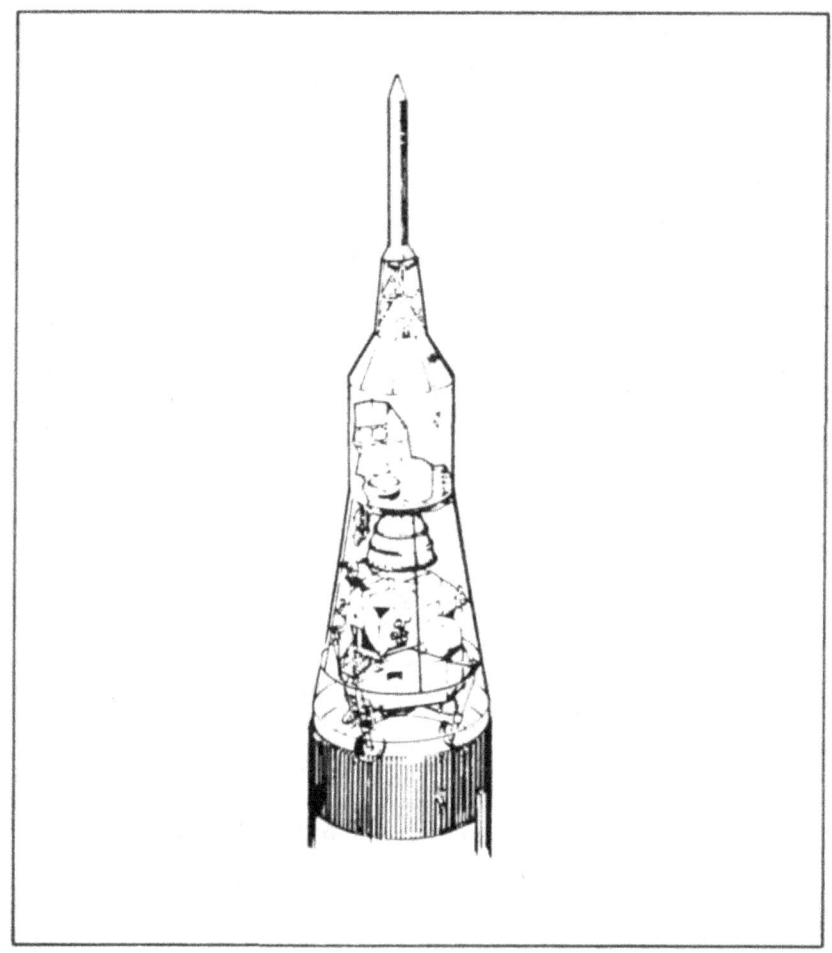

Figure 4. Configuration of the Apollo spacecraft at launch.

The Command Module (CM) was essentially a conical pressure vessel encased in a heat shield, as shown in the top half of Figure 5 (Johnston & Hull, 1975). It was 3.5 meters long, with a base diameter of 3.9 meters. There were three sections to the CM: a forward compartment contained reaction control engines and recovery parachutes; the middle section contained crew accommodations, controls and displays, and other spacecraft systems; the aft compartment housed additional reaction control engines and fuel, gas, and water storage tanks. Overall habitable volume was 5.95 cubic meters, more than four times as large as the Gemini vehicle. The three crew couches were positioned so that the seated crew faced the display console, toward the apex of the cone. The interior of the CM was divided into nine different equipment bays for compartmentalized stowage of the large amount of equipment, materials, and provisions. Two hatches were present, one at the side and one at the top of the capsule. After the Apollo 204 fire, the side access hatch was reconfigured to allow easier opening, particularly under one gravity. Five observation windows permitted extensive outside viewing and photography.

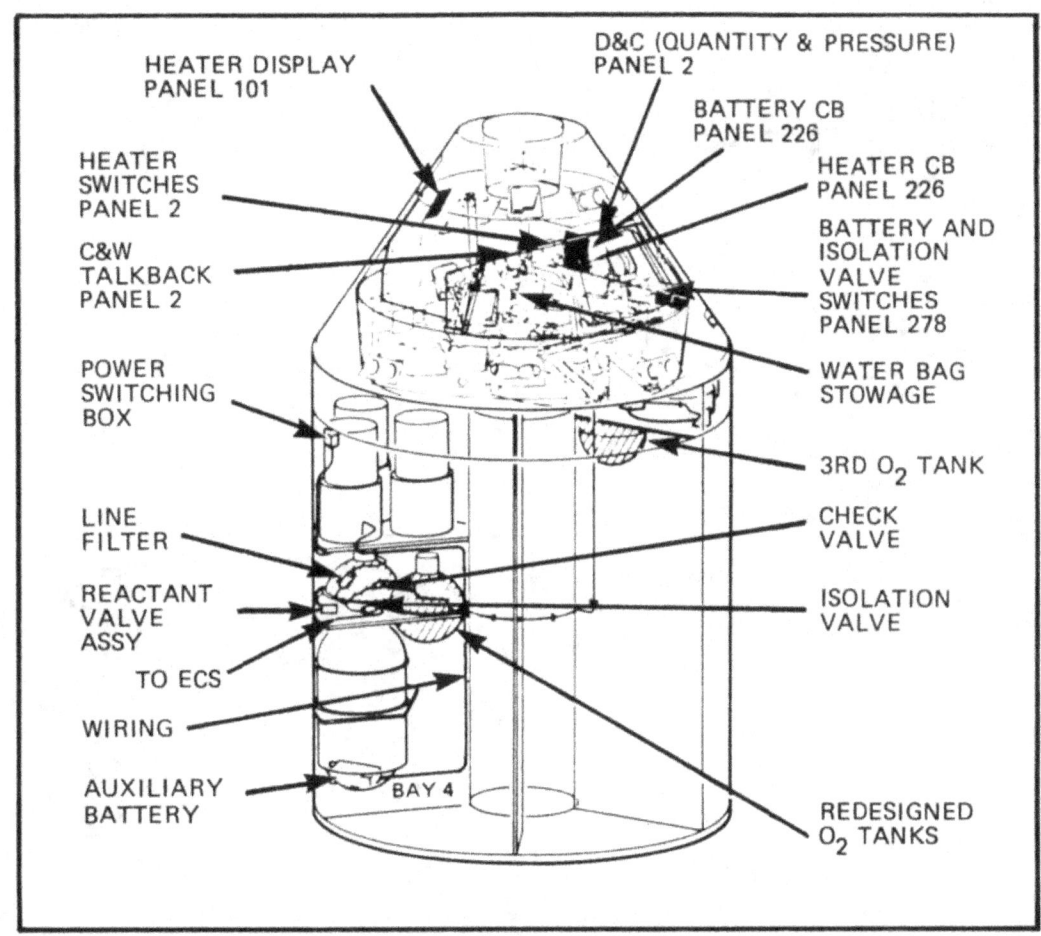

HEATER DISPLAY PANEL 101

D&C (QUANTITY & PRESSURE) PANEL 2

BATTERY CB PANEL 226

HEATER CB PANEL 226

HEATER SWITCHES PANEL 2

C&W TALKBACK PANEL 2

BATTERY AND ISOLATION VALVE SWITCHES PANEL 278

POWER SWITCHING BOX

WATER BAG STOWAGE

3RD O$_2$ TANK

LINE FILTER

CHECK VALVE

REACTANT VALVE ASSY

ISOLATION VALVE

TO ECS

WIRING

AUXILIARY BATTERY

BAY 4

REDESIGNED O$_2$ TANKS

Figure 5. Configuration of the Apollo Command and Service Module.

The Service Module (SM) was a cylindrical structure 3.9 meters in diameter by 6.9 meters long. Mounted below the CM (lower portion of Figure 5), this part of the spacecraft contained the main propulsion system and provided stowage for most of the consumable supplies. The SM remained attached to the CM from launch until just before Earth atmosphere reentry, when it was jettisoned. The service propulsion system was used for midtransit maneuvers and to reduce the velocity of the spacecraft before entering lunar orbit.

The Lunar Module (Figure 6) was a two-stage vehicle used to transport astronauts from the orbiting CSM to the lunar surface, provide living quarters and a base of operations on the moon, and return the crew to the CSM. It had an overall height of 7 meters and a diagonal width between landing gear of 9.5 meters. A Lunar Module Adapter provided an aerodynamic casing for the LM during launch, and was jettisoned shortly after the spacecraft left Earth orbit. Because it was designed to fly only in the vacuum of space, the LM itself was incapable of reentering Earth's atmosphere, and was returned in free-fall to the lunar surface.

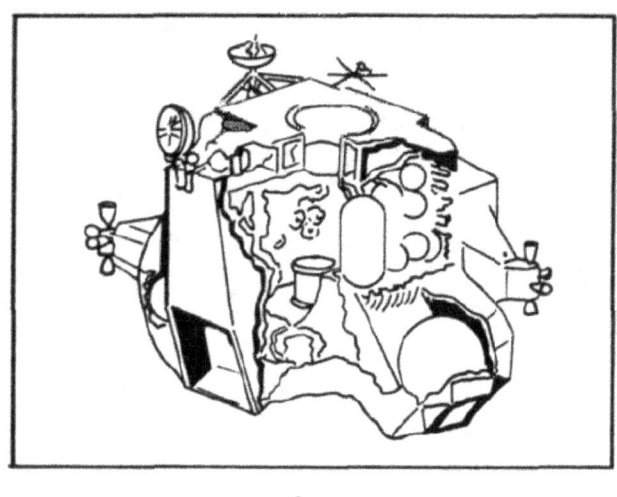

A

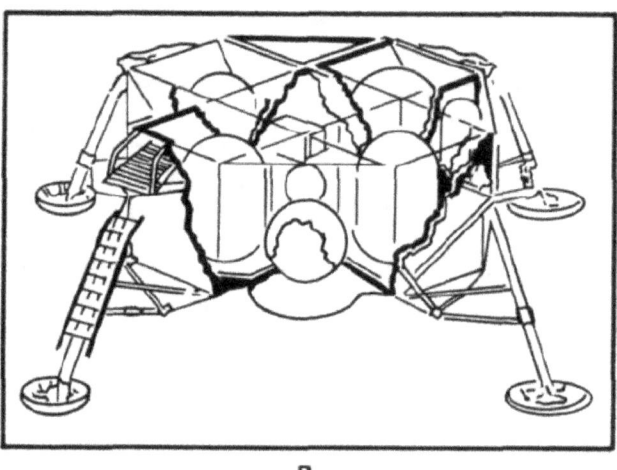

B

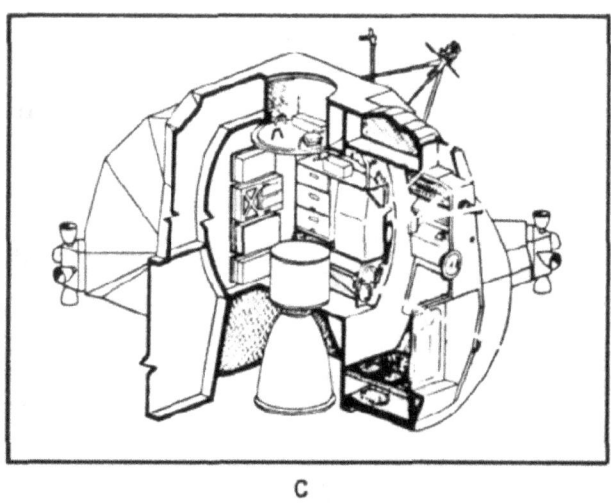

C

Figure 6. Lunar Module: A - ascent stage; B - descent stage; C - cabin interior.

The ascent stage of the LM (Figure 6-A) consisted of three sections: the crew compartment, the midsection, and the aft equipment bay. The crew compartment and midsection were pressurized. Habitable cabin volume was 4.5 cubic meters. The ascent stage was 3.8 meters long by 4.3 meters in diameter. Figure 6-C shows the interior of the Lunar Module cabin.

The descent stage (Figure 6-B) was the unmanned portion of the LM. It supported the ascent stage for the landing on the lunar surface, and contained the propulsion system used to slow the spacecraft for landing on the moon. During descent, four landing gear struts were unfolded to form the landing gear. Foot pads at the ends of the legs contained sensors which signaled the crew to shut down the descent engine upon contact with the lunar surface. Four bays surrounding the descent engine contained the propellant tanks, the Modularized Equipment Stowage Assembly (TV equipment, lunar sample containers, and portable life support systems), the Lunar Roving Vehicle (LRV), and the Apollo Lunar Surface Experiment Package (ALSEP).

Of these subassemblies, the LRV merits description here. Used for the first time on Apollo 15, the fourth lunar landing operation, this battery-operated vehicle doubled the traverse radius during lunar expeditions (Johnston & Hull, 1975). It was guided with a T-shaped hand controller, and was equipped with a television camera and a communications system which provided transmission of voice and biomedical/life support data. It had a lunar payload that was several times the vehicle's own Earth weight.

The only major operational problem encountered during any of the 11 manned missions that took place during this most ambitious of space programs was due to the explosion of an oxygen tank aboard Apollo 13, while enroute to the moon. A resulting loss of power in the CSM forced the crew to live in the LM while swinging around the moon to return home. The fact that no ships or crews were lost in space during this complex and accelerated program is a tribute to the design integrity of the vehicles as well as to the expertise of crews and controllers.

Skylab

As was mentioned earlier, one of the uses originally envisioned for the Apollo spacecraft was as an adjunct to advanced research and studies in Earth orbit. Various concepts for an orbital space station had been discussed in connection with this Apollo Applications Program. By 1969, the project had taken definitive shape: a Saturn IVB rocket stage would be outfitted as a workshop on the ground, with solar panels for power supply, and an external Apollo-Telescope Mount (ATM) for conducting solar observations. In 1970, as support for this and other NASA programs was scaled down, the Apollo Applications Program was renamed Skylab (Smith, 1981).

In manned orbital operation, Skylab consisted of five components: the Apollo ferry craft, the Orbital Workshop and, connecting the two, an Airlock Module and Multiple Docking Adapter; a fifth component was the Apollo Telescope Mount. These components are shown in launch configuration atop a Saturn 5 booster in the left-hand portion of Figure 7, and in fully deployed orbital configuration at the right (NASA, 1973).

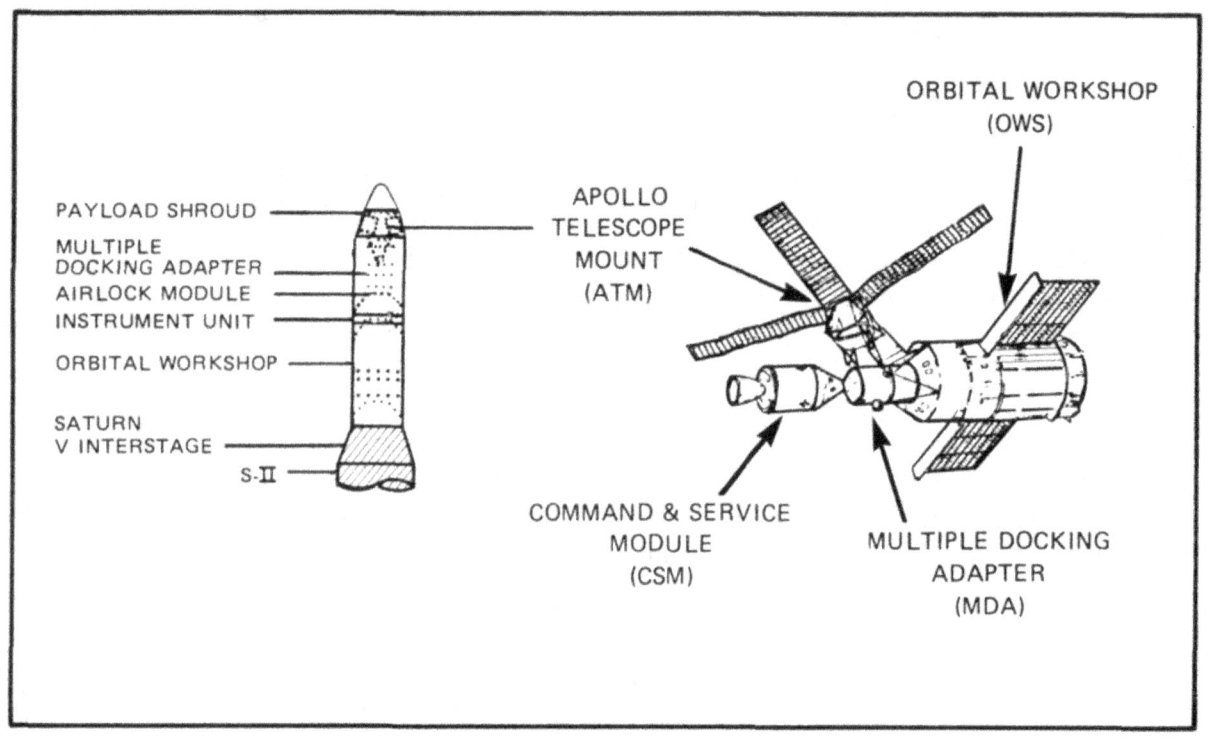

Figure 7. Skylab configuration at launch (left) and in Earth orbit (right).

The Apollo transport craft was identical to the lunar CSM described in the preceding section. The Skylab Orbital Workshop (SOW), which was the heart of Skylab, was 14.6 meters long and 6.7 meters wide, with a habitable volume of nearly 275 cubic meters (Smith, 1981). On Earth, the fully equipped workshop weighed 35,400 kilograms. Enveloping the SOW structure was a thin, aluminum meteoroid shield intended to absorb micrometeoroid impacts and to shade the workshop from direct solar radiation. (This shield broke off during launch, and was later replaced in orbit by an umbrella-like structure.) Internally, the SOW consisted of two major sections: an upper compartment where large-scale experiments were staged, and which contained two scientific airlocks; and a lower compartment containing areas for food preparation and eating, sleeping, waste management, and an experiment work area.

The habitation compartment's food system and provisions for waste management were highly inventive and efficient, and provided the most comfortable and normal living environment possible. The food system consisted of a wide variety of foods in frozen, thermostabilized, and freeze-dried form, and facilities for preparation and consumption. Approximately one ton of food was stored in the SOW at launch; packaged in six-day supply increments, it was moved as needed to the galley area for preparation and eating. The galley contained a freezer, a food chiller, hot and cold taps, and attachments for both trays and diners. Food trays had accessed openings (some of which had warmers) for holding food cans. Both food and water containers were designed for use in zero gravity (Johnston, 1977).

The Skylab Waste Management System included equipment for the collection, measurement, and processing of urine and feces as well as for the management of "household" trash and garbage. Feces were individually collected into a bag beneath a commode seat. The material was then weighed, labeled, processed, and stored by the crewman. Urine was collected in an individual 24-hour collection bag. The volume of the bag was regularly estimated, and every 24 hours a sample was removed and frozen for postflight analysis. Trash was discarded through an airlock into a holding tank. Other provisions for personal hygiene included a shower contained in a collapsible cloth bag; each crewman took one shower per week in this device (Johnston, 1977).

Three long-term three-man missions were flown to Skylab in 1973 and 1974, lasting, respectively, 28, 59, and 84 days. The third and longest mission set a spaceflight duration record that was not broken until 1978. Skylab's orbit eventually decayed, and the station reentered over western Australia in July 1979.

Space Shuttle

The Space Transportation System (STS) consists of three major components (shown in Figure 8): the Orbiter, an external tank (ET), and two solid rocket boosters (SRBs). The Orbiter is the actual spacecraft, and is designed to carry up to seven crewmembers and the payload into Earth orbit and return. The SRBs and the ET are part of the propulsion system which boosts the Orbiter into space (Luxenberg, 1981).

The Shuttle Orbiter

Figure 9 shows a more detailed diagram of the Orbiter vehicle. It is comparable to a DC-9 in size, and weighs approximately 68,000 kg. The major components and subsystems are described in Table 1.

The Orbiter is an extremely versatile vehicle. It is designed for vertical launch, but assumes horizontal flight for an unpowered aircraft-type approach and landing after reentering Earth's atmosphere. It contains crew accommodations for up to seven; although during an emergency the Orbiter can carry as many as ten persons. The orbital stay capability is for up to eight days, with the potential for extension to 30 days when special power modules are developed. The Orbiter can interface with a variety of different payloads and support many kinds of payload functions. For example, it can support Spacelab (the primary facility for life science experiments) and the space telescope, and it can accommodate up to five satellites. The payload bay contains one manipulator arm as standard equipment, which can be used to retrieve satellites or place them into orbit. A second arm can be installed and controlled if required by the mission.

Figure 8. The Space Transportation System (Space Shuttle).

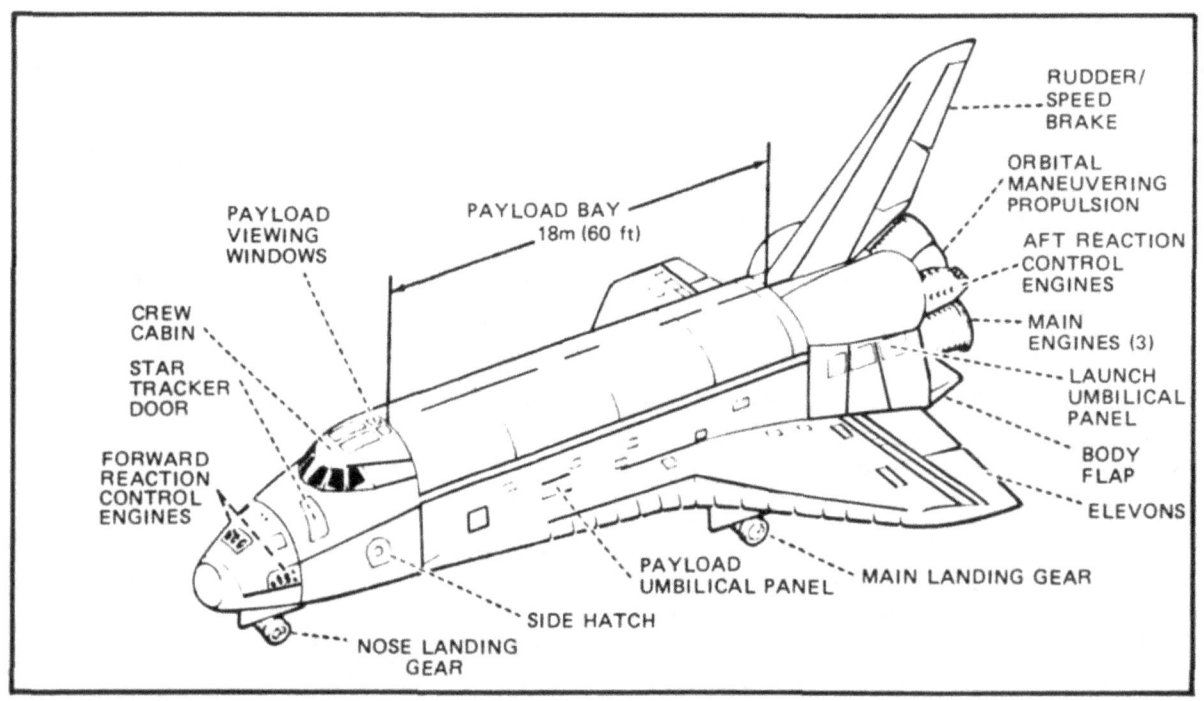

Figure 9. Configuration of Space Shuttle Orbiter.

Table 1
Components and Subsystems of the Orbiter

Orbiter Component	Principal Functions
Crew Cabin	A two-level cabin which provides seating and living accommodations for seven crew and passengers. During emergencies, the crew cabin can accommodate as many as ten people.
Payload bay	Designed to contain one of several payloads (such as the Spacelab, the space telescope, up to five individual satellites, or satellites with an additional propulsion stage for insertion into higher Earth orbit or deep space); payload bay doors are controlled by crew.
Orbital Maneuvering Subsystem (OMS)	Provides the thrust required to perform orbit insertion, orbit circularization, orbit transfer, rendezvous, and deorbit, after separation of the external tank. The propellants (MMH and N_2O_4) are contained in two pods, one on each side of the aft fuselage.
Reaction Control Subsystem (RCS)	Three modules (one in the forward fuselage and two in the aft) containing a total of 44 thrusters, fueled by MMH and N_2O_4 in separate RSC tanks. This propulsion subsystem provides attitude control of the spacecraft during orbit insertion, reentry, and on-orbit.

The aluminum hull of the Orbiter is covered with thermal materials to protect the spacecraft from solar radiation and the extreme heat of atmospheric reentry. Two types of reusable surface insulation—coated silica tiles and coated flexible sheets—cover the top and sides of the vehicle. The coatings on both types of insulation give the Orbiter an off-white color with optical properties that reflect solar radiation. Silica tiles with a high-temperature coating cover the bottom of the Orbiter and the leading edge of the tail. These tiles are glossy black in appearance, and provide protection against temperatures up to 1,260° C.

Fuel Tank and Boosters

The external tank contains the propellants for the Orbiter main engines. Fluid controls and valves for operation of the main propulsion system are located in the Orbiter. The ET is jettisoned when its fuel is exhausted, shortly after the separation of the SRBs, and is the only non-reusable portion of the Space Transportation System. The two solid rocket boosters are attached to the ET at launch; they provide additional initial ascent thrust. After their fuel is exhausted, they separate from the ET and Orbiter and deploy a system of self-contained parachutes. When the SRBs hit the water, the parachutes are jettisoned and the rockets are towed back to port by recovery ships.

Crew and Passenger Accommodations

The Orbiter cabin has a volume of about 71 cubic meters and consists of three levels. The upper level, or flight deck, contains the displays and controls used to pilot, monitor, and control the Shuttle vehicle and payload. The mid-deck contains payload specialist/passenger seating, a living area, accommodations for hygiene and sleeping, a galley, an airlock, and avionics equipment compartments. The lower deck contains the environmental control equipment, and is accessible from the mid-deck by removable floor panels.

Like Skylab, the Shuttle Orbiter is equipped with food, food storage, and food preparation and dining facilities sufficient to support the mission in progress. Facilities can accommodate missions ranging from two to seven crewmembers. Food is available in a choice of dehydrated, thermo-stabilized, irradiated, intermediate-moisture, natural, and beverage forms. The following food service facilities are provided: a water dispenser of ambient-temperature and chilled water for drinking and food reconstitution; a small, portable food warmer that can simultaneously warm meals for four crewmembers; and food trays with restraints for food items and accessories (Figure 10). A meal galley is carried on the majority of flights. The galley is a multi-purpose facility, mounted on the mid-deck floor and designed for airliner-like efficiency, which provides centralized food preparation facilities and storage of accessories. Figure 11 shows the food rehydration unit for operational Shuttle missions. The galley provides hot and cold water, and includes a pantry and an oven. It also includes a Personal Hygiene Station, used for washing up. A work/dining table doubles as a dining surface during meals and a work table during orbital operations.

The Orbiter's Waste Collection System (WCS) is, likewise, an integrated, multi-functional system designed to collect and process biowastes from male and female crewmembers in both zero gravity and one gravity. The system is used as a standard Earth-like facility, and is designed to perform the following general functions:

- Collecting, storing, and drying fecal wastes, toilet paper, and emesis collection bags

- Processing wash water from the Personal Hygiene Station

- Processing urine

- Processing water from the Extravehicular Mobility Unit in the airlock

- Transferring collected fluids to waste storage tanks in the waste management system

- Venting air and vapors from trash storage units

- Transfer of waste water overboard, as a contingency.

The WCS station is located on the mid-deck of the Orbiter in a small compartment. The unit itself is roughly cubical, with dimensions of 69x69x74 centimeters, and has separate assemblies for handling fluids (a cup-and-tube urinal) and solids (a commode). Various controls, filters, and fan separators are common to both assemblies.

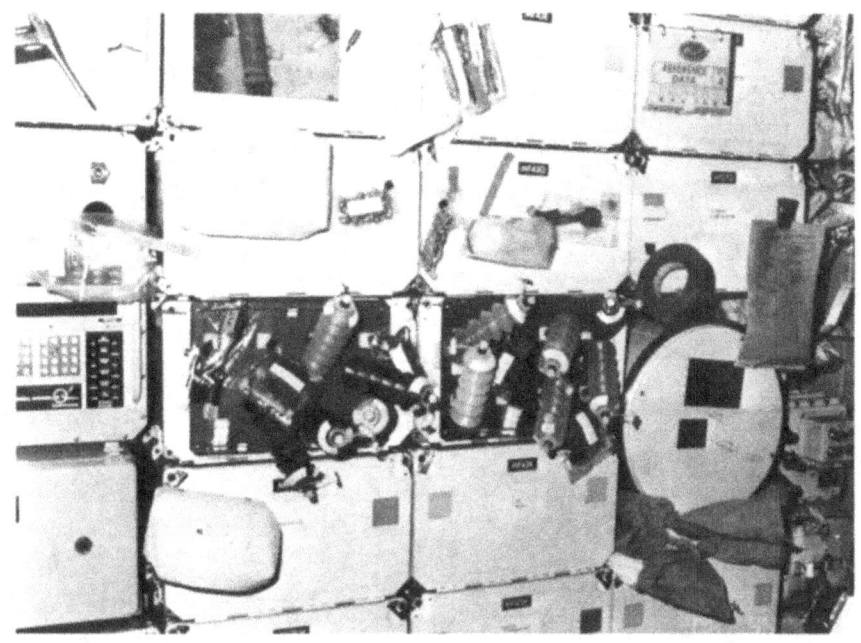

Figure 10. Fruit juice containers and other meal items are shown fastened to food trays and locker doors in the galley area of the Orbiter.

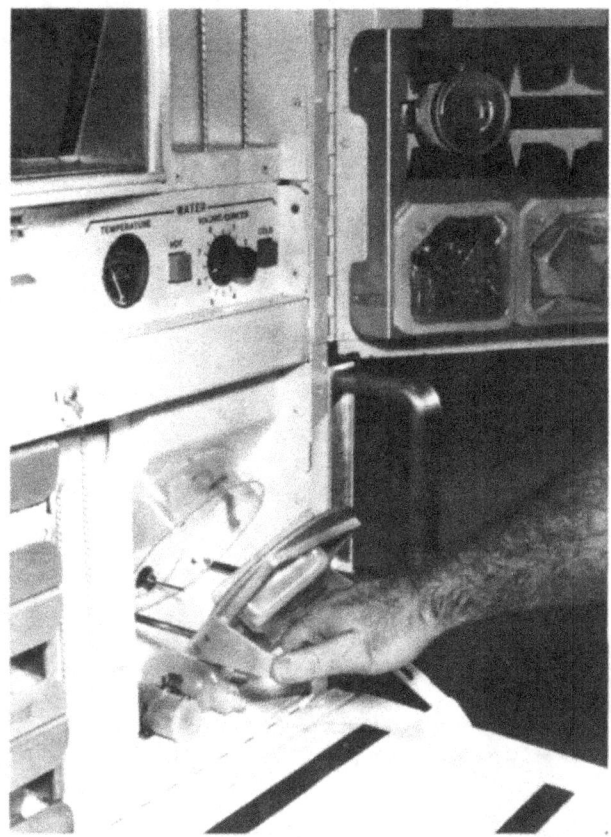

Figure 11. A view of the food rehydration unit used in meal preparation onboard the Orbiter.

Spacelab

The prime experimental facility for the Space Transportation System is Spacelab, constructed under the direction of the European Space Agency (ESA). Components of this modular facility can be carried in the payload bay of the Orbiter for experimental investigations in such fields as Earth observation, materials science, astrophysics, or life science. Spacelab has been designed for maximum flexibility to reduce the cost of performing experiments in space, and to meet the needs of scientists in many disciplines. Figure 12 shows a schematic drawing of one configuration of Spacelab contained within the payload bay of the Orbiter.

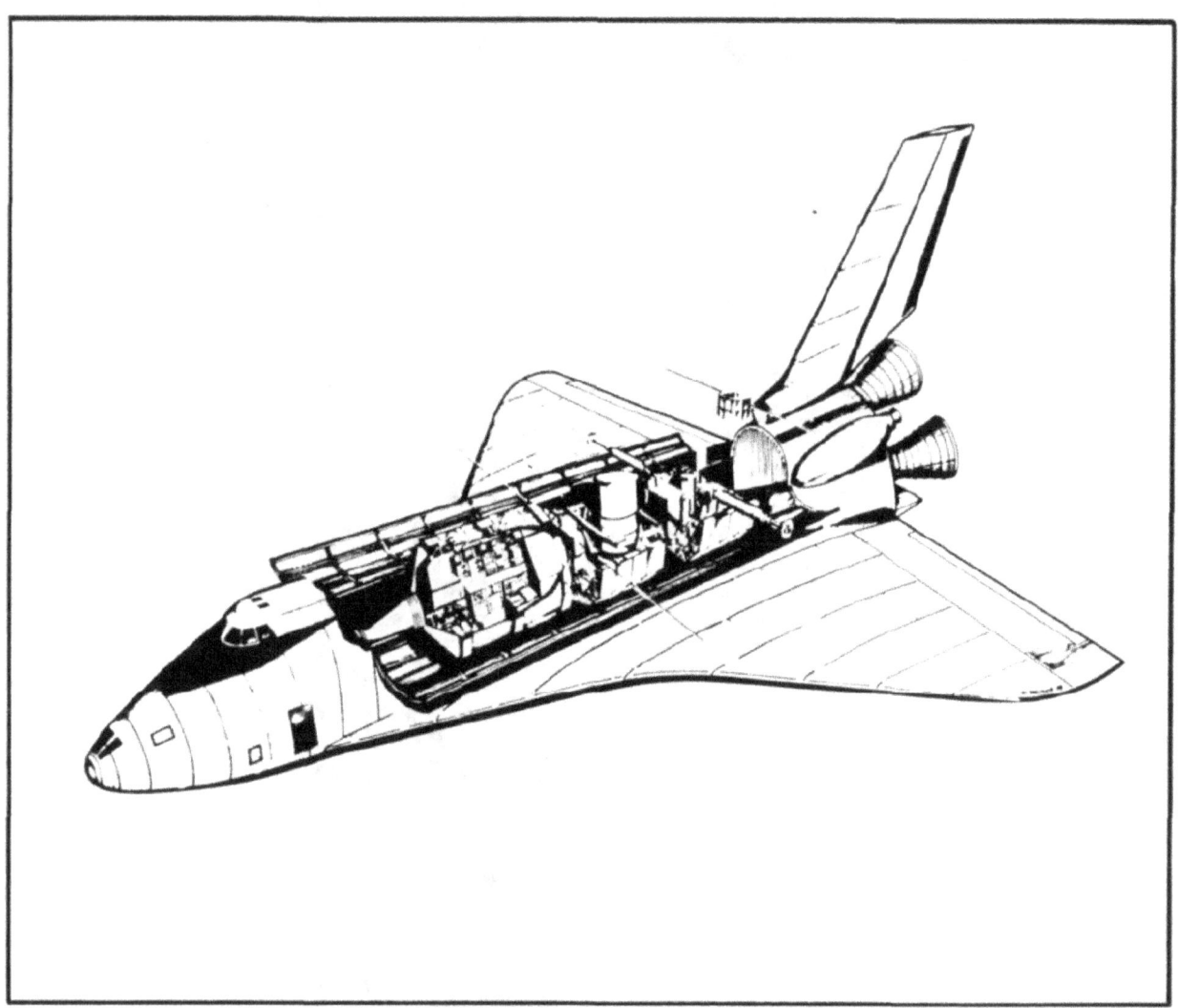

Figure 12. The Orbiter and Spacelab.

Description of Spacelab. The concept of Spacelab is to provide a versatile laboratory in which experiments can be conducted as they are on Earth. The two major components are called modules and pallets.

A module provides a habitable volume of variable size for the crew, biological specimens, and laboratory equipment. This area is accessible from the Orbiter cabin through a tunnel. Depending on the requirements of the mission, a module can consist of either one (short module) or two (long module) segments, as shown in Figure 13. The forward segment of the long module (and the only segment of the short module) is called the Core Segment, and contains the Spacelab electrical

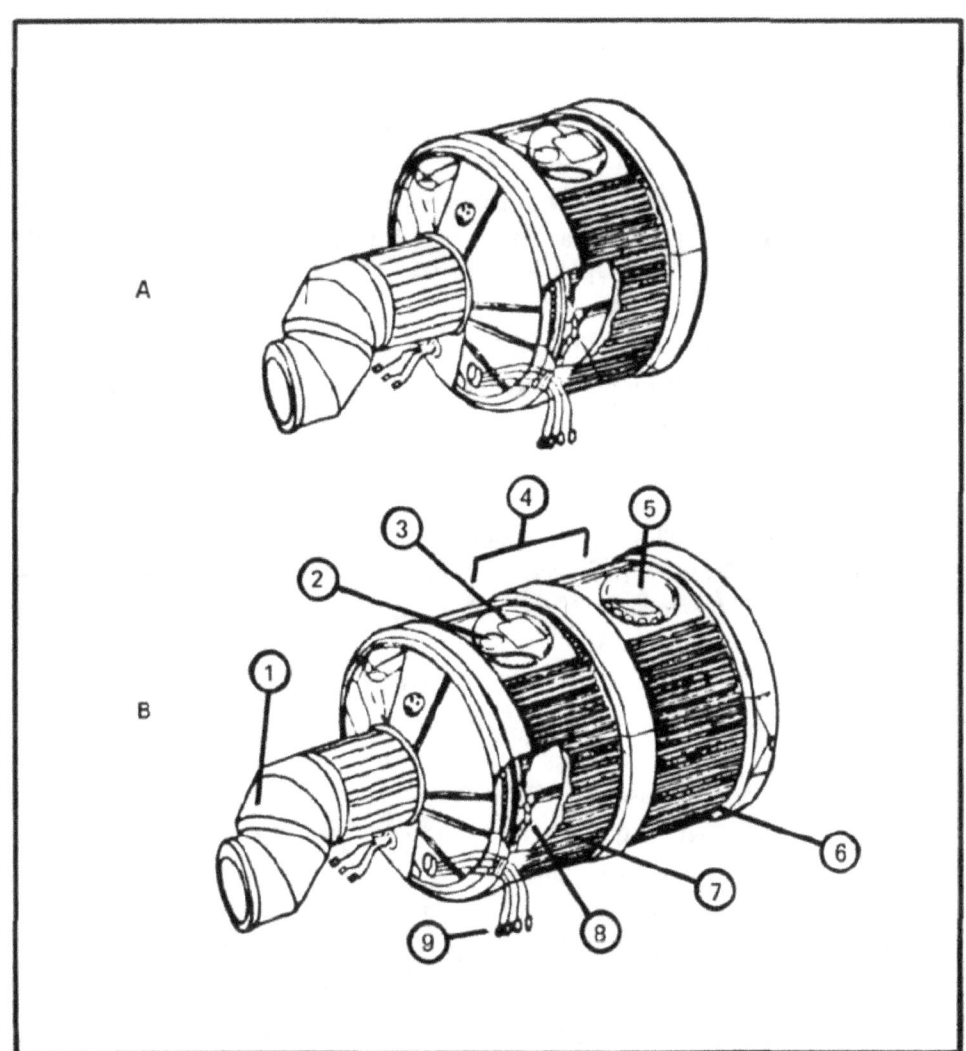

Figure 13. A. Short module (core segment only). B. Long module (core and experiment segments): 1—tunnel, 2—viewport, 3—optical window, 4—module, 5—airlock, 6—experiment segment, 7—core segment, 8—Orbiter attach fittings, 9—utility interface.

power, environmental control, and command and data management subsystems. Approximately 60 percent of the Core Segment volume is available for experimental equipment.

Pallets are rigid, U-shaped structures, 2.3 meters long, on which experiments and equipment can be mounted for direct exposure to the space environment. Pallet segments may be carried separately if no module is required for the mission. Figure 14 shows a diagram of a pallet mounted behind a long module.

When the Orbiter is carrying both a module and pallets, the Spacelab subsystems inside the Core Segment of the module provide the necessary power, data management, and other house-keeping equipment for the experiments mounted on the pallets. When only pallets are carried, these subsystems are contained in a cylindrical "igloo," shown in Figure 15.

The modular approach employed in the design of Spacelab permits a large number of configurations to meet the needs of specific missions (Figures 16 and 17). A long or short module can be carried with or without pallets, and pallet segments can be carried without a module. When the mission requires only a small number of Spacelab components, the remainder of the Orbiter's cargo bay can be used to carry other payloads, such as satellites.

The interior of the Spacelab module has been designed for maximum flexibility. It contains standard racks for experimental equipment, each of which is provided with power and data interfaces and cooling subsystems. Equipment can be stored in the ceiling containers and under the floor.

Possible Uses of Spacelab. Spacelab offers a radically new approach to space activities and can be used in a number of different ways in various scientific disciplines. From the first flights it can be used as an observation platform, a research laboratory, or a test bed for equipment designed for use in space. Eventually Spacelab may be used as a production facility to manufacture materials such as crystals, alloys with new properties, high quality optical supplies, and improved vaccines.

In order to reduce the number of conflicting requirements and maximize the efficiency of the time spent in orbit, most Spacelab missions will focus on a cluster of related disciplines that have similar operational and equipment requirements. For example, Spacelab micro-gravity missions dedicated exclusively to life science investigations are planned for launch every 18 months. For these missions, the Spacelab module is equipped with appropriate laboratory equipment and specialized support systems, and carries from 15 to 30 life science experiments. In addition to the dedicated Spacelab missions, life science experiments can be carried as mini-labs on multidisciplinary missions, or as self-contained carry-on experiments requiring a minimum of crew time.

The space environment is unique, and the STS offers enormous opportunities to perform investigations in this environment in a cost-effective way. As the prime experimental facility for the STS, Spacelab provides scientists with a valuable and versatile laboratory in which information may be acquired that is impossible to obtain on Earth.

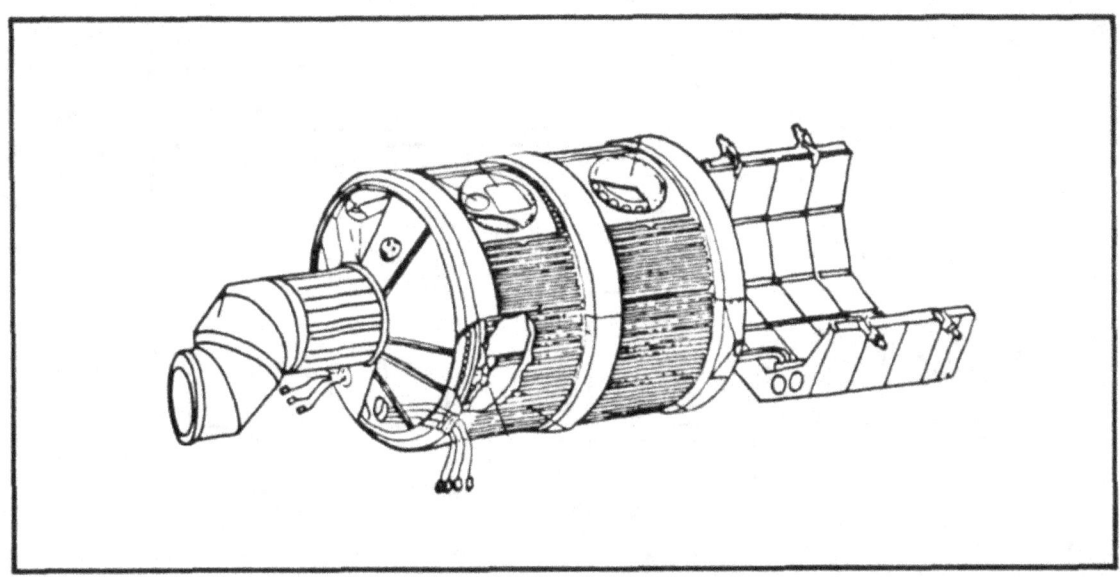

Figure 14. Long module with pallet.

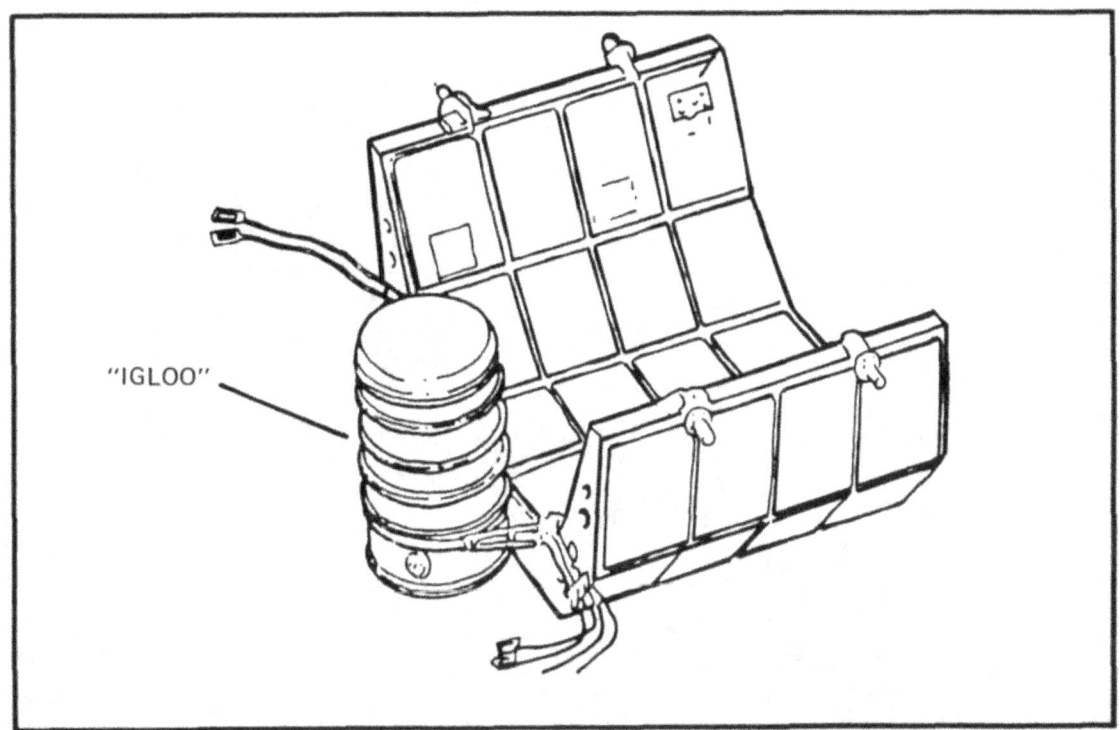

"IGLOO"

Figure 15. Pallet with cylindrical "igloo" housing subsystems.

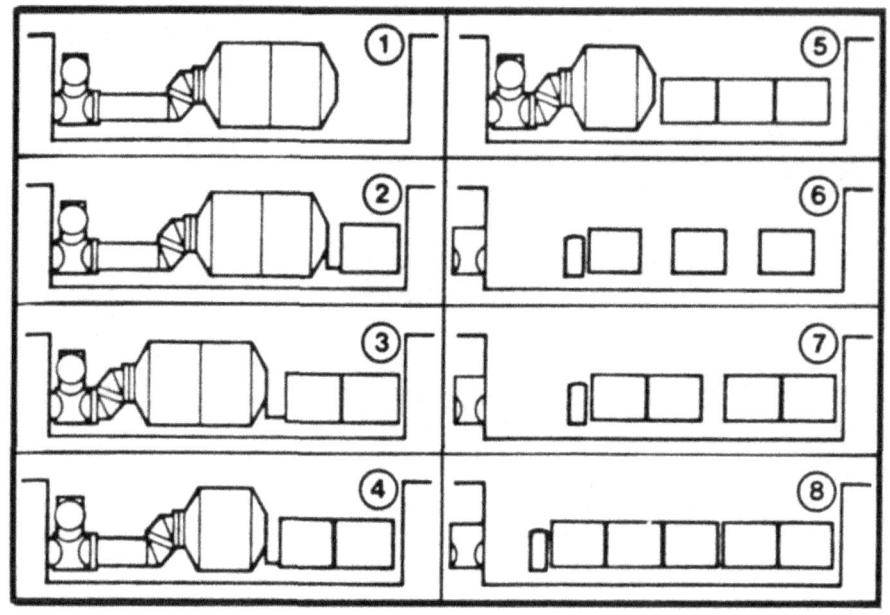

Figure 16. Basic configurations for Spacelab.

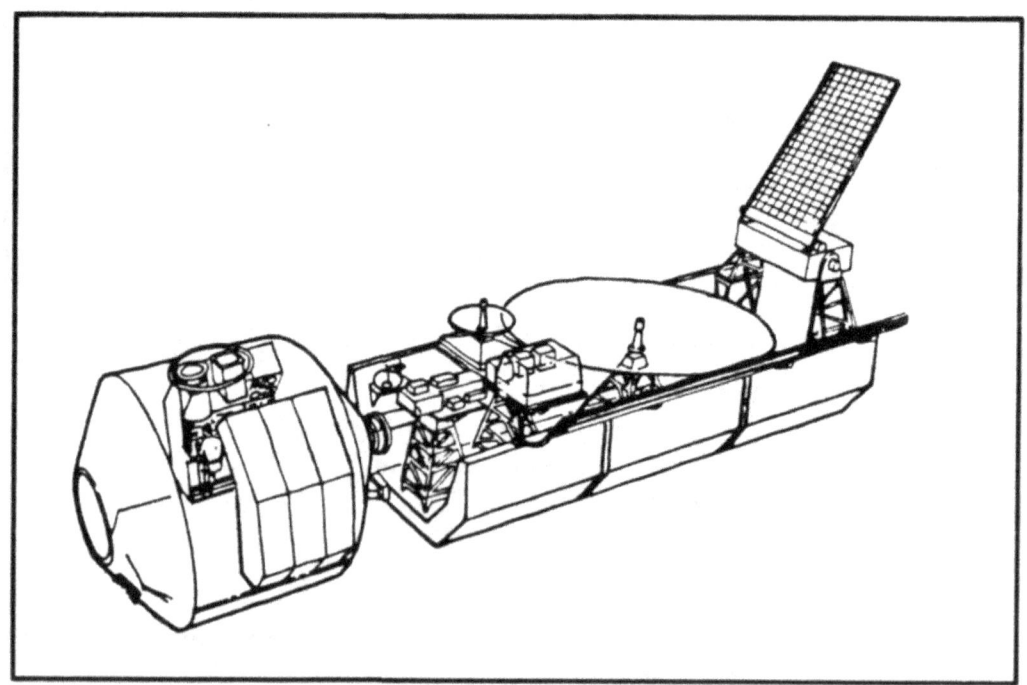

Figure 17. Typical configuration for an Earth observation mission.

The Soviet Union

Although their early achievements in space led many Western observers to conclude differently, the Soviet approach to space vehicle development and to the advancement of their space program in general has always been conservative, deliberate, and consistent. Space scientists and technicians have been able to convince the political leadership, which often has an engineering background, of the economic necessity of pursuing a large-scale program of space exploration and applications. The political advantages are self-evident to all concerned. Therefore, space milestones are represented as prominently in successive Five-Year Plans as are, for instance, agricultural goals. The economic investment in this program is quite substantial, and probably surpasses the U.S. expenditures by a considerable margin. The overall approach is toward the development of large space habitats utilizing the best standardized vehicle prototypes.

Vostok/Voskhod

These two programs of space flight, representing eight missions in all, constituted a "reconnaissance" of space. They served as trial runs for many of the basic activities to be conducted in space, and as a checkout of man's reactions to the space environment. The Vostok and Voskhod vehicles can be spoken of jointly, since the principal differences between them were the removal of an ejection seat to provide room in Voskhod for a three-man "shirtsleeves" crew and, in Voskhod 2, the addition of an airlock.

The Vostok/Voskhod spacecraft (Figure 18) consisted of two modules. A near-spherical manned cabin contained the crew couch(es), a life support system, radios, instrumentation and (in Vostok) an ejection seat apparatus for optional use. The capsule had three small portholes for exterior viewing, external radio antennas, and a covering of ablative material. The manned cabin was attached to a service module resembling two truncated cones base to base, with a ring of gas pressure bottles on the upper cone near the cabin. This module carried batteries, orientation rockets, and the retrorocket system.

In Earth orbit the spacecraft was allowed to tumble slowly to distribute the heat load evenly, but could be stabilized on command for operations such as retrofire. As in all succeeding Soviet manned missions until about 1980, the flight systems were almost exclusively automated, with the cosmonaut present as a contingency backup (and, of course, as a passenger performing research tasks). A key difference between this spacecraft and its American counterparts was that Vostok/Voskhod was primarily a land recovery operation. In Vostok, crews had the option of coming down with the parachute-slowed ship for a rather hard landing, or ejecting at 2,100 meters to come down independently by parachute. In Voskhod, the landing was slowed by a final braking rocket (Smith, 1976).

Soyuz

The workhorse of the Soviet manned space program has been the Soyuz vehicle, in almost continuous use between 1967 and 1981. If Vostok/Voskhod can be considered to correspond to

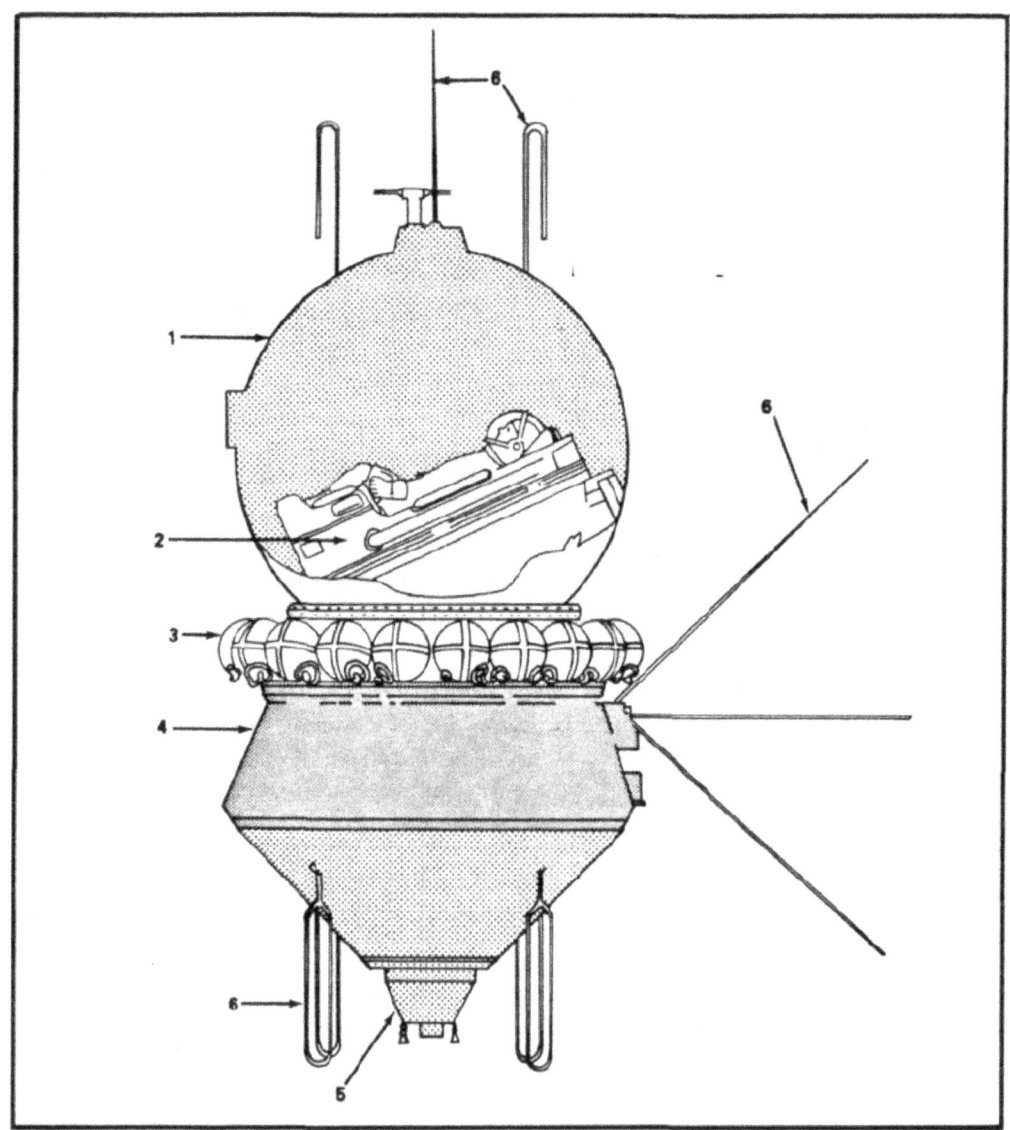

Figure 18. The configuration of the Vostok spacecraft (nearly identical to the later Voskhod vehicle). Labels denote: (1) ablative heat shield; (2) ejection seat; (3) oxygen and nitrogen pressure bottles; (4) service module; (5) retrorocket; and (6) antennas.

the American Mercury and Gemini spacecraft, then Soyuz corresponded to the Apollo ship: larger, more versatile, providing expanded capability for mission activities and duration, and incorporating many design improvements based on previous experience in space. The Soyuz craft had an overall length of 8 meters and a maximum diameter of about 2.5 meters. It consisted of two modules: an orbital module which served as a work compartment, and a command module, also termed the reentry module because this was the recoverable portion of the ship. The two compartments were separated by an airlock, and equipment and accessories are stowed in compartments aft of the command module, where a cruise engine was also located. An outer hatch in the orbital module permitted egress for EVA.

The total volume of the two modules was about 8.5 cubic meters. Appropriately outfitted, the ship had a 30-day stay time capability and, unlike Vostok/Voskhod, it was capable of large orbital path adjustments and maneuvering. Some versions could fly up to a distance of 1,300 km from Earth (Smith, 1976). Over the years of its use, the Soyuz craft evolved a number of variants to accommodate different needs. A ferry, or transport, version was used to take crews and some supplies back and forth from orbiting space stations (Figure 19). This type had a modest maneuvering capability and relied on batteries rather than solar panels for power. It could operate independently for about three days, but was sometimes left docked with a space station for up to 90 days awaiting further use. Another type of Soyuz craft was the "Progress" cargo ship (Figure 20) used to transport large quantities of fuel, food, equipment, and other necessities to space stations. This was essentially a stripped-down ferry vehicle outfitted with cargo bays and capable of carrying over 2,300 kilograms of payload. A third version of Soyuz had solar panels to enable it to carry out long-term, independent research missions. In this mode, the Soyuz functioned as a sort of small-scale space station.

Additional variations on these three types of Soyuz could be achieved by varying the number of crew seats (Soyuz could accommodate from one to three people in coveralls, and either one or two in pressure suits) and the type of docking gear. Different types of active and passive docking units were available (as well as vehicles without a docking unit).

Unlike its predecessors, which followed a ballistic trajectory on reentry, Soyuz had special aerodynamic features which permitted more controlled reentry and landing. Cosmonauts now remained in the cabin until they were on the ground. A drogue parachute was deployed at 9 km, and at one meter above the ground a gunpowder rocket fired, cushioning the impact.

Soyuz T

The newest model of the Soyuz line, first flown manned in 1980, is a sufficient departure from the old model to warrant independent discussion. Externally, this ship differs little from its predecessor. Internally the changes are substantial. The Soyuz T seats three crewmen rather than two. It includes an onboard computer complex that gives the cosmonaut much more autonomous control of docking and other activities (in this respect, it is similar to American spacecraft in use since the time of the Apollo program). The Soyuz T also has new instrumentation, improved radio communications and heat control systems, new life support and orientation and control systems, and new solar power panels (Malyshev, 1980).

Salyut Orbital Space Station

As of mid-1982, seven Salyut space stations had been placed in orbit (six of them successfully). This spacecraft is the base for long-term Soviet research missions in Earth orbit, and forms the basis for future Soviet plans regarding space manufacturing, materials processing, Earth resources studies, and space research. Although several successive versions have been developed, the basic

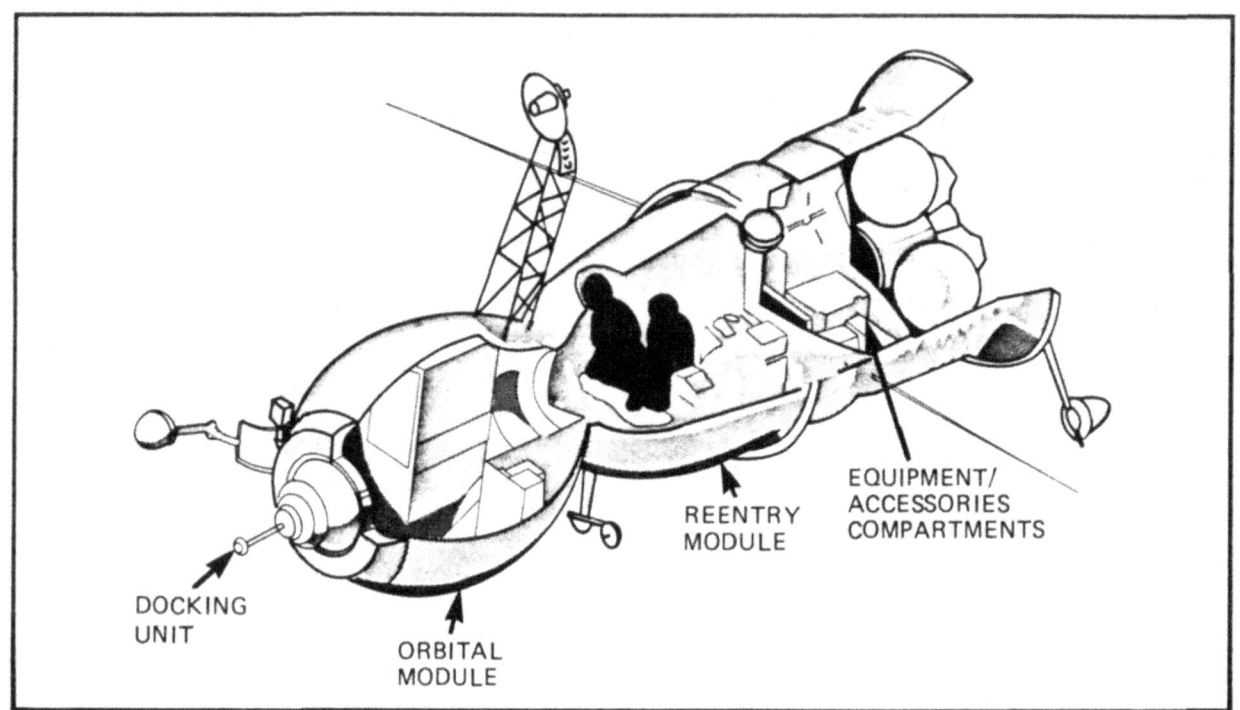

Figure 19. Configuration of the Soyuz used as a transport craft.

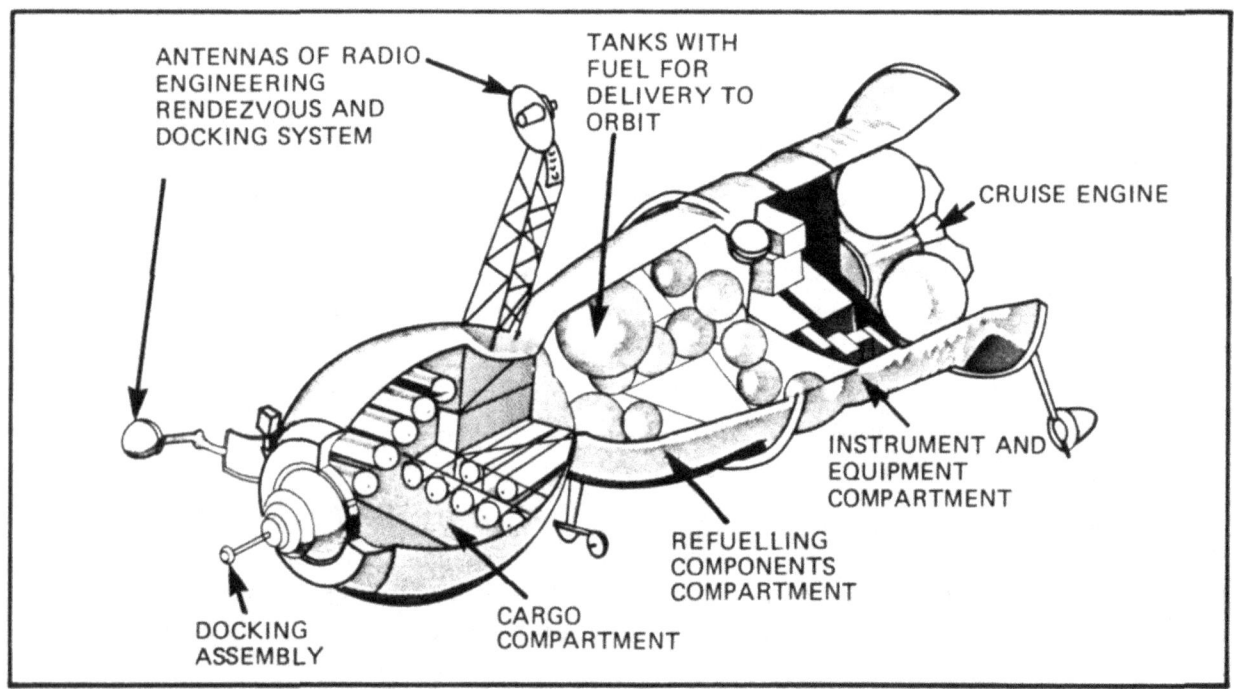

Figure 20. Configuration of the Soyuz "Progress" cargo craft.

configuration of the station does not vary substantially from one "model" to the next (Figure 21). The basic station is about 20 meters long, with a maximum diameter of 4.1 meters and a volume of 95 cubic meters. It is comprised of several compartments: a cylindrical transfer tunnel, a small and large cylinder connected by a conical section, and a cylindrical service module (Smith, 1976).

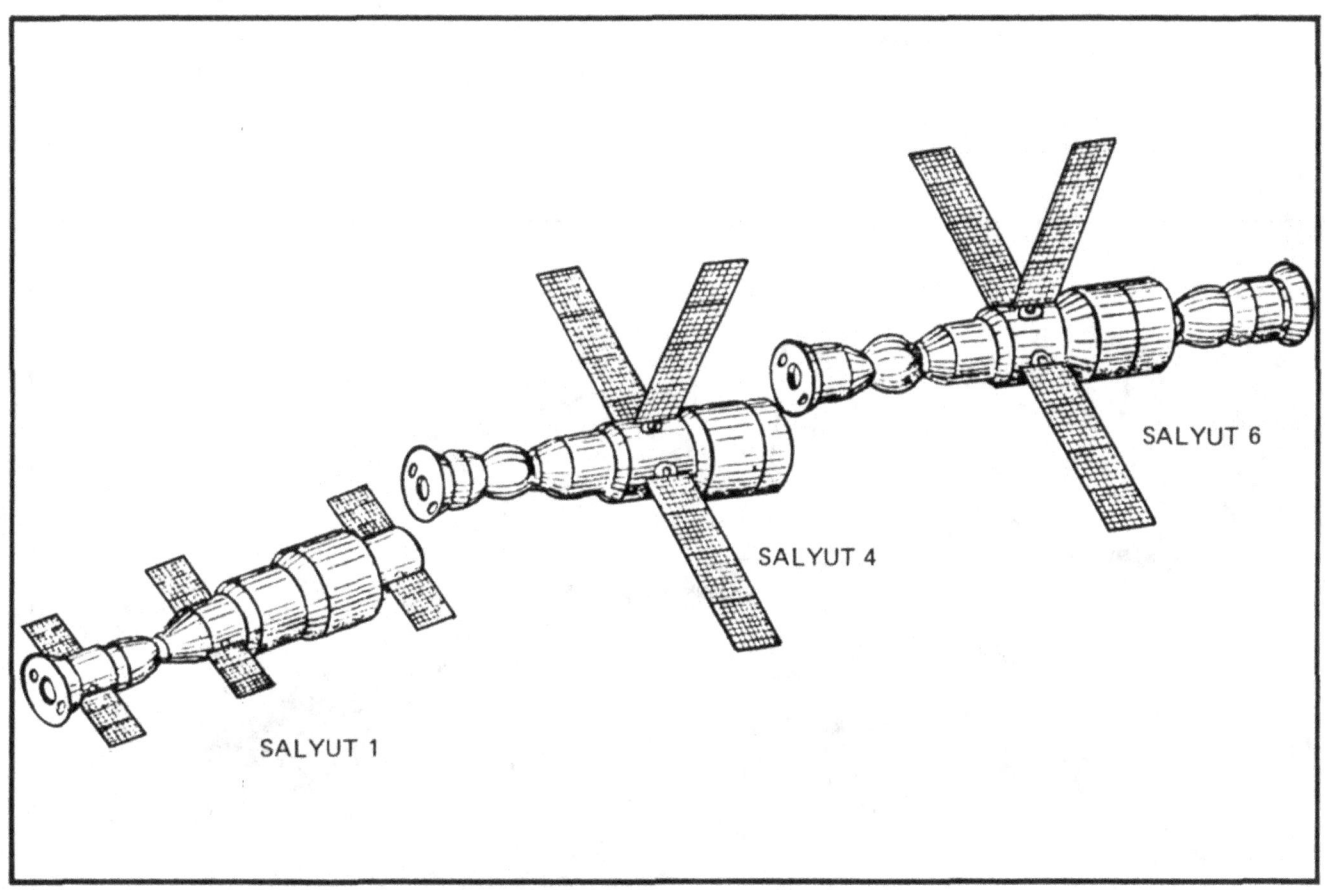

Figure 21. Salyut orbital stations of different generations. A Soyuz transport craft is shown docked to each station at the left. Salyut 6 also has a Progress cargo ship docked at the right.

Each station generates power by means of solar panels. The configuration and operation of these has changed during the development of the series. Salyut 1 had two short pairs of "wings" mounted transversely on the fore and aft modules of the station; Salyut 3 had three larger panels mounted around the perimeter of the central cylinder; these capable of being rotated 180°. Salyut 4 had a similar configuration, but the panels were now fully and independently rotatable by automatic means. In addition to the solar panels, other external features include the heat regulation system's radiators, orientation and control devices, and some of the scientific equipment. There are 20 portholes for viewing.

Information about the internal features and systems aboard Salyut was scanty for many years. Figure 22 shows the interior layout of Salyut 6, the first station about which substantial information was released. Emphasis is placed on comfort and "homey-ness" in the furnishings and decorations (for example, the interior is painted in soft pastel colors). Large, comfortable chairs are placed at seven different work stations. A permanent, rigid shower enclosure is provided, as well as a toilet. All sections of the working compartment (control, working, and living sections) are connected by a corridor and served by the same life support system. A galley area includes hot and cold water sources and a dining table. There are separate storage compartments for clothes, linens, and entertainment equipment (including stereo equipment and a small library).

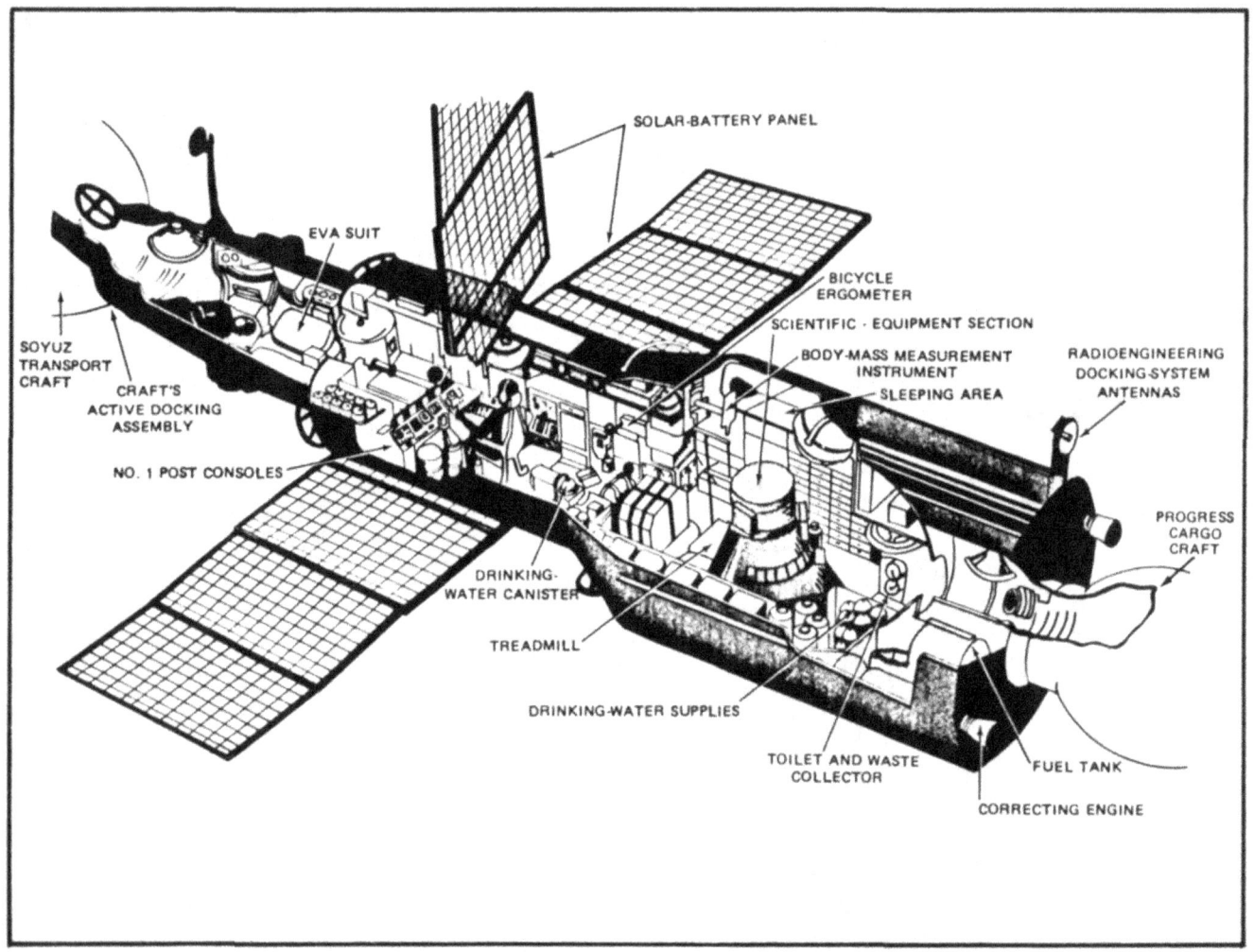

Figure 22. Schematic diagram of the Salyut 6 orbital station.

Transport of crews to the station is accomplished via the Soyuz ferry craft. Problems with adequate provisioning prompted a substantial reconfiguration in Salyut 6, so that two docking ports were provided instead of one: a Progress cargo ship could be docked to the second port while a Soyuz craft occupied the other. Salyut 6 was thus considered the first of a second generation of Salyuts.

This basic Salyut configuration is expected to be used by the Soviets as the base for space operations for many years. The next major change in design will probably be in the direction of a multi-manned station with multiple docking ports, to which will be docked modules dedicated to specific research programs (Oberg, 1982). Such a ship might also be used as the staging point for manned interplanetary expeditions. In any event, Soviet space vehicle development will continue to follow a conservative path, responding to developed capabilities, available economic resources, and a logical expansion of research and technological needs.

References

Grimwood, J.M. *Project Mercury: A chronology* (NASA SP-4001). Washington, D.C.: National Aeronautics and Space Administration, 1963.

Johnston, R.S. Skylab Medical Program overview. In R.S. Johnston and C.F. Dietlein (Eds.), *Biomedical results from Skylab* (NASA SP-377). Washington, D.C.: National Aeronautics and Space Administration, 1977.

Johnston, R.S., & Hull, W.E. Apollo missions. In R.S. Johnston, L.F. Dietlein, & C.A. Berry (Eds.), *Biomedical results of Apollo* (NASA SP-368). Washington, D.C.: National Aeronautics and Space Administration, 1975.

Luxenberg, B.A. Space Transportation System (Ch. 7). In *United States civilian space programs, 1958-1978*. Washington, D.C.: Congressional Research Service, Science Policy Research Division, January 1981.

Malyshev, Yu. Evolution of the Soyuz spacecraft. *Aviatsiya i Kosmonavtika*, 1980, No. 10: 38–39.

Mueller, G.E. Introduction. In *Gemini summary conference* (NASA SP-138). Washington, D.C.: National Aeronautics and Space Administration, 1967.

NASA. Pocket statistics. Washington, D.C.: National Aeronautics and Space Administration, March 1973.

Smith, M.S. Program details of man-related flights (Ch. 3). In *Soviet space programs, 1971-1975*. Washington, D.C.: Congressional Research Service, Service Policy Research Division, August 1976.

Smith, M.S. Manned spaceflight through 1975 (Ch. 6). In *United States civilian space programs, 1958-1978*. Washington, D.C.: Congressional Research Service, Science Policy Research Division, January 1981.

CHAPTER 5

SPACECRAFT ATMOSPHERES AND LIFE SUPPORT

Environmental Parameters

The atmospheric environment that we take for granted on Earth, a certain combination of gas pressure, composition, and temperature, is not present at survivable levels in the space environment. On Earth the atmospheric pressure is 760 torr (14.7 psi) at the surface. The composition of this gaseous medium is 20.9 percent O_2, 78.0 percent N_2, and 0.04 percent CO_2, with trace amounts of other gases. Over much of the Earth's surface the average temperature of the atmosphere ranges from 22°C to 27°C (72°-81°F), the ideal temperature for a lightly clothed man. By contrast, in the space environment the pressure approaches that of a perfect vacuum. Here there is no question of gas composition, and thermal exchange occurs solely by radiation, either from the unshielded sun or to the cold blackness of space. For man to survive in space, he must have available a volume of living space in which an atmosphere is provided at the proper pressure, gas concentrations, and temperature.

Pressure

A minimal atmospheric pressure (about 0.9 psi) is required to keep body fluids in the fluid state, and man's tolerance to extremely high pressures is limited. For all practical purposes, however, and particularly in the context of spacecraft, acceptable static pressures for a habitable environment are determined by the required partial pressures of the component gases, by their combined ability to support combustion, and by the necessity for change in pressure in the course of a mission.

Relative to changes in pressure, there are several potential physiological problems that must be considered in providing an onboard atmosphere. The three most significant concerns are barotrauma, explosive decompression syndrome, and decompression sickness.

Barotrauma occurs when gas is temporarily trapped in the middle ear or the sinuses, in teeth if a gas pocket has formed below a tooth restoration or in a decayed tooth, or in the gut. If trapped gas pockets exist, a change in the external pressure will produce pressure differences across the walls of these cavities, resulting in pain and tissue injury. Barotrauma is most likely (1) when swollen mucous membranes associated with a respiratory infection have obstructed passages that normally permit pressure equilibration of the ears and sinuses, (2) when poor dental care has resulted in air cavities in teeth, and (3) when the diet prior to pressure change has allowed large quantities of gas to form in the gut.

Barotrauma can be avoided by controlling the predisposing factors and by operationally limiting the rate of change in pressure. In the Shuttle, for example, cabin pressure is changed from the normal 14.7 psi only prior to an extravehicular activity, when pressure is reduced in the airlock to that of the pressure suit. For a nominal decompression or recompression, the rate of pressure is limited by specification to 0.1 psi/second and, in practice, to even slower rates. During an emergency recompression, the rate is limited to 1 psi/second.

Explosive decompression syndrome occurs when external pressure drops so rapidly that a transient overpressure develops in the lungs and other air cavities. The lungs may rupture at a pressure differential as low as 80 torr (1.6 psi). Should a tear or rupture of the lungs occur under these conditions, blood vessels will be severed and the positive pressure in the lungs is likely to force large quantities of gas into the bloodstream, resulting in a fatal air embolism.

The outcome of a rapid decompression or an explosive decompression depends on the following factors:

1) The rate of change of pressure
2) The absolute change in pressure
3) The absolute pressure prior to decompression
4) The ratio of initial pressure to final pressure
5) The ratio of lung volume at the time of decompression to maximum lung capacity
6) The ratio of the cabin orifice over the cabin volume compared to the ratio of airway orifice over the lung volume.

To understand the significance of some of these factors, it may help to walk our way through a scenario of an explosive decompression. Let us assume that a man is contained in a cabin at 12 psi surrounded by an ambient pressure of 5 psi. A large wall of the cabin blows away, so that the area available for gas to leave the cabin is very large relative to the cabin volume. In 0.01 sec the cabin reaches 5 psi. The man has his glottis open at the time of decompression and he has 3 liters of air in his lungs, which have a maximum capacity of 6 liters.

In this example the rate of cabin decompression is extremely high because of the large cabin orifice. The rate of decompression will depend on the area of the orifice, the volume of the cabin, the ratio of the pressure inside and out, and the final absolute pressure.

The rate of cabin pressure change in this example is so high that no appreciable gas volume will be able to escape through the open glottis. The initial pressure in the lungs is 12 psi. During the decompression the lung will expand to its maximal volume, doubling its volume and lowering the pressure by half to 6 psi. At this point, there will be a 1-psi (50-torr) differential pressure across the distended lung. As this pressure is below the 80-torr level at which the lung may rupture, this simple analysis (which does not consider the inertia of the expanding lung) would indicate that the decompression would be survivable.

An excellent review of rapid decompression written by Roth (1968) should be referred to for more rigorous analytical techniques and empirical data.

Explosive decompression was an important consideration for the Shuttle Orbital Flight Test missions. The ejection escape system involved a very rapid decompression of the cabin from

97

14.7 psi to ambient pressure. In certain contingency situations involving a shirtsleeve crewman with an oxygen supply, the maximum altitude for ejection was limited by considerations of rapid decompression.

Decompression sickness occurs when the pressure of dissolved gases in the tissues exceeds the ambient pressure. Under these conditions, bubbles may form in tissues and be carried by the bloodstream throughout the body. Decompression sickness can manifest itself by bubbles underneath the skin, by classic "bends" pain in the joints and muscles, by "chokes" or pain in the area of the lungs, by neurological manifestations, and by circulatory collapse and shock. Decompression is not normally a problem when the pressure of the diluent gas in the atmosphere does not exceed the final decompression pressure by more than a ratio of 1.5 to 1.8. When changes in pressure will result in conditions exceeding these limits, it is necessary to lower the pressure of dissolved gases in the tissues prior to decompression. Because of the high rate of tissue utilization of oxygen, this gas will not contribute significantly to the formation or growth of a bubble in the tissue. Therefore, an effective means to protect against decompression sickness is to breathe 100 percent O_2 prior to decompression, displacing nitrogen from the tissues, or to lower the concentration of N_2 in the breathing gas, thus reducing N_2 pressure in the tissue.

In American space missions prior to the Shuttle, decompression sickness was a problem primarily on liftoff, when there was a pressure change from 14.7 psi to 5 psi. The possibility of decompression sickness at this time was protected against by breathing pure oxygen for three hours prior to launch. This prebreathing period also protected the crew against a cabin decompression early in the mission, which would leave the crew in their pressure suits at 3.7 psi. When extravehicular activities (EVAs) were carried out on subsequent days, the decompression from the 5.0 psi cabin pressure to the 3.7 psi suit pressure involved no real hazard of decompression sickness.

The Shuttle vehicle has a cabin pressure of 14.7 psi, so that decompression sickness is not a concern on liftoff but becomes a concern if crewmembers perform EVA in a pressure suit at a lower pressure.

The method of protection that has been adopted for Shuttle missions involves a cabin decompression from 14.7 psi to 10.2 psi at least 12 hours prior to EVA (to lower tissue nitrogen), followed by a 40-minute period in the pressure suit at 10.2 psi breathing O_2, then decompression in the airlock to a suit pressure of 4.3 psi.

Gas Concentrations

Oxygen. In terms of the well-being of the crew, the most significant gas component of the atmosphere is oxygen partial pressure. The partial pressure of O_2 at sea level on Earth is 158 torr (3.06 psi). As the atmosphere is breathed, its components are diluted in the lungs by the addition of CO_2 and water vapor so that at the alveoli of the lungs, where O_2 transfer to the blood takes place, the O_2 partial pressure is 104 torr (2.01 psi). This is the physiologically important pressure, and it can be calculated for any atmosphere using the following equation:

$$P_A O_2 = F_i O_2 (P_B - 47) - PCO_2 \times \left[F_i O_2 + \frac{1 - F_i O_2}{0.85} \right]$$

where
$P_A O_2$ = alveolar partial pressure of oxygen

$F_i O_2$ = oxygen fraction in breathing atmosphere

P_B = barometric pressure of the breathing mixture

0.85 = an assumed respiratory exchange ratio

PCO_2 = partial pressure of CO_2.

On Earth the total pressure and the O_2 partial pressure vary as a function of altitude. People can live continuously at 3,660 meters (12,000 ft) with an alveolar O_2 partial pressure of 54 torr (1.05 psi). However, living at this altitude requires extensive physiological acclimation, and the acclimation is not complete. Even with acclimation, an individual living at this altitude cannot perform as well as a sea-level individual. Nearly complete acclimation (the exception being maximum oxygen uptake during hard work) can be achieved up to about 1,830 meters (6,000 ft) with an alveolar O_2 pressure of 77 torr (1.50 psi).

At sea-level, man, without acclimation, will show some measurable effects of hypoxia at an alveolar O_2 pressure of 85 torr (1.65 psi); at this level, some aspects of vision (low-illumination color vision threshold) begin to be affected. At alveolar O_2 pressures between 81 and 69 torr (1.57-1.33 psi), certain discrete types of mental performance (e.g., learning a new task) begin to be affected while others are not affected. As alveolar O_2 is dropped below this level, the scope and severity of visual, mental, and finally motor impairment increases, until at an alveolar O_2 of about 34 torr (0.67 psi), time of consciousness begins to be affected as shown in Table 1.

Table 1

Time of Useful Consciousness after Acute Exposure to Reduced Oxygen Levels

Altitude, km (ft)	Alveolar O_2 Partial Pressure (torr)	Time of useful consciousness, sec	
		Moderate Activity	Sitting Quietly
6.7 (22 000)	32.8	300	600
7.6 (25 000)	30.4	120	180
8.5 (28 000)	<30	60	90
9.1 (30 000)	<30	45	75
10.7 (35 000)	<30	30	45
12.2 (40 000)	<30	18	30
19.8 (65 000)	<30	12	12

(Adapted from Horrigan, 1979)

In the normal Earth environment, man is not exposed to alveolar O_2 levels in excess of 104 torr (2.01 psi). Oxygen at high partial pressures can be toxic. Subjects who breathe 100 percent oxygen at sea level for 6 to 24 hours complain of substernal distress and show a diminution of vital capacity of 500 to 800 milliliters (West, 1974). This loss is probably due to atelectasis, which occurs when all the O_2 in a poorly ventilated alveolus is absorbed by the blood. When this happens, the alveolus collapses. Surface tension tends to prevent reopening of a collapsed alveolus. Astronauts and test subjects who were exposed to the Apollo spacecraft atmosphere for periods of up to two weeks showed no acute effects of O_2 toxicity. They did evidence some changes in blood forming tissue, however (Kimzey et al., 1975). Similar effects on blood forming tissue have been reported after eight hours/day exposure to 8.00 psi O_2 (Hendler, 1974).

The Shuttle environmental control system (ECS) will control O_2 at 3.2 ±0.25 psi (165 ±12.9 torr), or essentially Earth-normal O_2 pressure. Mission rules require that O_2 masks be donned should O_2 pressure fall below 2.34 psi (121 torr). Hyperoxia will not be a concern; but it can be a consideration in altitude chamber training where N_2 elimination is accomplished by breathing 100 percent O_2 at 14.7 psi. If a test is delayed once prebreathe has begun, the prebreathe cannot be extended beyond six hours.

Carbon Dioxide. Carbon dioxide is normally present in the outdoor Earth environment at a concentration of 0.04 percent. Carbon dioxide is a product of respiration, so its concentration increases in indoor environments that are crowded or have poor ventilation. Its production by the crewman presents a particular problem in the closed-loop environmental control systems of the space cabin and pressure suit.

The effects of increased carbon dioxide in the atmosphere depend on the concentration and duration of exposure. Acute responses to increased CO_2 are increases in heart rate, respiration rate, and minute volume (Figure 1). Chronic exposure to CO_2 disturbs the acid-base balance of the body.

The PCO_2 limit for nominal Shuttle operations is 7.6 torr (0.15 psi). If this limit is exceeded, a mission contingency exists and every effort is taken to correct the situation. Breathing masks are donned if the PCO_2 exceeds 20 torr (0.39 psi). In the pressure suit, PCO_2 is limited to 7.6 torr for metabolic rates up to 1,600 Btu/m and to 15 torr (0.20 psi) for higher work rates.

Water Vapor (Humidity). Water vapor is a normal constituent of the Earth atmosphere. The partial pressure of water in the atmosphere is a function of (1) the exposure of the atmosphere to free water and (2) the temperature of the atmosphere, which may limit the water vapor pressure. Percent relative humidity is a measure of the water vapor pressure relative to the maximum water vapor pressure that the atmosphere will hold at that temperature. High relative humidity is associated with condensation in some areas and, in general, is conducive to microbial and fungal growth. Absolute humidity, the actual partial pressure of water vapor, is the more important physiological measure of humidity. Low humidity, encompassing levels that are common on Earth in the winter, causes drying of mucous membranes of the nose and throat as well as chapping of the lips and drying of the eyes and skin. Inactivation of cilia protecting the respiratory tract leads to an increased incidence of respiratory infections under these conditions. A water vapor pressure of 10 torr (0.19 psi) is optimum for habitability. The Shuttle ECS will control water vapor pressure between 6 torr and 14 torr (0.12-0.27 psi).

In addition to the direct physiological effects of humidity, the humidity is also significant in its influence on heat loss and heat balance.

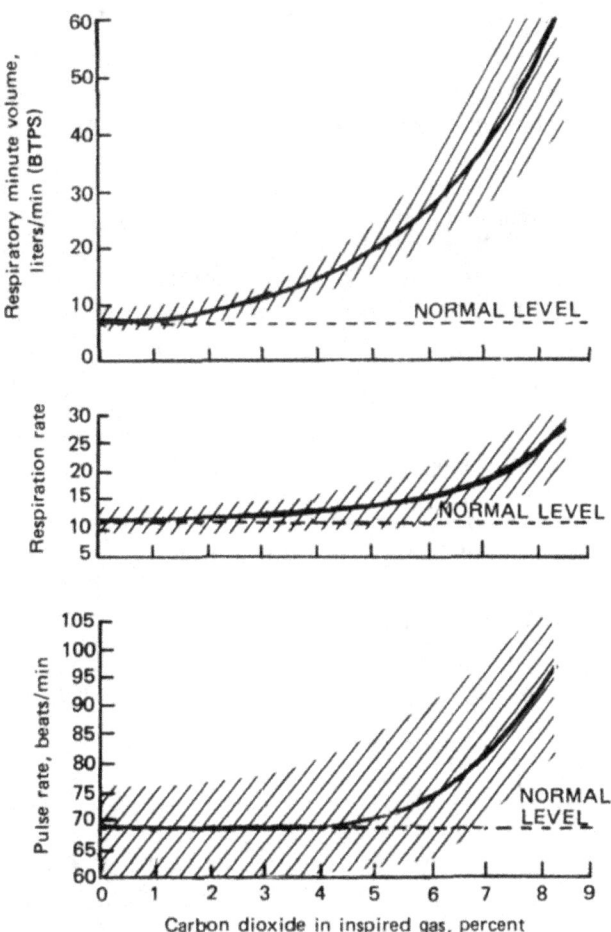

Figure 1. Immediate effects of increased CO_2 on pulse rate, respiration rate, and respiratory minute volume (BTPS = body temperature and pressure, saturated with water) for subjects at rest. Hatched areas represent one standard deviation on each side of the mean. To convert percentage CO_2 to partial pressure, multiply percent value by 1.013 for kilonewtons per square meter or by 7.6 for torr. (From Horrigan, 1979)

Temperature

The temperature of the atmosphere is an important component of the heat balance that must be maintained in the bodies of the crewmembers. Body temperature is closely guarded by physiological responses and sensory-behavioral responses. In addition, the thermal mass of the body resists change in temperature due to temporary imbalances in heat gain and heat loss. Physiologic thermo-regulatory control responses include shivering and vasoconstriction in the cold, and sweating and vasodilation in the heat. The element of behavioral control in temperature regulation is dependent on a strong and compelling sense of temperature discomfort when the environment is too hot or too cold. Discomfort in the heat or the cold has a deleterious effect on performance. Thermal comfort is, then, a critical factor if optimum performance is to be maintained. The observation has been made that at least five percent of a large group will find any one temperature uncomfortable (Fanger, 1970). An inclusive and practical approach to achieving comfort is to provide thermostatic temperature control around an optimum point, and to also provide some means for individuals to modify their heat balance, either by clothing selection or by individual air motion control.

The Shuttle ECS provides for temperature control within a range of $18^{\circ}C$ to $27^{\circ}C$ (64°-$81^{\circ}F$). Under conditions in which a comfortable heat balance is maintained, variations in humidity do not have a strong effect on comfort. However, when this heat balance can only be maintained at the upper limit of a comfort band or outside a comfort band, the humidity becomes very significant. To preserve a strong and effective thermoregulation response to overheating, particularly during short transients that may be encountered during exercise, an upper value of 14 torr (0.27 psi) PH_2O is a component of the Space Shuttle Orbiter specification.

Spacecraft Control Systems

A man-made and controlled spacecraft atmosphere is necessary to sustain life in the spacecraft. A large number of options are available in providing a viable spacecraft atmosphere; the choice of options has a major impact on crew comfort and safety, as well as on vehicle cost, weight, complexity, and reliability.

Evolutionary System Development

The Mercury spacecraft had a 5.0 psi O_2 atmosphere supplied from a store of pressurized oxygen. CO_2 was controlled by a lithium hydroxide absorber in the environmental control loop. Temperature control was accomplished through cooling provided by a sublimator heat exchanger, which was supplied by a tank of cooling water augmented by condensate water from the ECS water separator. The sublimator vented water vapor overboard and cooling resulted from the change of state. The crew stayed in their pressure suits and the ECS system supplied both the pressure suit and the cabin (Mercury Project Summary, 1963).

The Gemini spacecraft retained the 5.0 psi O_2 atmosphere used in Mercury; however, the primary source of oxygen was a liquid O_2 tank, with secondary O_2 supplies stored as high-pressure

102

O_2. CO_2 was again controlled by the use of lithium hydroxide absorber. The primary means of heat rejection in the Gemini vehicle was a spacecraft radiator which radiated the heat to space. Heat loss was controlled by the flow of coolant to the radiator and to ECS heat exchangers. A secondary system permitted sublimation of water to the space vacuum if required for additional cooling. In the later Gemini flights, extensive periods were spent outside the pressure suit. Gemini was, incidentally, the first American spacecraft to support a pressure suit in space. The ECS system supplied the pressure suit through an umbilical connection (Gemini Midprogram Conference, 1966).

The Apollo spacecraft atmosphere control systems were similar to the Gemini systems, but improved and more elaborate. The Apollo spacecraft system involved three modules: a Command Module ECS system, a Lunar Landing Module ECS system, and a Service Module which carried consumables to support the Command Module system. The atmosphere was 5.0 psi O_2 supplied from a cryogenic O_2 supply in the service module. Lithium hydroxide was used to absorb CO_2. Cabin temperature was maintained at $24^{o}C \pm 2.8^{o}$ ($75^{o}F \pm 5^{o}$), with relative humidity limited to the range of 40-70 percent. The primary system for heat rejection was again a space radiator. The use of the radiators was facilitated by a slow, controlled roll of the Command and Service Module called the passive thermal control flight mode. An evaporator was installed to provide additional cooling, but it was not used after Apollo 11 except for launch, Earth orbit, and entry (Brady et al., 1975).

The Skylab atmosphere control system incorporated some significant changes. Cabin pressure was 5.0 psi, but to avoid minor chronic effects of hyperoxia over a long mission and possible interference of such effects with medical experiments, a two-gas environment was used. The atmospheric composition was 70 percent O_2 and 30 percent N_2, to provide an O_2 partial pressure just slightly higher than Earth-normal. There was automatic control of the two gases, but in practice much of the control was accomplished manually to provide constant O_2 pressure during certain medical experiments. CO_2 absorption was accomplished with a regenerable molecular sieve system. This system allowed CO_2 to be flushed out of one bed by vacuum and heat and vented overboard while a second bed was absorbing CO_2. This regenerable system has obvious advantages for a long mission. However, a characteristics of this system was that it operated at a nominal CO_2 level of about 5 torr, so that although it met the same 7.6-torr CO_2 limit as earlier lithium hydroxide systems, the average CO_2 level was higher than on earlier spacecraft. The earlier systems had kept CO_2 near 1 torr most of the time. The thermal control system for Skylab was primarily a passive system. The vehicle was carefully painted with paints of varying emissivity in specific patterns so that very little active control with radiators or evaporators was necessary. Several large panels of the exterior surface of Skylab were lost during launch, and later two separate types of shades were deployed to shade this bare area (Belew, 1977).

As was noted earlier, the Shuttle is the first American spacecraft to use a 14.7-psi atmospheric pressure. Gas composition is 80 percent N_2 and 20 percent O_2, as on Earth. CO_2 absorption is accomplished with disposable lithium hydroxide cartridges, as in pre-Skylab flights, and thermal control is accomplished using radiators on the insides of the cargo bay doors.

Factors Impacting Pressure and Gas Control

A number of competing requirements influence the selection of cabin pressure and gas composition. Some of these factors are listed below:

- An O_2 partial pressure that is neither hyperoxic nor hypoxic
- An O_2 concentration that minimizes flammability of materials
- A gas density that will provide adequate cooling to gas-cooled electronics
- A total cabin/pressure suit pressure differential that is not conducive to dysbarism
- Structure strength sufficient to contain the cabin pressure
- Compensation for gas leakage.

To demonstrate the way in which these factors interact, and the tradeoffs that are involved in emphasizing certain system parameters at the expense of others, let us examine a current pressure/gas control system. The Shuttle cabin atmosphere, at 14.7 psi with 20 percent O_2 and 80 percent N_2, incorporates most of these factors; however, it does present a dysbarism problem when considered in combination with a 4.1-psi pressure suit. This pressure differential requires three to four hours prebreathing with O_2 prior to decompression in the suit. The lengthy prebreathing requirement reduces the time available for EVA during a reasonable work day. To arrive at a more acceptable combination of cabin and pressure suit pressures, either suit pressure must be increased or cabin pressure must be reduced. If cabin pressure is reduced and O_2 pressure is held constant, the percentage concentration of O_2 increases, and thus flammability is increased. If cabin pressure is reduced and O_2 concentration is held constant, then O_2 partial pressure is reduced and hypoxia becomes a problem. If suit pressure is increased, then suit mobility becomes a problem. The ideal solution to this dilemma may ultimately be provided by the development of a high-pressure suit (8.0 psi) with good mobility. In the meantime, the compromise solution for Shuttle is to use a 10.2-psi decompression stop for 12 hours, breathe O_2 for 40 minutes, and then decompress to a 4.3-psi pressure suit. This protocol involves a degree of reduction in gas cooling that is acceptable, an increase in flammability that is acceptable, some reduction in O_2 pressure (to 1,830 meters equivalent) that will not result in clinical hypoxia, a prebreathing requirement that reduces EVA time, and some increase in suit pressure.

Emergency Systems

Space suits were used in Mercury, Gemini, and Apollo test flights as an emergency pressure refuge. In operational Apollo orbital flights, the emergency pressure system was an emergency gas supply that would sustain cabin pressure in the face of a cabin leak through a hole of a given size, and sustain this pressure for sufficient time to allow an emergency return to Earth. In lunar Apollo flights, the Lunar Module (LM) served as a potential pressure refuge. During the Apollo 13 incident, after the Service Module cryogenic oxygen supply was lost and the Command Module was left without its main source of supply for oxygen, water, and electrical power, the LM's ECS system was activated to sustain the crew. The LM was not utilized as a pressure refuge during this incident, as the LM was able to maintain normal cabin pressure in both the Command Module and the LM.

The Shuttle vehicle has an emergency gas supply that will maintain the cabin at 8.0 psi in the face of a specified maximum leak (1.1 cm, or 0.45 in., diameter orifice) for 165 minutes—sufficient time for an emergency landing. However, the O_2 pressure at 8.0 psi would not be sufficient to avoid substantial levels of hypoxia. For this reason, the use of a mask breathing system providing supplemental O_2 is required for the emergency cabin pressure of 8.0 psi.

Biomedical Issues for Future Flight Operations

We have been very successful in maintaining a viable atmosphere for the crew in spacecraft and pressure suits in all the flight programs to date. Biomedical research has been directed primarily at studying the effects of environmental parameters, such as microgravity, that are not easy to control or to maintain at levels that do not have significant physiological effects.

The advent of long duration space flights and the extension of the population that may be exposed to space flight will present special problems and new considerations in spacecraft atmospheric control.

Recycling of Consumables

Permanent stations in space will present new problems in atmospheric provision and control, particularly in minimizing the use of consumables requiring resupply from the Earth. The consumables in present environmental control systems consist of the gas supplies of O_2 and N_2, carbon dioxide-absorbent material, and water for cooling the atmosphere by change of state. The regenerable molecular sieve CO_2 absorbers used on Skylab would eliminate the need for resupply of an absorber like lithium hydroxide; however, in regenerating the molecular sieve CO_2 absorbers, the CO_2 itself is vented to space and power is required to heat the sieve to drive the CO_2 off. Consumables could be further reduced if O_2 could be reclaimed from the CO_2. This might be accomplished by a biological conversion system or by power-costly direct chemical conversion. Important determinants of the quantity of O_2 and N_2 consumables required are the tightness of the pressure containment structure and the pressure in the pressurized area. The choice of a total pressure lower than Earth normal would tend to reduce the loss of gas consumables; however, since a near Earth-normal oxygen pressure is required for life support, lowering total pressure increases the percentage concentration of O_2, and thus increases flammability. The use of a total pressure less than sea-level atmospheric pressure would also reduce the cooling that could be applied to air-cooled electrical devices. This means that electric or electronic equipment might have to be tested or modified for use at a lower total pressure.

If extravehicular capability is planned in conjunction with a permanent station—and it almost surely will be—then a cabin/pressure suit pressure combination that will protect against decompression sickness and not require nitrogen washout must also be considered as a determinant of cabin pressure.

At the present time, the optimum atmosphere for a permanent station in space would appear to be a 14.7-psi atmosphere with 20 percent O_2 and 80 percent N_2. EVA would require transition

105

from this cabin pressure to an 8.0 psi mixed-gas pressure suit (to avoid hyperoxia) designed to provide good mobility at pressure.

The last consumable to be considered relative to the atmosphere is water for use in maintaining atmosphere temperature control. In all but the earliest spacecraft, evaporation of water for cooling has been used essentially as a back-up to the controlled radiation of heat to space. It seems very likely that, through configuration of station surface emissivities and the use of radiators, consumable water required for cooling will be minimal with the possible exception of the pressure suit cooling system.

Diverse Space Crews

The population flying in space in the 1980s and 90s will be more diverse than the astronauts who have flown to date. The impact of providing a suitable nominal atmosphere for the more diverse population should be minimal because our present guidelines are to provide a non-stressful, habitable environment for nominal operations. Consideration of a more diverse population will be more significant when we consider appropriate contingency or emergency environments to support the crew in the event of a system failure. The approach has been to ensure that environmental stress during contingency operations is limited to the point that the resulting physiological stress does not affect performance. The basis for these limits has been tolerance studies for each of the environmental factors. Most of these tolerance studies have been done on young men—usually college students and military recruits. Tolerance to environmental stresses will be more limited for some individuals in the more diverse flight population. Studies will have to be done to identify these limits, and action will have to be taken to provide more nominal environmental levels during contingency operations.

References

Belew, L.F. (Ed.). *Skylab, our first space station* (NASA SP-400). Washington, D.C.: U.S. Government Printing Office, 1977.

Brady, J.C., Hughes, D.F., Samonski, F.H., Jr., Young, R.W., & Browne, D.M. Apollo command and service module and lunar module environmental control systems. In R.S. Johnston, L.F. Dietlein, & C.A. Berry (Eds.), *Biomedical results of Apollo* (NASA SP-368). Washington, D.C.: U.S. Government Printing Office, 1975.

Fanger, P.O. *Thermal comfort*. Copenhagen: Danish Technical Press, 1970.

Gemini Midprogram Conference (NASA SP-121). Washington, D.C.: U.S. Government Printing Office, 1966.

Hendler, E. Physiological responses to intermittent oxygen and exercise exposures (NADC-74 241-40; AD B000 348L). November 1974.

Horrigan, D.J. Atmospheres. In *The physiological basis for spacecraft* (NASA RP-1045). Washington, D.C.: U.S. Government Printing Office, 1979.

Kimzey, S.L., Fischer, C.L., Johnson, P.C., Ritzmann, S.E., & Mengel, C.E. Hematology and immunology studies. In R.S. Johnston, L.F. Dietlein, & C.A. Berry (Eds.), *Biomedical results of Apollo* (NASA SP-368). Washington, D.C.: U.S. Government Printing Office, 1975.

Mercury project summary including results of the fourth manned orbital flight (NASA SP-45). Washington, D.C.: U.S. Government Printing Office, 1963.

Roth, E.M. *Rapid (explosive) decompression emergencies in pressure-suited subjects* (NASA CR-1223). Washington, D.C.: U.S. Government Printing Office, 1968.

West, J.B. *Respiratory physiology—The essentials*. Baltimore: The Williams and Wilkins Co., 1974.

CHAPTER 6
EXTRAVEHICULAR ACTIVITIES

Since the time of the Gemini missions, extravehicular activity (EVA) has been one of the most dramatic and challenging aspects of space flight. Although various kinds of problems were encountered in the five Gemini EVAs, those early missions demonstrated the feasibility of placing a human into free space, outside the protective confines of the space vehicle. The practicality of EVA as a useful mode of functioning in space was firmly established in subsequent manned missions; lunar surface activities during the Apollo program were highly successful in the deployment and conduct of numerous experiments; and a series of EVAs in Skylab permitted the carrying out of repairs that were crucial to the successful operation of the station.

During the Apollo program, 12 crewmen spent a total of 160 hours (the equivalent of four 40-hour work weeks) in space suits on the 1/6-gravity lunar surface. The Apollo crews also spent a total of almost eight hours in zero-gravity extravehicular activities (Waligora & Horrigan, 1975). The planned EVAs for Skylab were originally to require a total of 18 to 20 hours and to involve the replacement of film used in Apollo Telescope Mount (ATM) cameras. These plans were changed when the Orbital Workshop was damaged during the launch of Skylab 1. First, a set of EVAs was conducted to deploy the solar power panel. After it was demonstrated that a considerable amount of work could be performed successfully in zero-gravity, the number and duration of EVAs was extended, and additional EVAs were carried out to erect a solar canopy, repair an Earth Resources antenna, replace a gyro power-pack, and perform other vehicle and experiment repairs (Waligora & Horrigan, 1977).

Paramount among the planning considerations relative to EVA is the need to meet essential physiological requirements. The pressurized suit must supply oxygen and remove carbon dioxide while protecting the body from temperature extremes and micrometeoroids. But physiology is not the only factor involved. Operational and engineering considerations create separate sets of requirements that must be effectively integrated with the physiological requirements in the planning of EVA operations. One example, encountered during preparations for a contingency EVA in the initial flight tests of the Shuttle, involved the use of enriched onboard oxygen concentrations (Horrigan & Waligora, 1980). Increased flammability is an operational consideration which sets an upper limit to the O_2 concentration that can be used, while the danger of hypoxia and the need for nitrogen elimination prior to decompression are physiological considerations which set a lower limit on O_2 and an upper limit on nitrogen. If suit pressure is raised to decrease the extent of the decompression and thereby reduce the required time of preoxygenation, then mobility within the suit is lessened. These factors must be balanced off to find the ideal "window" for O_2 concentration aboard the spacecraft.

As our operational experience has increased throughout the manned space programs, our ability to understand and integrate life support, operational, and engineering requirements such as these has also increased. However, features introduced by future EVAs in conjunction with the

Space Shuttle and Spacelab present a number of new requirements that will have to be met. Thus it is pertinent to examine the current status of EVA technology and the basic problems that drive its development.

Adequacy of Life Support

The central element of EVA is the pressurized space suit. This system is actually a small spacecraft or space cabin that must supply all the essentials of life support for as long as seven or eight hours. It must provide adequate pressure, a suitable O_2 partial pressure, and a means of removing the CO_2 generated by the crewman. In addition, enough cooling must be available to remove the metabolic heat generated by the crewman, the heat generated in absorbing CO_2, and the heat generated by pumps and fans.

Given these requirements, the greatest concern in terms of the space suit's life support capability is the question of workload. While the pressurized suit was being developed for the first EVAs during Gemini, planners had to make educated guesses as to the metabolic cost of work in zero gravity and on the lunar surface. As it happened, although the Gemini life support system was generally adequate for average metabolic rates seen during EVA, it was not able to handle short-term spikes in the work rate and metabolic rates. Table 1 suggests the problems that were encountered in the Gemini series, both with overheating and with the completion of EVA objectives. Although metabolic rates were not measured directly, it was obvious on several occasions that metabolic rates exceeded the thermal control and carbon dioxide washout capabilities of the life support system.

Table 1
Gemini EVA Experience

Flight	Experience	Duration (hours)	Heart rates (beats per minute)	
			Mean	Peak
Gemini 4	Overheating during hatch closing—objectives completed.	0.60	155	175
Gemini 9	Visor fogging—hot at ingress—objectives not completed.	2.11	155	180
Gemini 10	No problem with heat or work rate—objectives completed.	0.65	125	165
Gemini 11	Exhausting work—no specific mention of heat—objectives not completed.	0.55	140	170
Gemini 12	Good restraints—no problems—objectives completed.	2.10	110	155

From Waligora & Horrigan, 1977.

Liquid Cooling Garment

The answer to these problems was the development, in the Apollo program, of a liquid-cooled garment (LCG). This system could suppress sweating at work rates up to 400 kcal/hr, and allowed sustained operations at work rates as high as 500 kcal/hr without thermal stress. By comparison, the heat removal capacity of the improved gas cooling system used on the later Gemini missions was limited to about 250 kcal/hr. The Apollo LCG was able to handle without difficulty any of the peak workloads that were encountered on the lunar surface during tasks involving heavy lifting, lengthy walking, and vigorous calisthenic-like exercise (Table 2). Later, the suit also proved adequate for all the EVAs on Skylab.

Table 2

Metabolic Expenditures During Apollo
Lunar Surface EVAs

Apollo Mission	No. of EVAs	Crewmen	Metabolic rate (kcal/hr)					EVA Duration (hrs.)
			Experiment Deployment	Geological Station Activity	"Overhead"	Lunar Roving Vehicle Operations	All Activities	
11	1	CDR[a]	195	244	214		227	2.43
		LMP[b]	302	351	303		302	2.43
	1	CDR	206	243	294		246	3.90
		LMP	240	245	267		252	3.90
12	2	CDR		218	215		221	3.78
		LMP		253	248		252	3.78
	1	CDR	182	294	219		202	4.80
		LMP	226	174	259		234	4.80
14	2	CDR	118	238	213		229	3.58
		LMP	203	267	213		252	3.58
	1	CDR	282	275	338	152	277	6.53
		LMP	327	186	293	104	247	6.53
15	2	CDR	243	293	287	149	252	7.22
		LMP		189	266	99	204	7.22
	3	CDR	261	242	311	138	260	4.83
		LMP	230	188	234	106	204	4.83
	1	CDR	207	216	273	173	219	7.18
		LMP	258	268	275	159	255	7.18
16	2	CDR		223	249	112	197	7.38
		LMP		244	236	105	209	7.38
	3	CDR		231	235	124	204	5.67
		LMP		242	264	103	207	5.67
	1	CDR	285	261	302	121	275	7.20
		LMP	278	300	285	113	272	7.20
17	2	CDR		261	302	121	207	7.62
		LMP		300	285	113	209	7.62
	3	CDR		261	302	121	234	7.25
		LMP		300	285	113	237	7.25
Mean			244	244	270	123	234	
Total time (hrs)			28.18	52.47	52.83	25.28	158.74	

[a]CDR = Commander   [b]LMP = Lunar Module Pilot.

From Waligora & Horrigan, 1975.

Metabolic Rate

In addition to improved cooling, one of the singular advantages offered by the LCG, particularly in the early Apollo missions, was the fact that it provided almost a closed calorimeter with which to measure metabolic rate. From the point of view of those responsible for medical monitoring and physiological studies this was a great benefit. The difference (ΔT) between the temperature of the water going into the suit and the temperature of the water coming out could now be determined, and from this ΔT the metabolic rate per minute and per hour could be derived. Thus there were now three different sources of information from which metabolic rate could be calculated: (1) the ΔT of coolant in the LCG; (2) the drop in pressure in the oxygen supply bottle; and (3) heart rate (HR vs. metabolic rate had been calculated preflight for each crewman).

Each one of these methods had certain drawbacks. With the LCG, very high workloads produced perspiration, which introduced a lag time into the ΔT of coolant because heat was lost into the garment's internal atmosphere. In addition, the lack of head cooling meant that heat lost in that area had to be estimated and factored in. With the drop in pressure in the oxygen bottle, the fidelity of the signal was such that it was more accurate to use larger decrements in pressure to determine average metabolic rates across a task rather than a minute-by-minute analysis. Psychogenic effects (excitement, anxiety) reduced the correlation between heart rate and metabolic rate, although the correlation was found to be reliable for high heart rates.

Notwithstanding the various sources of error, by using these methods in combination it was possible to determine metabolic rate during the various phases of EVA with considerable accuracy. As can be seen in Table 2, the average metabolic rate over the entire Apollo program during EVA was 234 kcal/hr, with the lowest rates seen while crewmembers were driving the Lunar Rover across the terrain (these rates were roughly equivalent to those experienced while driving a car). In Skylab, metabolic rates during EVA were essentially the same (230 kcal/hr) (Table 3). Thus, the effect of workload on life support requirements and on metabolic output became quite predictable.

Table 3
Metabolic Rates During Skylab EVA

Mission		Duration (hours)	Metabolic rate (kcal/hr)		
			CDR	PLT	SPT
Skylab 2	EVA-1 (Gas cooling only)	0.61		330	260
	EVA-2	3.38	315		265
	EVA-3	1.56	280		
Skylab 3	EVA-1	6.51		265	240
	EVA-2	4.51		310	250
	EVA-3 (Gas cooling only)	2.68	225		180
Skylab 4	EVA-1	6.56		230	250
	EVA-2	6.90	155	205	
	EVA-3	3.46	145		220
	EVA-4	5.31	220		185
Total Time		83.6	$\bar{X}$: 230 kcal/hr		

From Waligora & Horrigan, 1977.

111

Decompression Sickness

Apart from the problems of cooling and oxygen supply, another important concern relative to EVA is the danger of decompression sickness. The basis of this problem is the fact that, because of limitations on mobility of the limbs and hands as suit pressure is increased, the astronaut is subjected to lower pressures in the suit than in the space cabin. The difference between these two pressures is, generally speaking, proportional to the risk of developing decompression sickness.

Etiology

Decompression sickness occurs when the partial pressure of dissolved gases in bodily tissues exceeds the ambient atmospheric pressure. Typically this is the result of a rapid decompression, such as when a diver ascends quickly to the surface or when the astronaut dons a space suit which is then rapidly decompressed. Under these conditions, bubbles may form in tissues and blood and be carried by the bloodstream throughout the body. Decompression sickness can manifest itself by bubbles underneath the skin, by classic "bends" pain in the joints and muscles, by "chokes" or pain in the area of the lungs, by neurological manifestations such as a skin rash or numbness, and, in extreme cases, by paralysis, circulatory collapse, and shock. Various ways of categorizing the manifestations of decompression sickness have been devised. Most commonly, the symptoms are divided into two main categories: Type I and Type II (Kidd & Elliott, 1969). Type I includes those cases in which pain is the only symptom, or in which cutaneous and lymphatic involvement are seen either alone or with joint pain. Type II includes cases of a more serious nature, with nervous system and respiratory involvement. Other authors further subdivide these categories. Table 4 presents a more detailed listing of symptoms, drawing on the paradigm of Hills (1982).

In space flight, the main concern is with Type I bends. The expectation is that preventing Type I will also prevent the more severe symptoms. However, evidence described by Hills (1982) suggests that spinal, cerebral, and vestibular symptoms can be elicited by choice of conditions and may have different and independent etiologies. Thus there is not necessarily a progression from Type I to Type II. They may work in parallel, or Type II may appear without Type I, depending on the site of bubble formation. Since intra-arterial bubbles appear to be a cause of Type II, one concern has been the possibility of transpulmonary passage of venous emboli. Work by Butler and Hills (1979) has shown that the lung is an efficient bubble filter; and unless there is a pathological condition such as a patent foramen ovale, we would not expect to find transmission of bubbles from the venous to arterial side under levels of decompression found in normal space operations. Another factor that could impair the bubble-trap action of the lungs would be excessive oxygen toxicity (Butler & Hills, 1981), which could interfere with the functioning of pulmonary surfactant. Moreover, changes in pulmonary surfactant have also been reported to be involved in the pathophysiology of transpulmonary passage of venous gas emboli (Hills & Butler, 1981). Neither of these conditions is associated with atmospheres designed for use in space.

Table 4

Table 4
Decompression Sickness Manifestations for the
Six Types of Dysbarism

Organs Involved	Type	Symptoms					
		I	II	III	IV	V	VI
1. Bends							
1.a Skin		Rash Urticaria Pruritus					
1.b Extremities		Localized Joint Pains					
2. Central Nervous System							
2.a Cerebral			Unconscious-ness, Visual Disturbances, Convulsions Headache Collapse Paresthesia				
2.b Spinal Cord				Hemiplegia Paresthesia Abdominal Pain Muscular Weakness			
2.c Vestibular/ Cochlear					Vertigo Nausea/ Vomiting Nystagmus Incoordination Deafness Tinnitus		
3. Lungs						"Chokes": Cough Irritation	
4. Bones							Type A Osteonecrosis (juxta-articular) Type B Osteonecrosis (head, neck, shaft) Pain Neurologic Manifestations

Based on the paradigm developed by Hills, 1982.

Prevention

Decompression is not normally a problem when the original partial pressure of the diluent gas (usually nitrogen) in the atmosphere, and thus in the tissues, does not exceed the final decompression pressure by more than a ratio of 1.5 to 1.8. The actual rate varies from individual to individual and from tissue to tissue (for example, muscle absorbs and expels nitrogen much more quickly than does fatty tissue). But when changes in pressure will result in conditions exceeding these limits, it is necessary to lower the pressure of dissolved gases in the tissues prior to decompression. There are at least two basic ways of accomplishing this denitrogenation of tissues.

Preoxygenation. Because of the high rate of tissue utilization of oxygen, this gas will not contribute significantly to the formation or growth of a bubble in tissue. Therefore, an effective way to protect against decompression sickness is to breathe 100 percent O_2 prior to decompression, displacing nitrogen from the tissues, or to lower the concentration of N_2 in the breathing gas, thus reducing N_2 pressure in the tissue to a point at which the ratio is within the "safe" zone.

In Gemini and Apollo missions, the potential for decompression sickness was significant only on liftoff, when there was a pressure change from 14.7 psi to 5 psi. To protect against decompression sickness in this phase of the flight, crews breathed pure oxygen for three hours prior to launch, flushing nitrogen from their tissues. This procedure is also known as "whole body washout." It brought the ratio of tissue nitrogen pressure to cabin pressure at launch down to about 1.6, which was acceptable (Figure 1). The preoxygenation also protected the crew against either sudden cabin decompression or the need for an immediate ("contingency") EVA, which would have exposed the crew to 3.7 psi in their space suits. In either event, the N_2/suit pressure ratio would have been considerably higher—about 2.1 (Figure 1)—but probably not dangerous. However, neither problem occurred, and by the time crews embarked on their planned EVAs several days later, N_2 pressure in the tissues had been lowered sufficiently so that the 3.7-psi suit presented little danger of bends.

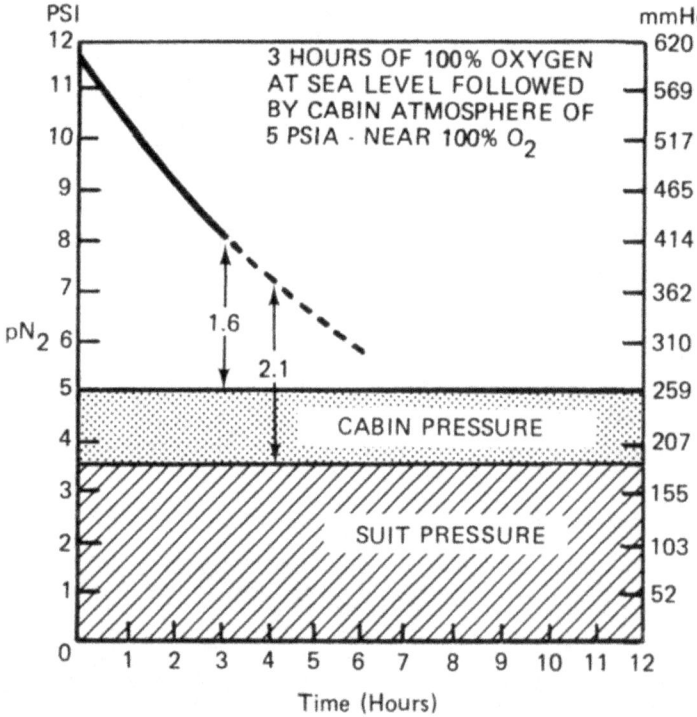

Figure 1. Estimated relationships of N_2 tension* to cabin and suit pressures in early U.S. space flights.

---

*The curve is for N_2 tension in a 360-minute tissue—that is, a tissue with an N_2-elimination half-time of 360 minutes.

The situation in Skylab was similar, except that the Skylab cabin contained a two-gas atmosphere (70% O_2, 30% N_2, at 5 psi). However, delivery to Skylab was accomplished via the 100-percent O_2 Apollo vehicle, so that an additional period of nitrogen washout was undergone. No cases of decompression sickness developed. Apollo-Soyuz presented an interesting variation, in that there were repetitive transfers of personnel between the 5-psi Apollo craft and the 10-psi Soyuz vehicle. The scenario was tested on the ground beforehand, and, as expected, was safe. Again, no cases of the bends were reported.

Space Shuttle: N_2 Equilibration. The Space Shuttle presents a very different set of concerns in the prevention of decompression sickness during EVA. In contrast to all preceding U.S. spacecraft, the Shuttle cabin has an "Earth-normal" atmosphere of 14.7 psi, with 20 percent O_2 and 80 percent N_2. Consequently, decompression sickness is a possibility not at liftoff but at the point of preparation for EVA. Initially a three-hour denitrogenation on 100 percent O_2 was contemplated. The assumption was that this would be adequate for individuals with a low fat-to-lean body mass ratio who were also bends-free during the spacesuit certification trials, and assuming EVA work rates similar to or less than those experienced on Apollo EVAs.

However, when planners examined the results of O_2-prebreathing studies in which similar final decompression pressures were obtained, they were disturbed by some of the findings (Table 5).

Table 5
Percent Protection from Bends Provided by
Selected Periods of O_2 Prebreathing

Final Pressure (psi)	Duration of Prebreathing at Sea Level (hours)						
	0.5	1.0	1.5	2.0	3.0	3.5	4.0
2.8	24	—	—	—	—	—	96
3.5	26	45	—	70	83	—	91
3.5	—	—	50	—	60	77	93
3.7	—	—	—	—	100	—	—
3.8	—	—	—	—	58	—	—
3.0	—	30.4	—	—	—	—	—
4.4	—	4.2	—	—	—	—	—
3.8	—	—	—	—	100[a]	—	—

[a]This represents space suit training conducted at Johnson Space Center. The policy was to provide at least three hours' prebreathing; some individuals may have had substantially more than this.

Studies from different laboratories; data compiled by NASA.

Although some researchers had achieved total protection from bends on a three-hour preoxygenation regimen, other results had ranged as low as 58 percent protection. Four hours of prebreathing appeared to offer consistently good protection, but from an operational standpoint this is a prohibitively long period of time to spend in a nonproductive mode.

Another problem with onboard prebreathing was a physical, or technical, one. When the crewman dons the EVA suit, he must not break the prebreathing regimen. As shown in Figure 2, this presents considerable difficulty. A nose clip and mouthpiece are provided to permit hands-free

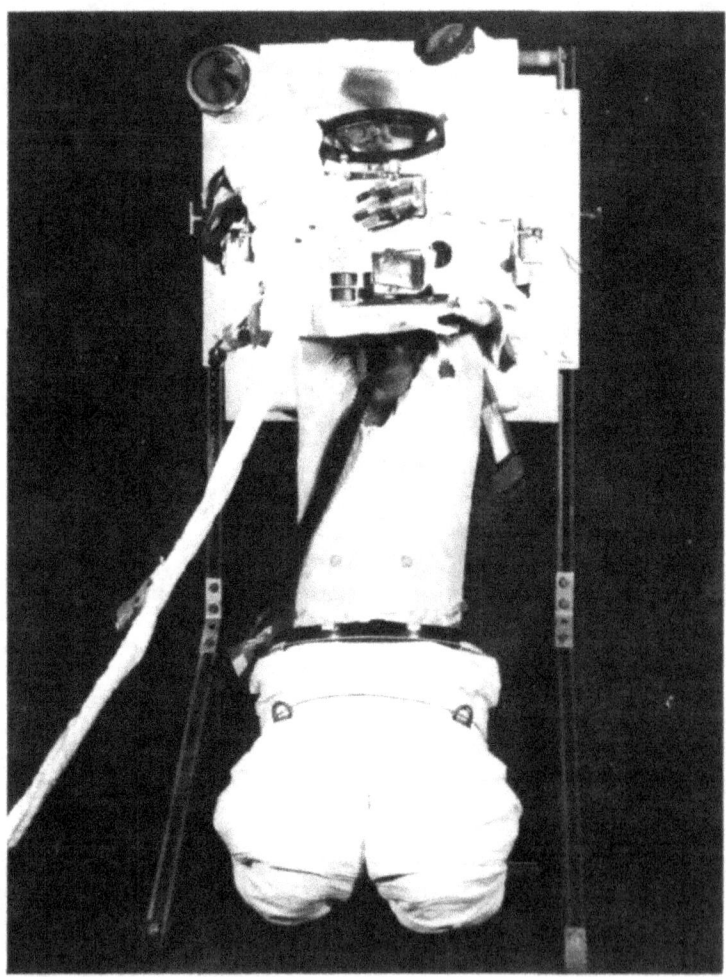

Figure 2. Donning the space suit while prebreathing was difficult, and could have compromised the prebreathing process.

manipulation, but to get completely into the suit the subject must remove the mouthpiece and hold his breath, then resume prebreathing once in the suit. When the helmet is attached, another 30 seconds of breathholding is required before the helmet is flushed of nitrogen. Considering these two operations, it is difficult to verify that some of the 80-percent nitrogen cabin air has not been inhaled.

This is an important concern because of the rate at which tissues become resaturated with nitrogen during washout (Figure 3). Air Force studies (Adams, Theis, & Stevens, 1977) indicated that a one-minute interruption, or "air break," would require 34 minutes of additional prebreathing to compensate for the renitrogenation of tissues. A human study conducted at the Shrine Burn Hospital in Galveston (Horrigan et al., 1979a, 1979b), using 80 percent argon and 20 percent oxygen for a normoxic denitrogenation, revealed a sharp slope of renitrogenation in quadriceps muscle tissue (Figure 4). The relatively rapid subsequent denitrogenation when exposed to 30 minutes of 100 percent O_2 prebreathing is probably due to the fact that muscle tissue is a "fast" (ratio of N_2 elimination) tissue.

These various problems with onboard preoxygenation led planners to search for alternative ways of accomplishing proper decompression prior to EVA in the Space Shuttle. The search centered

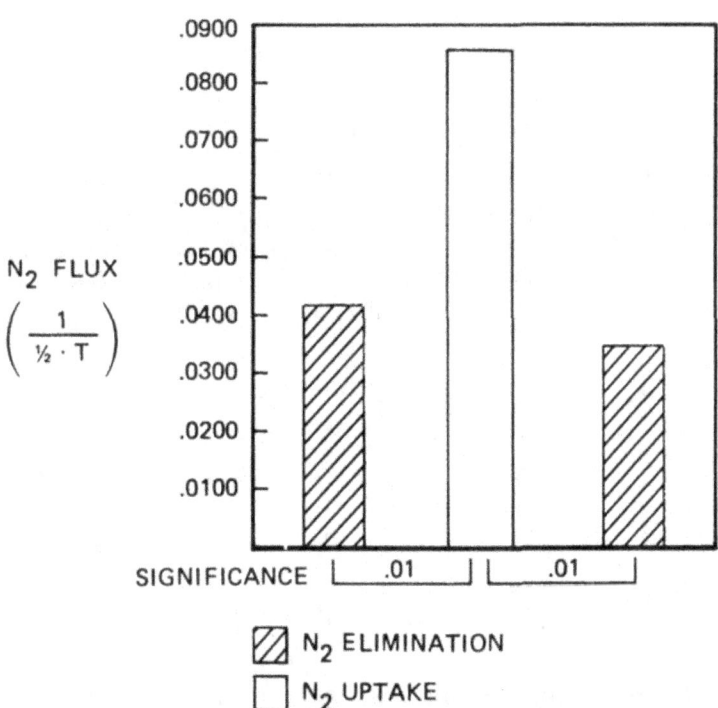

Figure 3. The uptake of nitrogen in muscle during an air break in nitrogen washout.

117

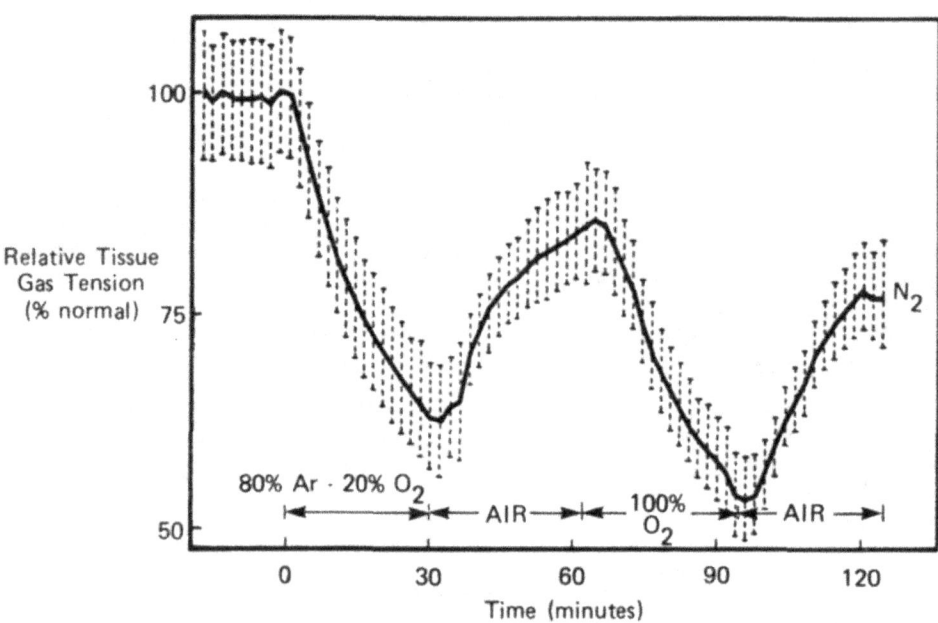

Figure 4. Nitrogen tension measured in the quadriceps muscle of human subjects exposed to selected atmospheres.

on a diving physiology technique, the Haldane method of using N_2 tissue/total pressure ratios (Boycott, Damant, & Haldane, 1908) to determine optimum combinations of pressures. Initially, the "equilibration" to 11 psi and 12 psi cabin pressure for 12 hours was considered. But in order to achieve final ratios comparable to those obtained with sea-level oxygenation, suit pressure would have to be raised from 4 psi to 5 psi (Figure 5). As this was not feasible, due to suit engineering, the procedure selected for the operational flight tests (OFT) of the Shuttle was to decompress the cabin to 9 psi at 28 percent O_2 for 12 hours, then don the suit for 30-45 minutes of 100 percent oxygen breathing at that pressure before decompressing the suit to 4 psi for EVA. Tests showed an approximately 6 percent incidence of Type I bends with this procedure (Adams et al., 1981), which was deemed a satisfactory risk in the event of contingency EVA during the OFTs. Since no EVAs were required during these missions, the procedure was not put to the test in space.

On the basis of decompression parameters alone, this method compared favorably with pre-oxygenation while allowing freedom of movement and activity. At the end of 12 hours of equilibration to this atmosphere, the pressure ratio would be 1.9—comparable to about 3½ hours of preoxygenation. The final 30-40 minutes on oxygen reduces the ratio to a still lower and safer number. Comparing these figures to the previous results of decompression studies (Figure 6), it is clear that as the ratio drops below 2.0 into the 1.6-1.8 range, the incidence of bends decreases markedly to an acceptable rate. Space suit training runs conducted at NASA Johnson Space Center confirmed these data. There were no cases of bends reported in 50 subject trials where the ratios averaged 1.80 for 360-minute tissue and 1.47 for 240-minute tissue.

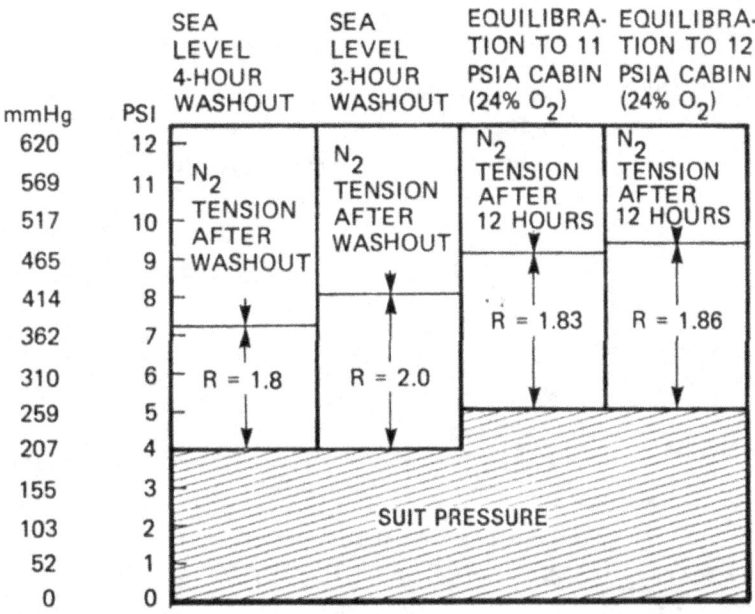

Figure 5. Comparison of sea-level nitrogen washout with suit/cabin pressure combinations which permit equilibrium to lower N$_2$ tension without oxygen breathing by mask. R = ratio of N$_2$ partial pressure before decompression to suit pressure after decompression.

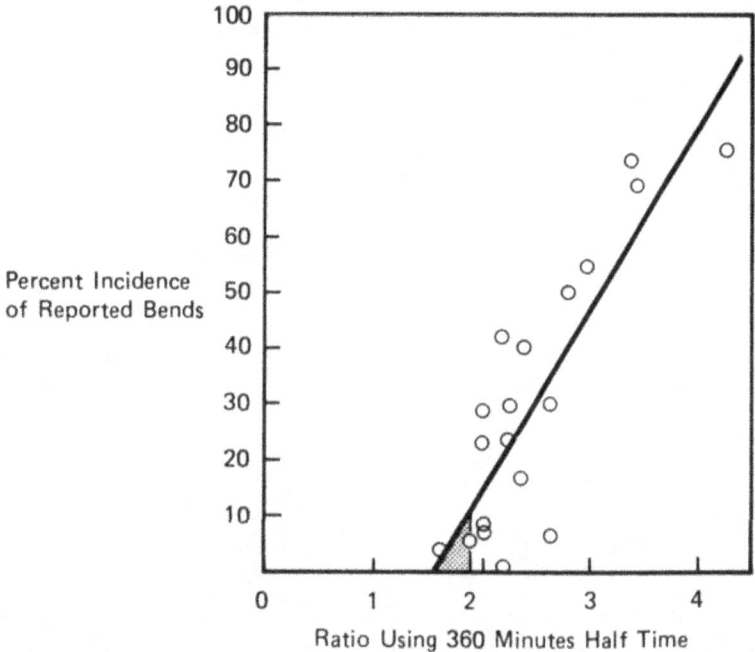

Figure 6. Comparison of Tissue N$_2$/final pressure vs. bends incidence in 500 subject exposures (1942-1981), using relatively slow (360-minute) tissues.

However, one potential problem emerged with the 9-psi equilibration. To minimize flammability, which is a consideration when using certain spacecraft cabin materials, a limitation of 30 percent had been placed on oxygen concentration. At 9 psi, when fluctuations of the gas pressure controller are taken into account, the worst-case oxygen level during the 12-hour equilibration period would approach the pO_2 found at 2,500 meters. Normally (on Earth) this is a safe altitude. But considering that the effects of pulmonary function changes and fluid redistribution in zero gravity are not yet completely known, the question of potential hypoxia had to be faced. Although bedrest studies suggest that oxygenation of the blood is not impaired at that pO_2 in zero gravity (Waligora et al., 1982), definitive assurance of the procedure's safety was not possible until pulmonary function data were obtained from Shuttle and Spacelab flights. Nevertheless, the 9 psi cabin was adopted for the Shuttle OFT flights, in the event that an EVA was required.

For the "mature" Shuttle runs (STS-5 and beyond), a compromise equilibration procedure was considered which alleviates the problem of potential hypoxia. It involves reducing cabin pressure to 10.2 psi at 27 percent O_2. Although this is a slightly lower concentration of oxygen, the pO_2 is higher (now equivalent to about 1,200 meters, well below hypoxic limits). In this procedure, the suit pressure is increased to 4.3 psi—an acceptable level—to keep the pressure ratio within the desirable limits. Equilibration still requires 12 hours, and 100 percent O_2 is still breathed for 40 minutes in the suit before final depressurization. Rapid EVA would still require a three-hour prebreathing of O_2 at sea level, followed by depressurization of the airlock to 10.2 psi (still on pure oxygen), donning of the suit for 40 minutes on O_2, and final depressurization to 4.3 psi. The above procedures will require further validation in both ground and flight tests to establish nitrogen washout times. In any event, emergency EVAs in early operational Shuttle flights can be conducted with minimal risk of bends by prebreathing O_2 at 14.7 psi in the suit for 3.5 hours prior to EVA.

Shuttle EVA Hardware

The Shuttle space suit (Figure 7) is called the Extravehicular Mobility Unit (EMU). Like previous space suits, the unit includes the pressure containment itself and an integrated life support system. It differs from prior suits in that the suit is not custom-made for each astronaut. Instead, it consists of ten separate components such as the hard upper torso, the arm assemblies, and lower torso assemblies. These are all "off-the-shelf" items that can be pieced together to form a suit appropriate for each individual's anthropometric proportions. The crewman in Figure 7 is seen wearing the mini work system equipped with tool caddies for performing EVA repairs and other work tasks.

The EMU is designed to provide a 4.3 psi nominal pressure. The suit is purged to a minimum O_2 concentration of 95 percent prior to EVA and all makeup gas is O_2. CO_2 is absorbed by a lithium hydroxide canister which will maintain CO_2 at 0.15 psi or less at metabolic rates up to 1,600 Btu/hr, and below 0.29 psi at higher metabolic rates. A liquid cooling vent garment removes heat from the body by a combination of conduction, convection, and evaporation. Coolant is pumped through the cooling garment and to a heat exchanger in contact with a sublimator which dissipates the heat to the vacuum of space through controlled sublimation of a water supply.

120

Figure 7. The Space Shuttle Extravehicular Mobility Unit. Crewman is wearing the mini work system equipped with caddies for holding hand tools.

The EMU also includes an in-suit drink bag to provide drinking water during a seven-hour EVA, and a urine collection device. It has a communications system to allow voice communication with the spacecraft and the ground, and a display and control module which will allow the crewman to control cooling temperature and select one of several operational modes. This module will provide the crewman with status of consumables through a visual display, as well as warnings of possible system failures. The Extravehicular Visor Assembly allows the crewman to see out of the space suit, and multiple visors allow the crewman to select the appropriate degree of protection from glare and ultraviolet radiation. Finally, a Secondary Oxygen Pack provides emergency pressure, O_2, CO_2 washout, and cooling for a 30-minute period by controlled release of a high-pressure O_2 supply in an open loop mode (not recirculated).

Outlook for EVA Operations

Early Shuttle EVAs

Shuttle EVAs are carried out primarily in the orbiter's payload bay. Crewmembers use a slide wire to translate between the forward and aft portions of the bay. On the starboard side next to the forward bulkhead is a Task Simulation Device (TSD), used in early missions as a testbed for representative tasks. TSD simulations include tasks typical for satellite servicing, such as cutting a thermal blanket, removing screws, disconnecting and reconnecting electrical contacts, and cutting grounding wires. Flight foot restraints and handholds have been developed for Shuttle EVA operations using experience gained in earlier missions.

Future EVA Requirements

As was pointed out earlier, the feasibility and usefulness of EVA has been amply demonstrated. The space suit life support system has protected astronauts against the space vacuum, against temperature extremes of -118° to $+132^\circ$ C (-180° to $+277^\circ$ F), and against the impact of micro-meteoroids while allowing useful work to be performed. In the future, a wider variety of tasks and a more variable crew population including females will require a continuing examination of the physiological limits, procedures, and hardware systems that have been established for EVA.

If a long-term space operations center is established at some point in the future, we may find a large number of hardhat workers of different physiological susceptibilities working in space. Such a scenario would require quantum improvements in the flexibility, durability, and reliability of EVA hardware. One such improvement would be the development of a one-half atmosphere (8 psi) suit, which would make EVA operations much more efficient. If engineering difficulties can be successfully overcome, this may in fact be the next big step to be made in the direction of simplifying and "normalizing" man's activity in space.

References

Adams, J.D., Theis, C.F., & Stevens, K.W. Denitrogenation/renitrogenation profiles: Interruption of oxygen prebreathing. Proceedings of the Aerospace Medical Association Meeting, Las Vegas, Nevada, May 1977.

Adams, J.D., Dixon, G.A., Olson, R.M., Bassett, B.E., & Fitzpatrick, E.L. Prevention of bends during Space Shuttle EVAs using stage decompression. Proceedings of the Aerospace Medical Association Meeting, San Antonio, Texas, 1981.

Boycott, A.E., Damant, G.C.C., & Haldane, J.S. The prevention of compressed air illness. *Journal of Hygiene*, 1908, *8*:342-443.

Butler, B.D., & Hills, B.A. The lung as a filter for microbubbles. *Journal of Applied Physiology*, 1979, *47*:537-543.

Butler, B.D., & Hills, B.A. Effect of excessive oxygen upon the capability of the lungs to filter gas emboli. In A.J. Bachrach and M.M. Matzen (Eds.), *Underwater Physiology VII*, Bethesda, MD: Undersea Medical Society, 1981.

Hills, B.A. Basic issues in prescribing preventive decompression (ltr. to the Editor). *Undersea Biomedical Research*, September 1982.

Hills, B.A., & Butler, B.D. Migration of lung surfactant to pulmonary air emboli. In A.J. Bachrach and M.M. Matzen (Eds.), *Underwater Physiology VII*, Bethesda, MD: Underwater Medical Society, 1981.

Horrigan, D.J. & Waligora, J.M. The development of effective procedures for the protection of Space Shuttle crewmembers against decompression sickness during extravehicular activities. Proceedings of the Aerospace Medical Association Meeting, Anaheim, California, May 1980.

Horrigan, D.J., Wells, C.H., Guest, M.M., Hart, G.B., & Goodpasture, J.E. Tissue gas and blood analyses of human subjects breathing 80% argon and 20% oxygen. *Aviation, Space, and Environmental Health*, 1979a, *50*(4):357-362.

Horrigan, D.J., Wells, C.H., Hart, G.B., & Goodpasture, J.E. Elimination and uptake of inert gases in muscle and subcutaneous tissue of human subjects. Proceedings of the 50th Scientific Meeting of the Aerospace Medical Association, Washington, D.C., May 13-17, 1979b.

Kidd, D.J., & Elliott, D.H. Clinical manifestations and treatment of decompression sickness in divers. In P.B. Bennett and D.H. Elliott (Eds.), *The physiology and medicine of diving and compressed air work*. Baltimore: Williams and Wilkins Company, 1969.

Waligora, J.M., & Horrigan, D.J. Metabolism and heat dissipation during Apollo EVA periods. In R.S. Johnston, L.F. Dietlein, and C.A. Berry (Eds.), *Biomedical results of Apollo* (NASA SP-368). Washington, D.C.: NASA, 1975.

Waligora, J.M., & Horrigan, D.J. Metabolic cost of extravehicular activities. In R.S. Johnston and L.F. Dietlein (Eds.), *Biomedical results from Skylab* (NASA SP-377). Washington, D.C.: NASA, 1977.

Waligora, J.M., Horrigan, D.J., Jr., Bungo, M.W., & Conkin, J. Investigation of the combined effects of bedrest and mild hypoxia. *Aviation, Space, and Environmental Medicine*, 1982, 53(7): 643-646.

SECTION IV

Physiological Adaptation to Space Flight

CHAPTER 7

OVERALL PHYSIOLOGICAL RESPONSE TO SPACE FLIGHT

Understanding the physiology of human response to extreme environments requires that the sequence of underlying mechanisms be identified. Knowledge of these mechanisms is of great importance to the development of rational and appropriate prophylactic or therapeutic treatments. In obtaining this knowledge, it is advantageous to observe the process of change in its entirety, without intervention. However, it has been difficult to avoid intervention in studies of human response to space stresses (principally weightlessness), since medical personnel are also responsible for insuring the health, well-being, and optimum performance of the crews under observation. This invariably requires some type of intervention during the period of exposure, further complicating the study of physiological processes.

There are a number of additional factors that complicate attempts to delineate the time course of response to weightlessness and the overall acclimation process in space. These include:

- **Sample Size.** To date, fewer than 150 astronauts/cosmonauts have flown in space. This small sample size makes it difficult to generalize to a larger population.
- **Limited Capabilities for Scientific Observations.** Biomedical observations have been restricted by the operational constraints imposed on most space missions, and also by time, since the longest missions flown so far have not exceeded six months.
- **Extensive Use of Countermeasures.** Prophylactic and therapeutic use of countermeasures, such as those for neurovestibular symptoms, cardiovascular deconditioning, and loss of lean body mass, has undoubtedly masked some of the direct effects attributable to weightlessness alone.

Despite these limitations, Skylab and Salyut missions have generated a wealth of biomedical data that points toward definite trends and lends itself to the formulation of hypotheses concerning acute responses in zero gravity and subsequent adjustments. Table 1 presents these trends for short missions (up to 14 days) and for longer missions (in excess of two weeks).

Time Course of Inflight Acclimation

Figure 1 presents a graphic summary of the time course of physiological shifts associated with space acclimation. It presupposes that all biological systems are in a state of homeostasis in the one-gravity environment at the beginning of the flight (one-gravity set point). Soon after orbital insertion, shifts occur in more susceptible physiological systems, although not necessarily simultaneously. For example, neurovestibular adjustments, with associated initial symptomatology, are likely to occur during the first few days in orbit, while marked changes in red blood cell (RBC) mass are detected only after a period of subclinical latency and peak at 60 days inflight. Other physiological functions do not exhibit detectable shifts early in flight, but are later shown to have undergone gradual and progressive changes. In particular, calcium loss, loss of lean body mass, and possible effects from cumulative radiation appear to increase continually regardless of flight duration or level of acclimation achieved by other body systems.

Table 1
Physiological Changes Associated with Short-Term and Long-Term Space Flight

Physiological Parameter	Short-Term Space Flights[a] (1-14 days)	Long-Term Space Flights (more than 2 weeks)[b]	
		Pre- vs. Inflight	Pre- vs. Postflight
Cardiopulmonary System			
Heart Rate (resting)	Increased postflight; peaks during launch and reentry, normal or decreased during mission. RPB:[c] up to 2 days.	Normal or slightly increased.	Increased. RPB:[c] 4-5 days.
Blood Pressure (resting)	Normal or decreased postflight.	Diastolic blood pressure reduced.	No change.
Orthostatic Tolerance	Decreased after flights longer than 5 hours. Exaggerated cardiovascular responses to tilt test, stand test, and LBNP postflight. RPB:[c] 3-14 days.	Highly exaggerated cardiovascular responses to inflight LBNP (especially during first 2 weeks), sometimes resulting in presyncope. Last inflight test comparable to R+0 (recovery day) test.	Exaggerated cardiovascular responses to LBNP. RPB:[c] up to 3 weeks.
Cardiac Size	Normal or slightly decreased cardiac/thoracic ratio postflight.		C/T ratio decreased postflight.
Stroke Volume and Cardiac Output	Decreased postflight. Gradual recovery after 5 days postflight.	Variable, usually increased during first month (impedance measurements).	Decreased postflight. Gradual recovery 5-21 days, depending on the level of exercise inflight.
ECG/VCG	Moderate rightward shift in QRS and T postflight.	Increased PR interval, QT_c interval, and QRS vector magnitude.	Slight increase in QRS duration and magnitude; increase in PR interval duration.
Systolic Time Intervals			Increase in resting and LBNP-stressed PEP/ET ratio. RPB:[c] 2 weeks.
Echocardiography			Decreased stroke volume and left end-diastolic volume. Ventricular function plots indicate no myocardial dysfunction postflight.
Arrhythmias	Usually premature atrial and ventricular beats (PABs, PVBs). Isolated cases of nodal tachycardia, ectopic beats, and supraventricular bigeminy inflight.	PVBs and occasional PABs; sinus or nodal arrhythmia at release of LBNP inflight.	Occasional unifocal PABs and PVBs.

[a]Compiled from biomedical data collected during the following space programs: Mercury, Gemini, Apollo, ASTP, Vostok, Voskhod, and Soyuz.

[b]Compiled from biomedical data collected during Skylab and Salyut missions.

[c]RPB: return to preflight baseline.

Table 1
Physiological Changes Associated with Short-Term and Long-Term Space Flight
(Continued)

Physiological Parameter	Short-Term Space Flights (1-14 days)	Long-Term Space Flights (more than 2 weeks)	
		Pre- vs. Inflight	Pre- vs. Postflight
Cardiopulmonary System *(Continued)*			
Exercise Capacity	No change or decreased postflight; increased HR for same O_2 consumption; no change in efficiency. RPB:[c] 3-8 days.	High exercise capacity inflight.	Decreased postflight; recovery time inversely related to amount of inflight exercise, rather than mission duration.
Lung Volume		Vital capacity decreased 10%.	No change.
Leg Volume	Decreased up to 3% postflight. Inflight, leg volume decreases exponentially during first 24 hours, and plateaus within 3 to 5 days.	Same as short missions.	Same as short missions.
Leg Blood Flow		Marked increase.	Normal or slightly increased.
Venous Compliance in Legs		Increased; continues to increase for 10 days or more; slow decrease later inflight.	Normal or slightly decreased.
Body Fluids			
Total Body Water	Decreased postflight.		Decreased postflight.
Plasma Volume	Decreased postflight (except Gemini 7 and 8).		Markedly decreased postflight. RPB: 2 weeks.
Hematocrit	Normal or slightly decreased postflight.		Normal R+0; decreased R+2 (hydration effect).
Hemoglobin	Normal or slightly increased postflight.	Increased first inflight sample; slowly declines later inflight.	Decreased postflight; RPB: 1-2 months.
Red Blood Cell (RBC) Mass	Decreased postflight; RPB: at least 2 weeks.	Decreased ~15% during first 2-3 weeks inflight; begins to recover after about 60 days; recovery of RBC mass is independent of the presence or absence of gravity.	Decreased postflight; RPB: 2 weeks to 2 months following landing.
Red Cell Half-Life (^{51}Cr)	No change.		No change.
Iron Turnover			No change.
Mean Corpuscular Volume (MCV)	Increased postflight; RPB: at least 2 weeks.		Variable, but within normal limits.
Mean Corpuscular Hemoglobin (MCH)	Increased postflight; RPB: 2 weeks.		Variable, but within normal limits.
Mean Corpuscular Hemoglobin Concentration (MCHC)	Increased postflight; RPB: at least 2 weeks.		Variable, but within normal limits.

Table 1
Physiological Changes Associated with Short-Term and Long-Term Space Flight
(Continued)

Physiological Parameter	Short-Term Space Flights (1-14 days)	Long-Term Space Flights (more than 2 weeks)	
		Pre- vs. Inflight	Pre- vs. Postflight
Body Fluids (Continued)			
Reticulocytes	Decreased postflight; RPB: 1 week.		Decreased postflight. In Skylab, RPB: 2–3 weeks for 28-day mission, 1 week for 59-day mission, and 1 day for 84-day mission.
White Blood Cells	Increased postflight, especially neutrophils; lymphocytes decreased; RPB: 1–2 days.		Increased, especially neutrophils; postflight reduction in number of T-cells and reduced T-cell function as measured by PHA* responsiveness; RPB: 3–7 days; transient postflight elevation in B-cells; RPB: 3 days.
Red Blood Cell Morphology	No significant changes observed postflight.	Increase in percentage of echinocytes; decrease in discocytes.	Rapid reversal of inflight changes in distribution of red cell shapes; significantly increased potassium influx; RPB: 3 days.
Plasma Proteins	Occasional postflight elevations in α_α-globulin, due to increases of haptoglobin, ceruloplasmin, and 2_α-macroglobulin; elevated IgA and C_3 factor.		No significant changes.
Red Cell Enzymes	No consistent postflight changes.	Decrease in phosphofructokinase; no evidence of lipid peroxidation and red blood cell damage.	No consistent postflight changes.
Serum/Plasma Electrolytes	Decreased K and Mg postflight.	Decreased Na, Cl, and osmolality; slight increase in K and PO_4.	Postflight decreases in Na, K, Cl, Mg; increase in PO_4 and osmolality.
Serum/Plasma Hormones	Postflight increases in HGH, thyroxine, insulin, angiotensin I, sometimes aldosterone.	Increases in cortisol. Decreases in ACTH, insulin.	Postflight increases in angiotensin, aldosterone, thyroxine, TSH and GH; decrease in ACTH.
Serum/Plasma Metabolites & Enzymes	Postflight increases in blood urea nitrogen, creatinine, and glucose; decreases in lactic acid dehydrogenase, creatinine phosphokinase, albumin, triglycerides, cholesterol, and uric acid.		Postflight decrease in cholesterol, uric acid.

*Phytohemagglutination.

130

Physiological Parameter	Short-Term Space Flights (1-14 days)	Long-Term Space Flights (more than 2 weeks)	
		Pre- vs. Inflight	Pre- vs. Postflight
Body Fluids *(Continued)*			
Urine Volume	Decreased postflight.	Decreased early inflight.	Normal or slightly increased.
Urine Electrolytes	Postflight increases in Ca, creatinine, PO_4, and osmolality, Decreases in Na, K, Cl, Mg.	Increased osmolality, Na, K, Cl, Mg, Ca, PO_4. Decrease in uric acid excretion.	Increase in Ca excretion; initial postflight decreases in Na, K, Cl, Mg, PO_4, uric acid, Na and Cl excretion increased in 2nd and 3rd week postflight.
Urinary Hormones	Inflight decreases in 17-OH-corticosteroids, increase in aldosterone; postflight increases in cortisol, aldosterone, ADH, and pregnanediol; decreases in epinephrine, 17-OH-corticosteroids, androsterone, and etiocholanolone.	Inflight increases in cortisol, aldosterone, and total 17-ketosteroids; decrease in ADH.	Increase in cortisol, aldosterone, norepinephrine; decrease in total 17-OH-corticosteroids, ADH.
Urinary Amino Acids	Postflight increases in taurine and β-alanine; decreases in glycine, alanine, and tyrosine.	Increased inflight.	Increased postflight.
Sensory Systems			
Audition	No change in thresholds postflight.		No change in thresholds postflight.
Gustation and Olfaction	Subjective and varied human experience. No impairments noted.	Same as shorter missions.	Same as shorter missions.
Somatosensory	Subjective and varied human experience. No impairments noted.	Subjective experiences (e.g., tingling of feet).	
Vision	Transitory postflight decrease in intraocular tension; postflight decreases in visual field; constriction of blood vessels in retina observed postflight; dark adapted crews reported light flashes with eyes open or closed; possible postflight changes in color vision. Decrease in visual motor task performance and contrast discrimination.	Light flashes reported by dark adapted subject; frequency related to latitude (highest in South Atlantic Anomaly, lowest over poles).	No significant changes except for transient decreases in intraocular pressures.

Table 1
Physiological Changes Associated with Short-Term and Long-Term Space Flight
(Continued)

Physiological Parameter	Short-Term Space Flights (1-14 days)	Long-Term Space Flights (more than 2 weeks)	
		Pre- vs. Inflight	Pre- vs. Postflight
Sensory Systems *(Continued)*			
Vestibular System	40-50% of astronauts/cosmonauts exhibit inflight neurovestibular effects including immediate reflex motor responses (postural illusions, sensations of tumbling or rotation, nystagmus, dizziness, vertigo) and space motion sickness (pallor, cold sweating, nausea, vomiting). Motion sickness symptoms appear early inflight, and subside or disappear in 2-7 days. Postflight difficulties in postural equilibrium with eyes closed, or other vestibular disturbances.	Inflight vestibular disturbances are same as for shorter missions; markedly decreased susceptibility to provocative motion stimuli (cross-coupled angular acceleration) after 2-7 days adaptation period. Cosmonauts have reported occasional reappearance of illusions during long-duration missions.	Immunity to provocative motion continues for for several days postflight. Marked postflight disturbances in postural equilibrium with eyes closed. Some cosmonauts exhibited additional vestibular disturbances postflight, including dizziness, nausea, and vomiting.
Musculoskeletal System and Anthropometry			
Height	Slight increase during first week inflight (~1.3 cm). RPB: 1 day.	Increased during first 2 weeks inflight (maximum 3-6 cm); stabilizes thereafter.	Height returns to normal on R+0.
Mass	Postflight weight losses average about 3.4%; about 2/3 of the loss is due to water loss, the remainder due to loss of lean body mass and fat.	Inflight weight losses average 3-4% during first 5 days; thereafter, weight gradually declines for the remainder of the mission. Early inflight losses are probably mainly due to loss of fluids; later losses are metabolic.	Rapid weight gain during first 5 days postflight, mainly due to replenishment of fluids. Slower weight gain from R+5* to R+2 or 3 weeks. Amount of postflight weight loss is inversely related to inflight caloric intake.
Body Composition		Large losses of water, protein and fat during first month inflight. Fat is probably regained. Muscle mass, depending on exercise regimens, is partially preserved.	

---

*Recovery day plus postflight days.

Table 1

Physiological Changes Associated with Short-Term and Long-Term Space Flight
(Continued)

Physiological Parameter	Short-Term Space Flights (1-14 days)	Long-Term Space Flights (more than 2 weeks)	
		Pre- vs. Inflight	Pre- vs. Postflight
Musculoskeletal System and Anthropometry *(Continued)*			
Total Body Volume	Decreased postflight.		Decreased postflight. Center of mass has shifted toward head.
Limb Volume	Inflight leg volume decreases exponentially during first mission day; thereafter, rate of decrease declines until reaching a plateau within 3-5 days. Postflight decrements in leg volume up to 3%; rapid increase immediately postflight, followed by slower RPB.	Early inflight period same as short missions. Leg volume may continue to decrease slightly throughout mission. Arm volume decreases slightly.	Rapid increase in leg volume immediately postflight, followed by slower RPB.
Muscle Strength	Decreased inflight and postflight; RPB: 1–2 weeks.		Postflight decrease in leg muscle strength, particularly extensors. Increased use of inflight exercise appears to reduce postflight strength losses, regardless of mission duration. Arm strength is normal or slightly decreased postflight.
EMG Analysis	Postflight EMGs from gastrocnemius suggest increased susceptibility to fatigue and reduced muscular efficiency. EMGs from arm muscles show no change.		Postflight EMGs from gastrocnemius showed shift to higher frequencies, suggesting deterioration of muscle tissue; EMGs indicated increased susceptibility to fatigue. RPB: about 4 days.
Reflexes (Achilles Tendon)	Reflex duration decreased postflight.		Reflex duration decreased postflight (by 30% or more). Reflex magnitude increased. Compensatory increase in reflex duration about 2 weeks postflight; RPB: about 1 month.
Nitrogen and Phosphorus Balance		Negative balances early inflight; less negative or slightly positive balances later inflight.	Rapid return to markedly positive balances postflight.

Physiological Parameter	Short-Term Space Flights (1-14 days)	Long-Term Space Flights (more than 2 weeks)	
		Pre- vs. Inflight	Pre- vs. Postflight
Musculoskeletal System and Anthropometry *(Continued)*			
Bone Density	Os calcis density decreased postflight. Radius and ulna show variable changes, depending upon method used to measure density.		Os calcis density decreased postflight; amount of loss is correlated with mission duration. Little or no loss from non-weightbearing bones. RPB is gradual; recovery time is about the same as mission duration.
Calcium Balance	Increasing negative calcium balance inflight.	Excretion of Ca in urine increases during 1st month inflight, then plateaus. Fecal Ca excretion declines until day 10, then increases continually throughout flight. Ca balance is positive preflight, becoming increasingly more negative throughout flight.	Urine Ca content drops below preflight baselines by day 10; fecal Ca content declines, but does not reach preflight baseline by day 20. Markedly negative Ca balance postflight, becoming much less negative by day 10. Ca balance still slightly negative on day 20. RPB: at least several weeks.

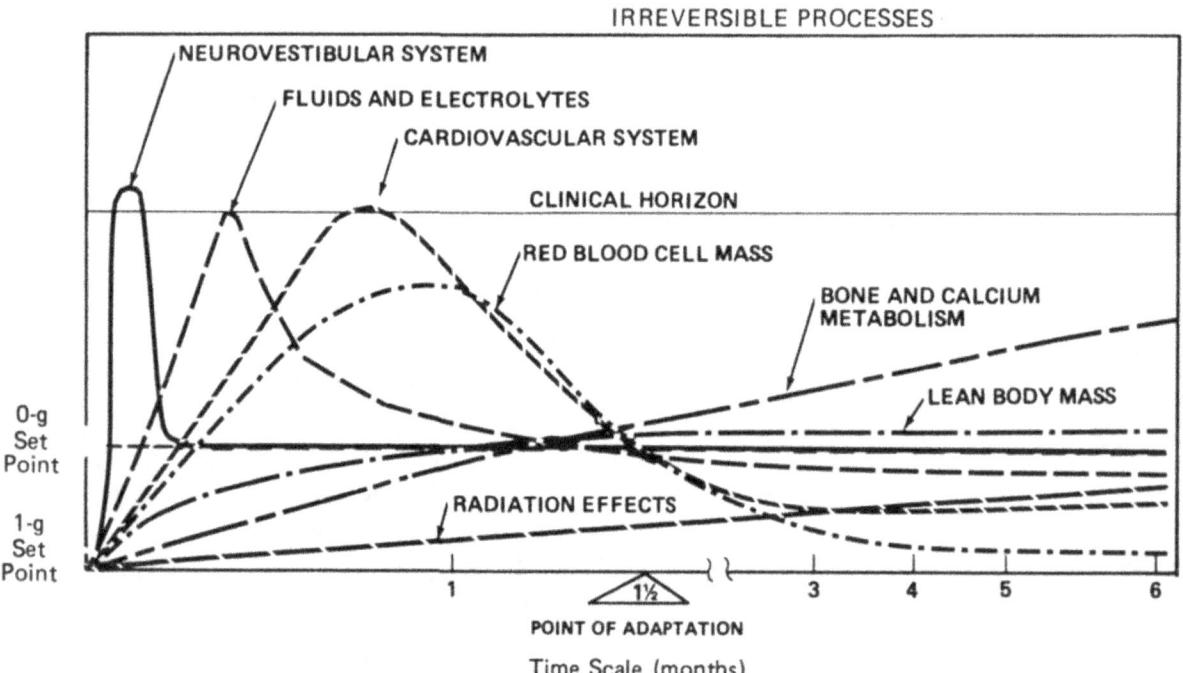

Figure 1. Time course of physiological shifts associated with acclimation to weightlessness.

Most physiological systems appear to reach a new steady state compatible with "normal" function in the space environment within four to six weeks. This acclimation process may not be complete for any physiological system, however, without a very long period of exposure to weightlessness.

Readaptation to Earth's Environment

The biomedical data collected on returning astronauts and cosmonauts indicate that a compensatory period of physiological readaptation to one gravity is required after each space flight. The amount of time necessary for readaptation and the characteristic features of the process exhibit large individual differences. Some differences may be attributed to variations in mission profile and duration, sample size, or the use of countermeasures. Furthermore, different physiological systems appear to achieve readaptation at varying rates. Nevertheless, it is possible to draw tentative conclusions concerning the time course of readaptation, especially for those systems that are minimally affected by mission duration. Figure 2 illustrates these trends.

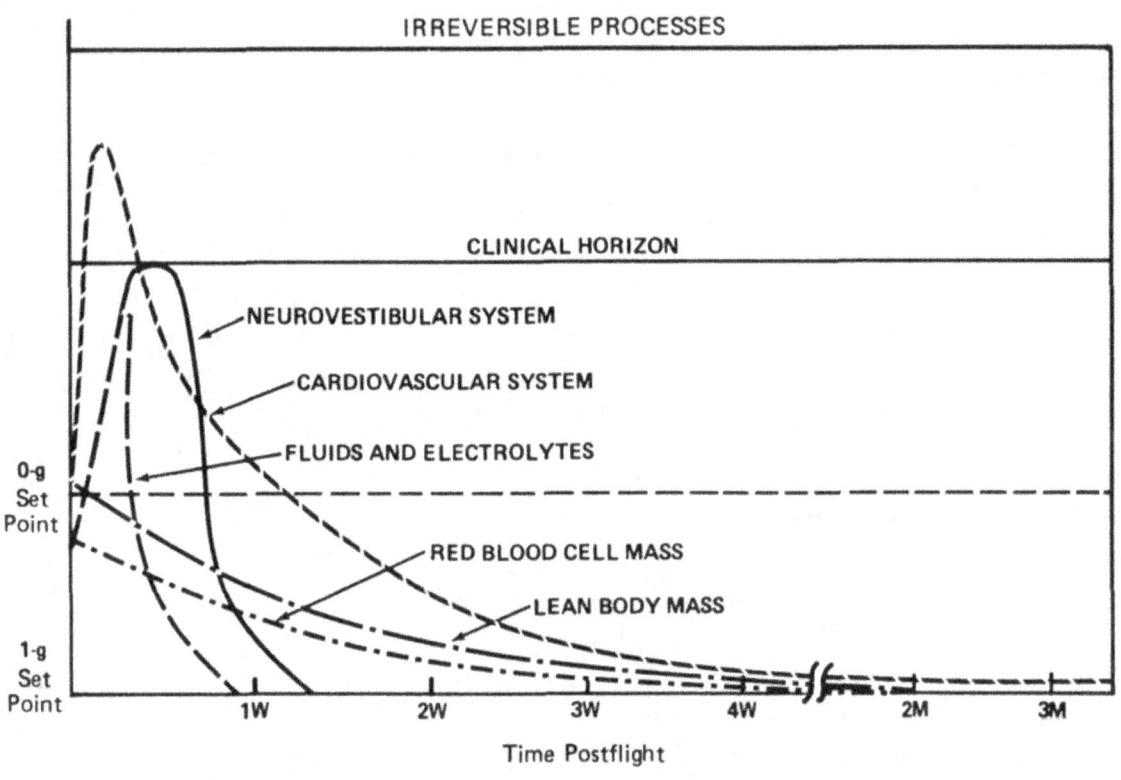

Figure 2. Time course of physiological shifts during readaptation to 1-g, in body systems in which the characteristics of readaptation are minimally dependent on flight duration.

The return to Earth's environment is invariably characterized by the reappearance of shifts in several physiological systems, sometimes leading to overt symptomatology. For example, astronauts and cosmonauts have consistently demonstrated postflight orthostatic intolerance, associated with major realignments due to body fluid shifts and associated reflex responses of the cardiopulmonary neuroreceptors. The postflight readaptation process in the neurovestibular system is usually heralded by difficulties in postural equilibrium; an array of symptoms ranging from mild to frank sickness has been observed in some individuals. However, most of the measured parameters do return to preflight baselines within one to three months postflight. Longer space missions usually require more time for readaptation, but for some parameters (e.g., RBC mass), the reverse has been observed. Again, concern remains regarding the recovery of bone mineral mass and of tissue damaged by radiation. Another unknown involves the effects of weightlessness on lean body mass, which raises the possibility of focal atrophy in the antigravity muscles even if countermeasures, such as vigorous exercise, are not employed.

Individual Differences in Acclimation Capacity

Despite these numerous physiological shifts, humans appear to acclimate quite well to environmental variations accompanying space flight and return to Earth, particularly with the help of adequate countermeasures and application of principles of preventive medicine. So far, astronauts have not demonstrated overt, progressive, or measurable residual pathology upon return to Earth. Nevertheless, there is probably wide variability in individual tolerance and ability to acclimate to weightlessness and to readapt to Earth's gravity. This leads to the hypothesis that there are two major groups of individuals that respond differently to the space environment.

The first group makes the transition from Earth to space and back with no apparent difficulty and no deterioration in performance or health. The second group responds with significant early physiological shifts associated with symptomatic responses during the first week of exposure; this group reaches a reasonably steady state later in flight. It is possible that there is a small third group of individuals who do not reach equilibrium in the space environment, and who will continue to show progressive pathophysiological deterioration despite extensive use of countermeasures.

Superimposed Effects of Combined Stresses

Although space flight is associated with numerous environmental changes that may influence biological systems, the most remarkable and consistent feature is the relative absence of gravity. The continuous gravitational pull has been a major factor in shaping the evolutionary process of all biological systems on Earth. Growth, development, structure, function, orientation, and motion have all evolved features to cope with and take advantage of gravity.

Gross structural changes have been observed in space. For example, astronauts typically demonstrate increases in height (Thornton, Hoffler, & Rummel, 1977). The weightless environment is also associated with a tendency to assume a fetal position, as shown in Figure 3. The well-documented fluid shifts provide additional substantiation for the profound effects of weightlessness (Gazenko, Grigor'yev, & Natochin, 1980; Hoffler, Bergman, & Nicogossian, 1977) and, by inference, for the reliance of structural architecture and functional dynamics on gravity.

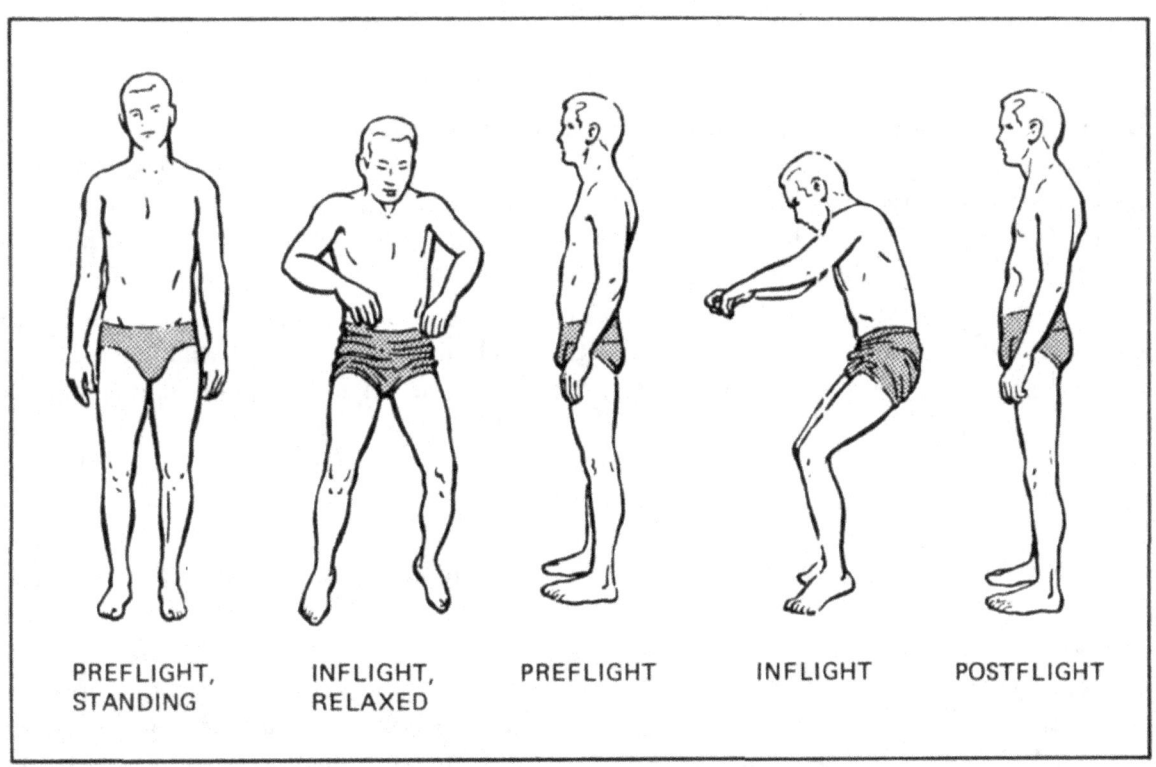

Figure 3. Postural changes inflight. (From Thornton, Hoffler & Rummel, 1977)

These changes in elastic forces in body tissues may be accompanied by changes at the cellular levels. Presumably, some of the endocrine and metabolic alterations observed inflight are related to the mechanical effects of weightlessness.

Although the foregoing discussion points to the fact that weightlessness is the primary contributing factor to the observed physiological changes, it is probably not the sole operant factor. Launch and reentry into Earth's atmosphere entail exposure to increased G forces and vibration which may have distinct physiological impacts (Hordinsky et al., 1981). Once in space, it is impossible for humans to exist without a spacecraft to provide protection and support for physiological and psychological needs. This kind of self-contained and confined habitat, which is rarely encountered on Earth, presents additional environmental stress factors.

On a physiological level, the craft must supply life support systems, such as atmospheric control and food and waste management, in the relative absence of gravity. On a psychological level, the crew must contend with the use of these life support systems, as well as with sustained workloads, sometimes altered work/rest cycles, isolation, confinement, and restricted quarters. The environment outside of Earth's atmosphere also exposes humans to more intense radiation and altered geomagnetic and electrical fields. More research is required to determine what physiological changes these conditions are responsible for, and especially how they interact with the influence of weightlessness.

During the Shuttle era, the nominal duration in space missions will be seven days, although some missions may last as long as 30 days. Opportunities to study the effects of long-term space flight on human physiology will thus be limited. However, Shuttle missions will provide ample opportunity to further define the characteristics and time course of the early acclimation phase and readaptation process postflight. In addition, the Shuttle era will permit investigations of the effects of frequent and repeated space flights on human physiology. It is possible that the acclimation process is more difficult for first-time astronauts. In addition, the determination of cumulative effects resulting from repeated space flights must be a focus of investigation. The completion of two 6-month-duration missions within a period of 18 months by a Soviet cosmonaut, with no apparent adverse effects or difficulties in acclimation (Gazenko, Genin, & Yegorov, 1981) allows some optimism in this regard.

The Space Shuttle will also provide opportunities to study countermeasures, particularly those designed to counteract weightlessness or to optimize the overall acclimation processes. As more information about these processes is obtained, the space environment will become more easily accessible, from a biomedical standpoint, for astronauts and members of the general population. In particular, the advent of space stations and platforms will necessitate a more complete understanding of basic physiological needs and human factors in space. Such data will be essential in, for instance, designing large-scale habitats and providing means to facilitate adaptation to weightlessness in a larger population. The following chapters describe the effects of weightlessness on individual physiological systems in greater detail.

References

Gazenko, O.G., Genin, A.M., & Yegorov, A.D. Major medical results of the Salyut-6/Soyuz 185-day space flight. NASA NDB 2747. Paper presented at the XXXII Congress of the International Astronautical Federation, Rome, Italy, 6-12 September 1981.

Gazenko, O.G., Grigor'yev, A.I., & Natochin, Yu.V. Fluid-electrolyte homeostasis and weightlessness. From JPRS, *USSR Report: Space Biology and Aerospace Medicine*, 30 October, 1980, *14*(5):1-11.

Hoffler, G.W., Bergman, S.A., & Nicogossian, A.E. In-flight lower limb volume measurement. In A.E. Nicogossian (Ed.), *The Apollo-Soyuz Test Project: Medical report* (NASA SP-411). Washington, D.C.: National Aeronautics and Space Administration, 1977.

Hordinsky, J.R., Gebhardt, U., Wegmann, H.M., & Schafer, G. Cardiovascular and biochemical response to simulated space flight entry. *Aviation, Space, & Environmental Medicine*, 1981, *52*(1):16–18.

Thornton, W.E., Hoffler, G.W., & Rummel, J.A. Anthropometric changes and fluid shifts. In R.S. Johnston and L.F. Dietlein (Eds.), *Biomedical results from Skylab* (NASA SP-377). Washington, D.C.: National Aeronautics and Space Administration, 1977.

References Used in Compiling Table 1

Akulinichev, I.T., et al. Results of physiological investigations on the space ships Vostok 3 and Vostok 4. In V.V. Parin (Ed.), *Aviation and space medicine* (NASA TT F-228). Washington, D.C.: National Aeronautics and Space Administration.

Gemini Midprogram Conference (NASA SP-121). Washington, D.C.: National Aeronautics and Space Administration, 1966.

Gemini Summary Conference (NASA SP-138). Washington, D.C.: National Aeronautics and Space Administration, 1967.

Johnston, R.S., & Dietlein, L.F. (Ed.). *Biomedical results from Skylab* (NASA SP-377). Washington, D.C.: National Aeronautics and Space Administration, 1977.

Johnston, R.S., Dietlein, L.F., & Berry, C.A. (Eds.). *Biomedical results of Apollo* (NASA SP-368). Washington, D.C.: National Aeronautics and Space Administration, 1975.

Kakurin, L.I. Medical research prepared on the flight program of the Soyuz-type spacecraft (NASA TTF-141026). Washington, D.C.: National Aeronautics and Space Administration, 1971.

Mercury project summary including results of the fourth manned orbital flight (NASA SP-45). Washington, D.C.: U.S. Government Printing Office, 1963.

Nicogossian, A.E. (Ed.). *The Apollo-Soyuz Test Project: Medical report* (NASA SP-411). Washington, D.C.: National Aeronautics and Space Administration, 1977.

Volynkin, Yu.M., & Vasil'yev, P.V. Some results of medical studies conducted during the flight of the Voskhod. In N.M. Sisakyan (Ed.), *The problems of space biology*, Vol. VI (NASA TT F-528). Washington, D.C.: National Aeronautics and Space Administration, 1969.

Vorob'yev, Ye.I., et al. Some results of medical investigations made during flights of the Soyuz 2-6, Soyuz 7, and Soyuz 8 space ships. *Space Biology and Medicine*, 1970, *4*(2):65-73.

Vorob'yev, Ye.I., Gazenko, O.G., Gurovskiy, N.N., Nefedov, Yu.G., Yegorov, B.B., Bayevsky, R.M., Bryanov, I.I., Genin, A.M., Degtyarev, V.A., Yegorov, A.D., Yeremin, A.V., & Pestov, I.D. Preliminary results of medical investigations carried out during the second mission of the Salyut-4 orbital station. *Space Biology and Aerospace Medicine*, September-October 1976, *10*(5):3-18.

Yegorov, A.D. Results of medical studies during long-term manned flights on the orbital Salyut-6 and Soyuz complex (NASA TM-76014). Moscow: Academy of Sciences USSR, Ministry of Public Health, Institute of Medical and Biological Problems, 1979.

Yegorov, A.D. Results of medical research during the 175-day flight of the third prime crew on the Salyut-6–Soyuz orbital complex (NASA TM-76450). Moscow: Academy of Sciences USSR, 1980.

CHAPTER 8
THE NEUROVESTIBULAR SYSTEM

The neurovestibular system provides information concerning the direction and magnitude of the net gravito-inertial forces acting on the body—information necessary to maintain equilibrium and spatial orientation. The gravitational component is an important part of this total force on Earth. During space flight, when the gravitational force is neutralized, significant realignments occur in the components of this sensory system, in particular in vestibular interactions with other sensory systems, resulting in an array of clinical symptoms. The most important disturbance associated with space flight is the so-called "space motion sickness," a condition that has affected 30 to 40 percent of individuals flying in space.

Vestibular System Structure

The principal components of the vestibular system consist of the otolith organ and the semicircular canals. The otolith organ includes the saccule and utricule, which provide information about gravity and linear acceleration. The receptor cells of the otolith are embedded in a gelatinous mass containing otoconia (crystals of calcium carbonate). The weight of the crystals causes the gelatinous mass to shift its position during changes in head orientation in one gravity, and during linear acceleration in both one and zero gravity. This produces a shearing force on sensory hair cells (cilia) and results in the transmission of neural impulses.

The three semicircular canals provide information about angular accelerations of the head. Each canal consists of a membranous tube filled with endolymph, floating within a bony canal filled with perilymph. At the end of each membranous canal is an enlargement called the ampulla, which contains the sensory receptors. The orientation of each canal roughly corresponds to one of the three anatomical planes of the head; each canal responds maximally to angular acceleration in its plane.

The two components of the vestibular system deal with independent aspects of orientation but do not function entirely independently of each other or of certain other body systems. According to Guedry (1978), the semicircular canals localize the angular acceleration vector relative to the head during head movement and contribute to sensory inputs for (1) appropriate reflex action relative to an anatomical axis and (2) perception of angular velocity about that axis. Perception of orientation relative to the Earth depends upon sensory inputs from the otolith and somatosensory systems. The otoliths provide both static and dynamic orientation information (relative to gravity) and contribute to the perception of tilt.

Space Motion Sickness

Graybiel et al. (1977) distinguish between two categories of vestibular side effects associated with space flight: (1) immediate reflex motor responses, which include postural illusions, sensations of rotation, nystagmus, dizziness, and vertigo; and (2) space motion sickness.

Symptoms

Table 1 presents the clinical symptoms and biochemical changes recorded with motion sickness occurrence in Earth environments, as compared to those in space environments. In general, the progressive cardinal symptoms consist of pallor, cold sweating, nausea, and vomiting. The key symptoms of space motion sickness appear to be similar to those found on Earth, although there have been instances in which vomiting occurred suddenly in space, without the prior occurrence of pallor or nausea (Homick & Miller, 1975). The diagnostic criteria used in Skylab for the evaluation of space motion sickness are shown in Table 2 (Graybiel et al., 1977). In Skylab, as subjects made experimental head movements, an observer in collaboration with the subject estimated the severity of each predesignated symptom and recorded it and any other symptom not mentioned in the table. This and similar techniques (Galle, 1981, for example) provide an approach to the quantitative description of vestibular responses.

Incidence

Table 3 lists the reported incidence of space motion sickness in the U.S. and USSR space missions. About one-half of all astronauts and cosmonauts have experienced at least some symptoms, which range from mild dizziness and stomach awareness to nausea and frank vomiting. Since many spacecrews took anti-motion sickness drugs prophylactically, it is possible that the severity of vestibular symptoms would otherwise have been greater.

It is interesting to note that of the 12 Apollo astronauts who walked on the moon, only three reported mild symptoms, such as stomach awareness or loss of appetite, prior to their moon walk. None reported symptoms while in the one-sixth gravity of the lunar surface. In no instances were symptoms noted upon return to weightlessness.

Susceptibility to Space Motion Sickness

A number of different research techniques have been used over the years to elicit motion sickness symptomatology in susceptible subjects. Examples of these include:

1. Vertical oscillators
2. Swings
3. Caloric irrigation
4. Roll and pitch rockers
5. Visual inversion or reversal techniques
6. Off-vertical rotation
7. Optokinetic stimulation
8. Z-axis recumbent rotation
9. Cross-coupling accelerations produced by head movements
10. Parabolic flights

Table 1
Motion Sickness Manifestations

Physiological System	Manifestations	Spaceflight Findings
Cardiovascular	Changes in pulse rate and/or blood pressure.	Not measured.
	↑ tone of arterial portion of capillaries in the fingernail bed.	
	↓ diameter of retinal vessels.	
	↓ peripheral circulation, especially in the skin of the head.	
	↑ muscle blood flow.	
Respiratory	Alterations in respiration rate.	Not measured.
	Sighing or yawning.	
Gastrointestinal	Inhibition of gastric intestinal tone and secretions.	Not measured.
	Salivation.	Reported.
	Gas or belching.	Reported.
	Epigastric discomfort or awareness.	Reported.
	Sudden relief from symptoms after vomiting.	Reported.
Body fluids, Blood	Changes in LDH concentrations.	Not measured.
	↑ hemoglobin concentration.	
	↑ pH and ↓ $paCO_2$ levels in arterial blood, presumably from hyperventilation.	
	↓ concentration of eosinophils.	
	↑ 17-hydroxycorticosteroids.	
	↑ plasma proteins.	
Urine	↑ 17-hydroxycorticosteroids.	Not measured.
	↑ catecholamines.	
Temperature	↓ body temperature.	Not measured.
	Coldness of extremities.	
Visual System	Ocular imbalance.	Not observed.
	Dilated pupils during emesis.	
	Small pupils.	
Behavioral	Apathy, lethargy, sleepiness, fatigue, weakness.	Reported.
	Depression and/or anxiety.	Reported.
	Mental confusion, spatial disorientation, dizziness, giddiness.	Variable.
	Anorexia, unusual sensitivity to repulsive sights or odors, or excessive discomfort from previously tolerable stimuli such as heat, cold, or tightness of clothing.	Reported.
	Headache, especially frontal headache.	Reported.
	↓ muscular coordination and psychomotor performance.	
	↓ time estimation.	
	↓ motivation.	

Adapted from Money, 1970; Reason and Brand, 1975.

Table 2
Diagnostic Categorization of Different Levels
of Severity of Acute Motion Sickness

Category	Pathognomonic 16 points	Major 8 points	Minor 4 points	Minimal 2 points	AQS[1] 1 point
Nausea syndrome	Nausea III[2], Vomiting or retching	Nausea II	Nausea I	Epigastric discomfort	Epigastric awareness
Skin		Pallor III	Pallor II	Pallor I	Flushing/Subjective warmth $\geqslant$ II
Cold sweating		III	II	I	
Increased salivation		III	II	I	
Drowsiness		III	II	I	
Pain					Headache (persistent) $\geqslant$ II
Central nervous system					Dizziness (persistent) Eyes closed $\geqslant$ II Eyes open III

Levels of Severity Identified by Total Points Scored

Frank sickness (FS) $\geqslant$ 16 points	Severe Malaise (M III) 8-15 points	Moderate Malaise A (M IIA) 5-7 points	Moderate Malaise B (M IIB) 3-4 points	Slight Malaise (M I) 1-2 points

[1]AQS = Additional qualifying symptoms.

[2]III, Severe or marked; II, moderate; I, slight.

From Graybiel et al., 1977

Table 3
Incidence of Space Motion Sickness[1]
In U.S. and USSR Space Missions

United States (as of November 1982)			Soviet Union (as of 1981)		
Program	Number of Crewmen[2]	Incidence of Motion Sickness	Program	Number of Crewmen[2]	Incidence of Motion Sickness
Mercury	6	0	Vostok	6	1
Gemini	20	0	Voskhod	5	3
Apollo	33	11	Soyuz	38	21
Skylab	9	5	ASTP	2	2
ASTP	3	0	Salyut 5	6	2
Shuttle	12	5	Salyut 6	27	12

[1]Reports of one or more symptoms of space motion sickness are included. No attempt was made to categorize by severity of symptoms.

[2]Includes some crewmen who flew more than once.

144

Although some of these tests (particularly the ones that involve exposure to cross-coupled stimulation) have been useful in predicting susceptibility in one-gravity motion environments, they have not successfully predicted space motion sickness susceptibility. Also, personal motion sickness histories, which document previous incidences of motion sickness in provocative one-gravity environments, are not useful in predicting susceptibility to space motion sickness.

There are several projected reasons for the failure of these tests to predict motion sickness susceptibility in space. First, individual susceptibility to different motion environments in one gravity often varies widely, and it is common for people to be susceptible under certain motion conditions without transfer to other motion environments. Second, an astronaut's susceptibility to space motion sickness probably involves three major features:

1. Initial "susceptibility" to the provocative motion environment
2. Rate of adaptation
3. Degree of adaptation.

Most of the tests that have been used to predict susceptibility to space motion sickness so far only measure the first component; however, an individual's capacity to adapt to a new motion environment may be more important than his initial susceptibility to an acute motion stimulus.

Other variables that may have interfered with the predictive accuracy of these tests include the prophylactic and therapeutic use of anti-motion sickness remedies, and the limited number of subjects.

Time Course of Adaptation

Susceptibility to space motion sickness varies not only among individuals, but within the same individual over time. For example, it appears that individuals who have participated in previous space missions are less likely to experience space motion sickness than first-time astronauts. On any particular flight, the symptoms of space motion sickness usually appear in susceptible individuals early after insertion into orbit, and are aggravated by head and body movement, especially if the eyes are open. In most cases, the symptoms disappear within two to four days, and do not recur for the remainder of the mission. According to Yakovleva et al. (1980), adaptation occurs in one of three patterns: (1) very little or no sensory discomfort, (2) intense discomfort for a short period, or (3) prolonged adaptation without severe symptoms.

On the Skylab missions, Graybiel et al. (1977) measured susceptibility to provocative motion stimuli preflight, inflight (beginning on or after Mission Day 5), and postflight to determine changes in thresholds for vestibular response. The stressful motion environment was produced by rotating subjects in a litter chair and requiring them to execute head movements with eyes covered. Symptoms were scored using the criteria in Table 2. In general, once acclimatized, the astronauts demonstrated a marked increase in the threshold of susceptibility to motion sickness during the inflight provocative tests. This "immunity" to cross-coupled angular acceleration stimuli continued in these subjects for a week or more postflight.

Most experiments and observations indicate that readaptation to one gravity involves a reappearance of vestibular side effects, particularly for space crews participating in longer missions. Tests of postural equilibrium with eyes closed revealed marked postflight deficits in one Apollo crewmember and all of the Skylab crews (Homick & Miller, 1975; Homick, Reschke, & Miller, 1977). Postural stability appeared to return to normal within ten days. Several of the returning astronauts from Skylab and Apollo reported a variety of additional postflight disturbances, which appear to be vestibular in origin and included dizziness and lightheadedness, vertigo during rapid head movements, and difficulty in turning corners.

More pronounced postflight vestibular disturbances have been experienced by Soviet cosmonauts. These include lasting postural instability; sweating while walking; dizziness; nausea; and vomiting, particularly during head movements (Gazenko, 1979). One of the crewmembers of the 175-day Salyut mission had, in addition to most of the foregoing, long-lasting postural difficulties. He also experienced pronounced illusory reactions (autokinetic illusions during fixation) postflight (Matsnev, in press). Some Soviet investigators (Kornilova et al., 1979), comparing postflight measures of otolith reflex and statokinetic disorders in several Salyut crews, concluded that the degree and duration of symptoms was proportional to mission length.

Mechanisms Underlying Space Motion Sickness

Two major theories that have been advanced to account for space motion sickness are the sensory conflict, or "sensory rearrangement," theory and the fluid shift hypothesis. A third theory suggests that a slight difference between the weights of the left and right otoconia may cause disturbances as the body adjusts to the loss of this difference.

Sensory Conflict Theory

The most likely hypothesis to explain the onset of space motion sickness, and motion sickness in general, is the sensory conflict or "sensory rearrangement" theory. Under one-gravity conditions, human orientation in three-dimensional space is based on four sensory inputs. These are (1) otolithic information on the gravity vector and linear accelerations, (2) angular acceleration data provided by the semicircular canals, (3) visual information concerning orientation of the body, and (4) pressure information from the proprioceptive receptors. In normal environments, information from these systems is compatible and complimentary, and matches that expected on the basis of previous experience. When the motion environment is altered in such a way that the information from the body sensory systems is not compatible and/or does not match previously stored neural patterns, motion sickness results. Reason and Brand (1975) distinguish between two different kinds of sensory rearrangements that can induce motion sickness:

1. Visual-inertial rearrangement, in which the sensory conflict arises between input from the visual system and vestibular apparatus.
2. Canal-otolith rearrangement, in which the conflict is between the input from the semicircular canals and the otolith organ.

146

Fluid Shift Hypothesis

According to this theory, the cephalad fluid shifts accompanying weightlessness might produce concomitant changes in cranial pressure, thereby altering the response properties of vestibular receptors. For example, engorgement of the blood vessels surrounding the endolymphatic duct might restrict the flow of endolymph from the cochlea to the endolymphatic sac, resulting in hydrops.

Parker and Money (1978) point out that most evidence does not favor this hypothesis. For example, Graybiel and Lackner (1977) did not find an increased susceptibility to provocative motion stimulation during head-down tilt. Furthermore, Soviet scientists have not been able to predict individual susceptibility to space motion sickness using this hypothesis (Homick, 1978).

Otolith Asymmetry

Von Baumgarten and Thumler (1979) have proposed a mechanism which complements the sensory conflict theory to explain the adaptation to weightlessness and readaptation to one-gravity, and the individual differences in space motion sickness susceptibility. They suggest that the central nervous system compensates for innate differences in the weights of right and left otoconial membranes by providing neural impulses to make up any deficit from the underweight membrane. This compensation is inappropriate in zero gravity, since the weight differential is nullified. The result is an imbalance producing rotary vertigo, eye movements, and posture changes until the central "compensating centers" adjust to the new situation. A similar imbalance would be produced upon return to one-gravity, resulting in postflight vestibular disturbances. Individuals with a greater degree of asymmetry in otolith morphology would thus be more susceptible to space motion sickness.

Prevention and Treatment of Space Motion Sickness

The disruptive nature of space motion sickness, occurring as it does during the early and critical stages of a mission, has led to a variety of approaches attempting to prevent or control this malady. Only limited success has been achieved to date.

Research directed toward the prevention of space motion sickness has proceeded on three broad lines of inquiry. Each of these research fields is described below in terms of its potential for control and problems associated with its use.

Training

The use of training procedures for the control of space motion sickness is based on the general physiological principle that increasing the levels of a stress agent will lead to a heightened level of adaptation. The problem, of course, is that training for space motion sickness prior to a mission cannot be conducted in a zero-gravity environment. Therefore, the training must be conducted in a one-gravity situation on Earth, recognizing that the transfer of the training effect to the space situation may be limited or might even be non-existent.

147

It is difficult to assess the training effectiveness of any of these procedures inasmuch as control groups—individuals not participating in the training—are not used. Therefore, there is no baseline against which to measure the amount of protection afforded. However, there is at least a basis in theory for each of these procedures.

One type of adaptation training consists of short exposures to weightlessness while flying in a parabolic trajectory. The brief duration of zero gravity makes this more a matter of indoctrination than of real training.

Aerobatic maneuvers in high performance aircraft also may be used to produce the stimulation known to cause motion sickness in susceptible individuals. Astronauts frequently have participated in flights of this type prior to missions. Based on anecdotal reports, one can assume that some protection is afforded. However, results from space missions, in which astronauts have suffered space motion sickness, make it clear that this protection is not total.

A third procedure for adaptation training places the subject in a rotating environment, such as a rotating chair or a slow rotation room. Work by Graybiel and his colleagues has shown that it is possible to lower an individual's susceptibility to motion sickness in a particular environment by exposing him to gradually increasing levels of stress intensity. In one study (Graybiel & Knepton, 1978), it was found that over-adaptation in one motion environment appears to provide protection in other motion environments. Using the slow rotating room, subjects executed standardized head and body movements leftward or rightward until either the motion sickness end point was reached or 1200 head movements were executed. Next, subjects executed head movements in the three unpracticed quadrants. A marked adaptation effect was measured under these conditions. This transfer of training effect now needs to be evaluated for possible benefit under weightless conditions.

Pharmacological Countermeasures

It has long been known that drug therapy offers some relief for many individuals who suffer from motion sickness in aircraft or in any moving vehicle. Logically, therefore, as soon as a motion sickness problem was identified in space flight, a search for effective pharmacological control began. Drugs selected for evaluation began with those known to offer some benefit for the motion sickness encountered in one gravity.

Graybiel et al. (1975) reported on a series of evaluations using drugs of known effectiveness with administration both singly and in certain combinations. The authors concluded that a fixed-dose combination of promethazine hydrochloride and ephedrine sulfate (25 mg each) proved to be outstanding as this combination exhibited a supra-summation effect. Substantial individual differences in response to the drugs also were noted, implying that individual assessments must be made for best protection. Finally, based on a few tests conducted using larger than usual doses, it was concluded that individuals may vary both with regard to the choice of drug and to the amount administered.

There has been recent interest in the Transdermal Therapeutic System (TTS), a topical method for delivering an anti-motion sickness drug (scopolamine). This allows the drug to be administered to the systemic circulation in a controlled manner for a predetermined period. Although claims have been made that transdermal application minimizes side effects, recent evaluations indicate this may not be so. Side effects may include dryness of the mouth, drowsiness, and blurred vision.

Wurster et al. (1981) compared the transdermal application of scopolamine with other drug combinations administered orally. Because of the great intra- and inter-individual variations, the responses to the drugs were not significant statistically. However, they revealed a strong tendency for both a promethazine-ephedrine combination and scopolamine, each administered orally, to reduce motion sickness. Scopolamine administered transdermally could not be distinguished from the placebo.

A more recent evaluation of transdermally administered scopolamine (Homick et al., 1982) reached many of the same conclusions as earlier studies. Here the particular interest was in a form of administration which would allow a drug such as scopolamine to be effective for a longer period than the four to six hours found with oral administration. Results of this study showed average improvements of 45, 32, 53, and 54 percent, relative to placebo, where measured at the 16-, 24-, 48-, and 72-hour test times. Although there was a consistent beneficial effect, it was not statistically significant since individual differences were large. The authors feel that one causal factor may be variations in rates of scopolamine absorption or metabolism. On the basis of a number of side effects noted and the large individual variations, it was concluded that operational use of TTS-scopolamine should be contingent on careful screening and further efficacy testing.

The current practice of NASA for the prevention of space motion sickness is to administer prophylactic therapy to those individuals about to fly for the first time in space and to those who have flown previously and experienced motion sickness. The general practice is to use an orally administered drug combination of scopolamine (0.4 mg) and dextroamphetamine (5.0 mg). The exact dosage is adjusted individually based on ground-based vestibular testing prior to flight.

Biofeedback Procedures

Use of autogenic biofeedback techniques as a means for training subjects to control the symptoms of space motion sickness is being considered, especially in the light of an Air Force program for rehabilitating crewmen grounded with supposedly disqualifying air sickness. The premise of the Air Force work is that air sickness (pallor, cold sweating, stomach awareness, nausea, etc.) is an autonomic nervous system response. Interruption of this response through voluntary control of the autonomic nervous system, therefore, should lead to prevention of air sickness. In the Air Force program (Levy, 1980), air crews are given relaxation training in a two-axis motion chair with the individual connected to physiological monitoring equipment. As rotation increases, patients recognize physiological responses prior to actual motion sickness symptoms. Through biofeedback conditioning, a student pilot rotating at 5 rpm, for example, who experiences a surface skin temperature drop of 8 to 10 degrees F, should at the conclusion of the treatment be able to control this to a 1- to 2-degree drop and even to reverse the trend. Notable success has been achieved in restoring crewmen to flight status.

Additional support for the feasibility of biofeedback techniques comes from research at the NASA Ames Research Center. Cowings and Toscano (1982) used biofeedback techniques with one group highly susceptible to motion sickness and another identified as moderately susceptible. These were compared with two control groups of matched susceptibility. Both treatment groups showed significantly improved performance on the Coriolis Sickness Susceptibility Index.

Although biofeedback procedures offer promise, investigators have identified certain problems with their use. Lackner (1978) notes that there are great individual differences in the extent to which subjects are able to gain control over the variable being "reinforced" and that laboratory training may not transfer to situations where the subject's active participation in other tasks is required. The effectiveness of biofeedback techniques in preventing motion sickness under operational conditions remains to be determined.

Mechanical Devices

It has been reported that space motion sickness is aggravated by movement of the body and, in particular, the head and neck. Soviet scientists, operating on the principle that a reduction in movement should result in a decrease in the severity of symptoms of space motion sickness, have developed a head restraint system. Crewmembers of the Salyut 6 mission used as a countermeasure this specially designed device, termed "A Neck Pneumatic Shock-Absorber" (Matsnev, in press). The device supplies a controlled load of known force to the cervical vertebrae and neck anti-gravitational muscles and restricts head movements during adaptation to weightlessness (Figures 1 and 2). The device consists of a soft cap with loop holes for rubber cords. With the cap on, a crewman must exert considerable force with his neck muscles in order to move his head from its natural erect position. Movement requires a force equal to the tensile strength of the two attached rubber cords. The device includes unstretchable straps attached to the right and left shoulders which restrict head tilt and turn. Flight test results from the Salyut missions indicate that use of the head restraint cap was of benefit in controlling the development of space motion sickness symptomatology. Matsnev concludes that the benefit of this head restraint system may be a result of its control of the vestibulo-cervical reflex, known to involve the labyrinth (semicircular canals and otolith organs) as a receptor and neck muscles as effectors.

In addition to the head-restraint cap, Soviet crewmen also have used specially tailored shoes which provide a pressure of up to 60 torrs to the soles of the feet (Matsnev, in press). A Soyuz 38 cosmonaut reported that use of this device reduced the severity of spatial illusions and motor disturbances.

Summary

The weightless environment of space produces disturbances in the functioning of the vestibular apparatus that result in a number of side effects, the most important of which is space motion sickness. The major symptoms of this disorder, which occurs in susceptible individuals soon after insertion into orbit, include nausea and vomiting. These symptoms typically subside or disappear

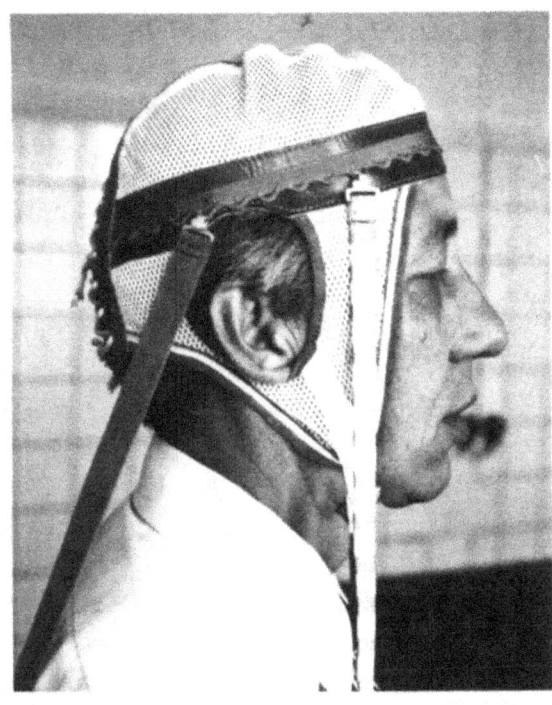

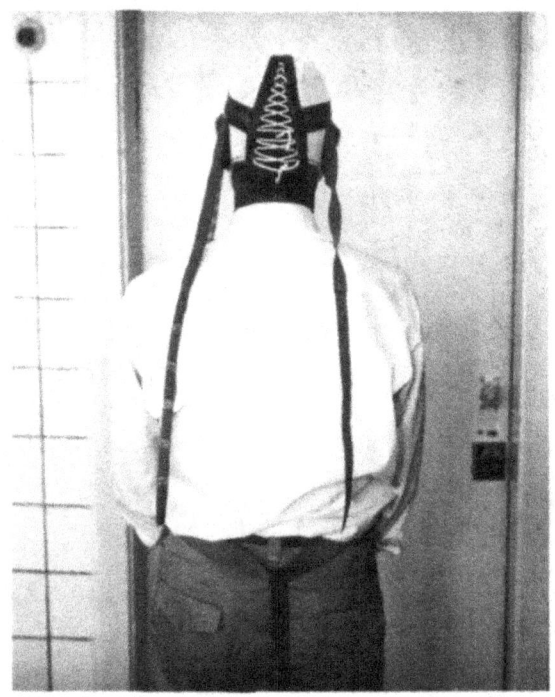

Figure 1. Head-restraint helmet developed for testing by Salyut 6 crewmembers.

Figure 2. Rear view of head-restraint helmet shows straps restricting head tilt.

within two to four days. Almost one-half of the astronaut and cosmonaut population has shown at least some symptoms of space motion sickness inflight. However, tests designed to measure vestibular function or susceptibility to motion sickness in one gravity have not successfully predicted which individuals will be susceptible inflight.

Dietlein and Johnston (1981) state that space motion sickness "represents the greatest research challenge facing life scientists in contemporary space medicine and physiology." A solution to the problem of space motion sickness will not come easily. Progress in all likelihood will come principally from experiments carried out in space. Investigations to be conducted on the Spacelab missions will study the function of the otolith apparatus in humans and the morphology of the otolith in animals. Of particular importance will be use of a human linear accelerator to measure neurovestibular responses associated with weightlessness. Through these programs, new knowledge concerning the functioning of the vestibular system, its interaction with other neurosensory components, its rate of adaptation, and the effectiveness of procedural and pharmacological preventive measures may be achieved.

References

Armstrong, H.S. Air sickness. In H.G. Armstrong (Ed.), *Aerospace Medicine*. Baltimore, MD: Williams & Wilkins Co., 1961.

Barrett, S.V., & Thornton, C.L. Relationship between perceptual style and simulator sickness. *Journal of Applied Psychology*, 1968, *52*: 304-308.

Chinn, H.I. Motion sickness in the military service. *Military Surgeon*, 1951, *108*: 20-29.

Chinn, H.I. Evaluation of drugs for protection against motion sickness aboard transport ships. *Journal of the American Medical Association*, 1956, *106*: 755-760.

Cowings, P.S., & Toscano, W.B. The relationship of motion sickness susceptibility to learned autonomic control for symptom suppression. *Aviation, Space, and Environmental Medicine*, 1982, *53*(6): 570-575.

Dietlein, L.F., & Johnston, R.S. U.S. manned space flight: The first twenty years. A biomedical status report. *Acta Astronautica*, 1981, *8*(9-10): 893-906.

Galle, R.R. Quantitative evaluation of clinical manifestations of motion sickness. *Space Biology and Aerospace Medicine*, 1981, *15*(3): 72-75.

Gazenko, O.G. (Ed.). *Summaries of reports of the 6th All-Soviet Union Conference on Space Biology and Medicine* (Vol. I & II). Kaluga, USSR, June 5-7, 1979.

Graybiel, A. Measurement of otolith function in man. In H.H. Kornhuber (Ed.), *Vestibular system—Part 2: Psychophysics, applied aspects and general interpretations*. Berlin: Springer-Verlag, 1974.

Graybiel, A. Prevention and treatment of space sickness in Shuttle-Orbiter missions. *Aviation, Space, and Environmental Medicine*, 1979, *50*(2): 171-176.

Graybiel, A., & Knepton, J. Bidirectional overadaptation achieved by executing leftward or rightward head movements during unidirectional rotation. *Aviation, Space, and Environmental Medicine*, 1978, *49*(1): 1-4.

Graybiel, A., & Lackner, J.R. Comparison of susceptibility to motion sickness during rotation at 30 rpm in the Earth-horizontal, 10° head-up, and 10° head-down positions. *Aviation, Space, and Environmental Medicine*, 1977, *48*(1): 7-11.

Graybiel, A., Miller, E.F., & Homick, J.L. Experiment M131: Human vestibular function. In R.S. Johnston and L.F. Dietlein (Eds.), *Biomedical results from Skylab* (NASA SP-377). Washington, D.C.: U.S. Government Printing Office, 1977.

Graybiel, A., Wood, C.D., Knepton, J., Hoche, J.P., & Perkins, G.F. Human assay of antimotion sickness drugs. *Aviation, Space, and Environmental Medicine*, 1975, *46*(9): 1107-1118.

Graybiel, A., Wood, C.D., Miller, E.F., II, & Cramer, D.B. Diagnostic criteria for grading the severity of acute motion sickness. *Aerospace Medicine*, 1968, *39*: 453-455.

Guedry, F.E., Jr. Vestibular function. In *U.S. naval flight surgeon's manual* (2nd ed.). Prepared under Office of Naval Research Contract N00014-76-C-1010 by the Naval Aerospace Medical Institute and BioTechnology, Inc. Washington, D.C.: U.S. Government Printing Office, 1978.

Hemingway, A. Airsickness during early flying training. *Journal of Aviation Medicine*, 1945, *16*: 409-416.

Hill, J. The care of the sea-sick. *British Medical Journal*, 1936, Oct.-Dec.: 802-807.

Homick, J.L. Special note: Space motion sickness in the Soviet manned spaceflight program. In J.L. Homick (Ed.), *Space motion sickness symposium proceedings*. Lyndon B. Johnson Space Center, Houston, TX, 15-17 November 1978. (Prepared by General Electric Company under Purchase Order T-1830G).

Homick, J.F., Degioanni, J., Reschke, M.F., Leach, C.S., Kohl, R.L., & Ryan, P.C. An evaluation of the time course of efficacy of transdermally administered scopolamine in the prevention of motion sickness. *Preprints of the 1982 Annual Scientific Meeting of the Aerospace Medical Association*, Bal Harbour, FL, 1982 (ISSN 0065-3764).

Homick, J.F., & Miller, E.F., II. Apollo flight crew vestibular assessment. In R.S. Johnston, L.F. Dietlein, and C.A. Berry (Managing Eds.), *Biomedical results of Apollo* (NASA SP-368). Washington, D.C.: U.S. Government Printing Office, 1975.

Homick, J.L., Reschke, M.F., & Miller, E.F., II. The effects of prolonged exposure to weightlessness on postural equilibrium. In R.S. Johnston and L.F. Dietlein (Eds.), *Biomedical results from Skylab* (NASA SP-377). Washington, D.C.: U.S. Government Printing Office, 1977.

Kaplan, I. Motion sickness on railroads. *Industrial Medicine and Surgery*, 1964, *33*: 648-651.

Kornilova, L.N., Syrykh, G.D., Tarasov, I.K., & Yakovleva, I.Ya. Results of the investigation of the otolith function in manned space flights (NASA TM-76103). Translated from *Vestnik Otorino-laringologii*, 1979, *6*: 21-24.

Lackner, J.R. Training countermeasures. In J.L. Homick (Ed.), *Space motion sickness symposium proceedings*. Lyndon B. Johnson Space Center, Houston, TX, 15-17 November 1978. (Prepared by General Electric Company under Purchase Order T-1830G).

Levy, R.A. Biofeedback rehabilitation of airsick aircrew. *Preprints of the 1980 Annual Scientific Meeting of the Aerospace Medical Association*, Anaheim, CA, 1980.

Matsnev, E.I. Space motion sickness: Phenomenology, countermeasures, mechanisms. Moscow, USSR: Institute of Biomedical Problems, USSR Ministry of Health, in press.

McDonough, F.E. NRC Committee on Aviation Medicine, Report No. 181, 1943.

Miller, E.F., II. Evaluation of otolith organ function by means of ocular counter-rolling. In J. Stahle (Ed.), *Vestibular function on Earth and in space*. Oxford: Pergamon Press, 1970.

Money, K.E. Motion sickness. *Physiology Review*, 1970, *50*: 1-39.

Parker, D.E., & Money, K.E. Vestibular/motion sickness mechanism. In J.L. Homick (Ed.), *Space motion sickness symposium proceedings*. Lyndon B. Johnson Space Center, Houston, TX, 15-17 November 1978. (Prepared by General Electric Company under Purchase Order T-1830G).

Reason, J.T. Relations between motion sickness susceptibility, the spiral after-effect and loudness estimation. *British Journal of Psychology*, 1968, *59*: 385-393.

Reason, J.T., & Brand, J.J. *Motion sickness*. London: Academic Press, 1975.

Reason, J.T., & Graybiel, A. An attempt to measure the degree of adaptation produced by differing amounts of Coriolis vestibular stimulation in the Slow Rotation Room (NAMI-1084, NASA Order R-93). Naval Aerospace Medical Institute, Pensacola, FL, 1969.

von Baumgarten, R.J., & Thumler, R.R. A model for vestibular function in altered gravitational states. In R. Holmquist (Ed.), *Life sciences and space research* (Vol. XVII). Oxford: Pergamon Press, 1979.

Whittingham, H.E. Medical aspects of air travel. I. Environment and immunization requirements. *British Medical Journal*, 1953, *1*: 556-558.

Wilding, J.M., & Meddis, R. Personality correlates of motion sickness. *British Journal of Psychology*, 1972, *63*: 619-620.

Wurster, W.H., Burchard, E.C., & von Restorff, W. Comparison of oral and TTS-scopolamine with respect to anti-motion sickness potency and psychomotor performance. *Preprints of the 1981 Annual Scientific Meeting of the Aerospace Medical Association* (ISSN 0065-3764), San Antonio, TX, 1981.

Yakovleva, I. Ya., Kornilova, L.N., Tarasov, I.K., & Alekseyev', V.N. Results of the study of the vestibular apparatus and the functions of the perception of space in cosmonauts (pre- and post-flight observations) (NASA TM-76485). Translated into English from *Report to the XI Joint Soviet-American Working Group on Space Biology and Medicine*, Moscow, October 1980.

Yegorov, A.D. Results of medical research during the 175-day flight of the third prime crew on the Salyut 6-Soyuz orbital complex (NASA TM-76450). Moscow, USSR: Academy of Sciences USSR, Ministry of Health, 1980.

CHAPTER 9
NEUROPHYSIOLOGY AND PERFORMANCE

Since the onset of the Mercury program in the U.S. and the Vostok flights in the USSR, attempts have been made to evaluate the neurophysiological manifestations associated with exposure to the space environment. There has been ongoing concern over the capability of a human operator to perform complex tasks in space such as docking maneuvers, extravehicular activities, and navigation assignments due to potential performance decrements associated with observed physiological changes. During the period of manned space flight in both nations, many attempts have been made to study the impact of neurophysiological change on different sensory-motor performance parameters. The information obtained has come through a combination of objective measurements and subjective reports. However, due to the complexities and time constraints associated with conducting research during a space mission, neither approach has proven completely satisfactory in defining with precision the nature and magnitude of neurophysiological change in weightlessness. Nonetheless, data from these experiments have been valuable in placing the different issues in a hierarchy of relative importance. As a result, where it has been found that the observed changes (for example, visual performance) did not impact mission success, research interest has declined. In other instances (for example, vestibular dysfunction), research interest has intensified.

The following sections describe the principal neurophysiological/performance issues which have been studied in manned space programs. While these issues must be addressed in planning for long-term space missions, they are not key factors underlying mission success. There is one exception, however. Vestibular dysfunction has been observed repeatedly in manned missions. "Space motion sickness" can seriously degrade mission performance during initial phases. Due to the importance of this disturbing phenomenon, therefore, vestibular neurophysiology is treated in a separate chapter.

Visual Performance

The visual system is the most critical of all the senses for orientation and adaptation to living and working conditions in space. Dr. Joseph Kerwin noted that, during his initial hours in Skylab 2, when he closed his eyes, "My instinct was to grab hold of whatever was nearest and just hang on, lest I fall." His orientation to the spacecraft was entirely vision-bound. Dr. Kerwin's experience clearly shows that the visual apparatus serves as the primary point of reference in space, just as it does on Earth.

Interest in visual performance in space was stimulated by knowledge that the visual environment would be different. First, the brightness of objects under direct solar illumination is higher. The atmosphere of the Earth absorbs at least 15 percent of the visible radiation. Water vapor, smog, and clouds can make this absorption considerably higher. In general, this means that the level of illumination in which astronauts work during daylight is about one-fourth higher than that on Earth. Second, on surfaces such as the moon, where there is no atmosphere, there is no scattering of

156

light (Figure 1). This causes areas not under direct solar illumination to appear much darker and results in a reordering of normal visual relationships. The extent to which these environmental differences might interact with subtle physiological changes in sensory receptor systems was an unknown at the time of early space missions.

Figure 1. Dark shadows from astronaut and rocks illustrate lack of light scattering in the lunar environment.

Tests of the visual acuity of Gemini 5 astronauts were obtained preflight, inflight, and postflight by the Visibility Laboratory of the Scripps Institute of Oceanography (Duntley et al., 1966). Measurements were obtained with an Inflight Vision Tester, a small, self-contained, binocular optical device containing a transilluminated array of high- and low-contrast rectangles. Astronauts judged the orientation of each rectangle and indicated their response by punching holes in a record card.

The second part of the Gemini 5 visual acuity measurement program used large rectangular patterns displayed at ground sites in Texas and Australia. The task of the astronauts was to report the orientation of the rectangles. Displays were changed in orientation between passes and adjustments for size were made in accordance with anticipated slant range, solar elevation, and the visual performance of astronauts on preceding passes.

Results of the Gemini 5 onboard measurement program indicated that the visual performance of the astronauts neither degraded nor improved during the eight-day mission. Meteorological conditions interfered in large measure with the viewing of ground panels, although success was achieved on one pass. Results confirmed that the visual performance of the astronauts was within the limits predicted by preflight measures of visual acuity.

In the Apollo program, interest turned more toward the visual apparatus per se. Photographic studies of the retinal vasculature showed a significant decrease in the size of both veins and arteries about three and one-half hours after flight for one crewmember, and a decrease in veins only for another crewmember after four hours. The degree of constriction of retinal vasculature in this crew was greater and persisted for a longer time than could be accounted for by the vasoconstrictive effect of oxygen alone (Hawkins & Zieglschmid, 1975).

Apollo astronauts also showed a postflight decrease in intraocular tension in all cases when compared with preflight tension measurements. The postflight intraocular tension reverted to its preflight value at a slower rate than expected, as had been the case during similar investigations in the Mercury and Gemini programs. The reason for the slow return to normal remains unknown.

Soviet investigators concluded after the single orbit of Yuri Gagarin that during a brief exposure to the space environment, the basic functions of the visual system do not undergo noticeable change (Lazarev et al., 1981). Since that flight, systematic investigations of visual acuity, contrast sensitivity, color vision, and general visual capability have been conducted. Soviet investigators conclude that, during the first days of flight, the main visual functions deteriorate by 5-30 percent, followed by a certain restoration of function until an approximation of preflight capability is achieved. Contrast sensitivity is subjected to the most pronounced change, suffering a ten percent loss immediately after entry into weightlessness and progressing to a 40 percent loss after five days. Even at these levels of change, it was concluded that the effect of spaceflight conditions on the principal visual functions under normal conditions of illumination are relatively small. The authors conclude that "In general, the vision of cosmonauts in flight is as reliable as that on Earth. This enables the extensive use of vision to carry out scientific research and control the space vehicle under normal levels of brightness and illumination."

Motor Performance

During normal earthly pursuits, all activities are accomplished under the constant application of a one-gravity force. During space flight, this force is removed and all actions, whether sleeping, eating, or simply moving about, must be done differently. Prior to the first manned space missions, there was concern that the absence of gravity could cause difficulty in motor performance, particularly in those activities requiring skilled and precise movement.

One of the first studies of perceptual-motor performance during weightlesness was conducted by Gerathewohl et al. in 1957. In this investigation, subjects experienced practical weightlessness for a minimum period of ten seconds during vertical dives in a jet aircraft. An eye-hand coordination test was used in which subjects were required to aim at and hit the center of a test chart with a metal stylus. The chart had a bull's eye center surrounded by six concentric circles, each one cm apart. Results of trials during weightlessness showed a consistent tendency of subjects to hit slightly above the target center. It was concluded that this type of performance is moderately disturbed by decreased gravity. However, it was found that subjects rapidly compensated for the deviation caused by decreased arm weight and readjusted their aiming response. Over the six trials during the weightless state, performance improved until it was comparable to that found under normal conditions.

The very first manned flights dispelled any real fears concerning the ability of an astronaut to carry out routine perceptual-motor activities during weightlessness. Even though movement was rather restricted in the tight confines of the Mercury capsule, it was clear that no motor difficulties were encountered when dealing with the internal management of the spacecraft. As the task requirements and room for movement both increased in the Gemini and Apollo programs, the capability of man to perform in the space environment with no significant loss in precision was confirmed.

Gross motor movements, such as involved in locomotion, appear to be aided by the absence of gravity. Less effort is involved and, once new procedures are learned, movement through a space cabin is done easily and efficiently. The absence of gravity appears to help rather than hinder this type of movement.

The Skylab program offered an opportunity to study astronaut performance in detail and during extended exposure to weightlessness (Figure 2). It was anticipated that inflight performance of some tasks would be slightly affected, while the performance of others in the zero-gravity environment would exhibit more pronounced changes in time and/or in the patterning of the elements comprising the tasks (Kubis et al., 1977). Photographic recordings were made of such activities as use of the Lower Body Negative Pressure Device, preparing and exercising on the ergometer, assembling and using photometer and camera systems, maintaining hardware assemblies, donning and doffing an EVA space suit, and accomplishing all the activities involved in the preparation of astronaut food. The tasks were categorized into classes requiring fine, medium, and gross motor dexterity. Inflight trials were compared with those done preflight, using for the most part time-to-completion as a score.

The Skylab studies produced no evidence of performance deterioration that could be attributed to the effects of long-duration exposure to the space environment. The first inflight performance of a task generally took longer than the last preflight performance. The investigators felt this to be the result of a number of factors—stress of last-minute flight preparations, change to the zero-gravity Skylab environment, use of greater care and caution in task performance, and some measure

159

of initial work overload. However, performance recovery was quite rapid. By the end of the second performance trial, approximately half of all tasks were completed within the time observed for the last preflight trial.

Figure 2. Astronaut Alan Bean demonstrates weightless acrobatics in the dome of the Skylab Orbital Workshop.

Sleep

The longest flight in the Mercury program lasted for slightly more than 34 hours. The results from this flight allowed the conclusion that "Sleep in flight was proved to be possible and subjectively normal" (Link, 1965). While it now was known that sleep in space was possible, there was no information as to its quality or any special arrangements which would be required for longer missions.

The first opportunity for any serious evaluation of sleep in space occurred with the four-day Gemini mission. A great deal of difficulty was encountered in obtaining satisfactory sleep periods during this flight. No long sleep period was obtained by either crewman, with four hours being the longest continuous period of sleep. The Command Pilot estimated that he did not get more than seven and one-half to eight hours of good sleep in the entire four days (Berry et al., 1966). Reasons for the sleep disturbance included noise from events such as thruster firings, communications from ground control, movement in the spacecraft, staggered sleep periods, altered diurnal cycles, and the "mission responsibility" felt by each crewman.

Attempts were made to improve sleeping conditions on the eight-day Gemini mission. However, due to conflicts with flight plan activities, sleep remained poor. On the 14-day Gemini flight, the flight plan was designed to allow the crew to sleep during hours corresponding to nighttime at Cape Kennedy. In addition, both crewmen slept at the same time, thus lessening noise in the spacecraft. As a result, sleep was much improved. The crew of the four-day flight was markedly fatigued following the mission. The eight-day crew was less so, and the 14-day crew the least fatigued of all (Berry et al., 1966).

The first attempt by the United States to obtain objective measures of sleep during a space mission was in the Gemini 7 flight. However, due to problems with the recording equipment, only limited information was obtained. In the Skylab program, more successful measurements were made of the quality of astronauts' sleep (Frost et al., 1977). Recordings were made of the electro-encephalogram, electrooculogram, and head-motion signals during sleep periods (Figure 3). One crewmember participated in the sleep monitoring activities during each Skylab flight.

Figure 3. A Skylab astronaut demonstrates use of the vertical "bed" which permitted monitoring of bioelectric functions during sleep (as well as affording better quality of sleep).

161

The results of the Skylab experiments do not show any major adverse changes in sleep as a result of prolonged space flight. Only during the 84-day flight did one subject experience any real difficulty in terms of sleep time. Even here, the problem diminished with time, although sleeping medication was required on occasion. The most significant changes occurred in the postflight period, with alterations more of sleep quality than quantity. It appears that readaptation to a one-gravity environment is more disruptive to sleep than the adaptation to zero gravity. In all, the Skylab investigators feel that adequate sleep can be obtained in a zero-gravity environment providing separate sleeping areas are used, noise levels are minimal, and a familiar time reference for the sleep period is used.

Postural and Illusion Problems

Postural equilibrium is a function of vestibular inputs, vision, kinesthesia, and touch. The operation of these four sensory channels represents a tightly bound system in which an unusual change in any one is capable of producing a significant disturbance in the complete system. It therefore is logical that in weightlessness, when otolithic stimulation is removed, some orientation problems should arise. One also might expect certain illusions, particularly those involving the visual sense, if other parts of the complete system provide aberrant inputs.

The occurrence of visual illusions during weightlessness was studied in a Skylab experiment (Graybiel et al., 1977). This experiment examined changes in the oculogyral illusion (a form of apparent motion observed following stimulation of the semicircular canals) for eight astronauts during the Skylab flights. Four exhibited some loss of ability to perceive the illusion and four showed no change. The evidence for any decreased sensitivity of the semicircular canals thus is inconclusive. At the same time, several reports were received of a spontaneous oscillatory target movement illusion during the course of testing. Since these reports were received when subjects were restrained and were drowsy, the illusory oscillations may be simply a function of reflexive eye movement prior to sleep.

During the Skylab program, postflight postural equilibrium tests were given to crew members to determine the extent of change. As an example, it was found (Homick et al., 1977) that the Skylab 3 Scientist Pilot and Pilot both showed a decrease in postural equilibrium performance with their eyes open when tested on the second day after splashdown. It is of interest, however, that a more pronounced decrement in ability to maintain upright posture was found with the eyes closed. Without vision, on the second day following recovery, the pilot experienced difficulty in standing, contrasting seriously with his excellent balance measured preflight.

The Soviet space program also has encountered postural as well as illusionary problems. Soviet investigators examined 24 cosmonauts who had made flights in craft of the Soyuz type and who had served in the Soyuz-Salyut orbital complex. The missions in which these cosmonauts participated ranged from as short as four days to as long as 175 days. It was found that illusory reactions occurred inflight more often than the more well-known vestibular dysfunctions. Twenty-one of the 24 cosmonauts reported illusory reactions of the inversion type ("hanging upside-down"),

162

occasionally in the form of displacement of surrounding objects (Yakovleva et al., 1982). This is the same type of illusion reported by many American astronauts, beginning with those in the Mercury program. The Soviets reported that reactions occurred immediately after weightlessness was experienced, in most cases, although in other instances the illusion did not appear until as long as two hours later. For some the illusion lasted only a few minutes while for others it lasted four or more hours. The illusion recurred sporadically during flight, most often at times of increased motor activity or while attempting some visual task. In some cases, the illusion could be suppressed if the cosmonaut fixed his eyes on some object, immobilized himself in the seat, and undertook a relaxation exercise.

Space Operations

Studies in both the American and Soviet space programs lead to the conclusion that no serious neurophysiological manifestations will occur which might degrade future advanced space operations. The neurophysiology underlying work performance adapts well to conditions in space missions. Complex motor activities, such as represented by operation of the remote-manipulator arm of the Space Shuttle (the space "teleoperator"), can be performed with precision and efficiency. Care must be taken, of course, to insure that the operator is rested and in good condition and that the man/machine interface between the operator and the system has been designed to account for use in the environment of space.

References

Berry, C.A., Coons, D.O., Catterson, A.D., & Kelly, G.F. Man's response to long-duration flight in the Gemini spacecraft. In *Gemini Midprogram Conference* (NASA SP-121). Washington, D.C.: U.S. Government Printing Office, 1966.

Duntley, S.Q., Austin, R.W., Taylor, J.H., & Harris, J.L. Experiment S-8/D-13, visual acuity and astronaut visibility. In *Gemini Midprogram Conference* (NASA SP-121). Washington, D.C.: U.S. Government Printing Office, 1966.

Frost, J.D., Jr., Shumate, W.H., Salamy, J.G., & Booher, C.R. Experiment M133. Sleep monitoring on Skylab. In R.S. Johnston & L.F. Dietlein (Eds.), *Biomedical results from Skylab* (NASA SP-377). Washington, D.C.: U.S. Government Printing Office, 1977.

Gerathewohl, S.J., Strughold, H., & Stallings, H.D. Sensomotor performance during weightlessness. In R.J. Benford (Ed.), *The Journal of Aviation Medicine* (Vol. 28). St. Paul: The Bruce Publishing Company, 1957.

Graybiel, A., Miller, E.F., & Homick, J.L. Experiment M131. Human vestibular function. In R.S. Johnston and L.F. Dietlein (Eds.), *Biomedical results from Skylab* (NASA SP-377). Washington, D.C.: U.S. Government Printing Office, 1977.

Hawkins, W.R., & Zieglschmid, J.F. Clinical aspects of crew health. In R.S. Johnston, L.F. Dietlein, & C.A. Berry (Managing Eds.), *Biomedical results of Apollo* (NASA SP-369). Washington, D.C.: U.S. Government Printing Office, 1975.

Homick, J.F., Reschke, M.F., & Miller, E.F., II. Effects of prolonged exposure to weightlessness on postural equilibrium. In R.S. Johnston & L.F. Dietlein (Eds.), *Biomedical results from Skylab* (NASA SP-377). Washington, D.C.: U.S. Government Printing Office, 1977.

Kubis, J.F., McLaughlin, E.J., Jackson, J.M., Rusnak, R., McBride, G.H., & Saxon, S.V. Task and work performance on Skylab missions 2, 3, and 4. Time and motion study—experiment M151. In R.J. Johnston & L.F. Dietlein (Eds.), *Biomedical results from Skylab* (NASA SP-377). Washington, D.C.: U.S. Government Printing Office, 1977.

Lazarev, A.I., Kovalenok, V.V., Ivanchenkov, A.S., & Avakyan, S.V. Atmosphere of Earth from Salyut-6. Translation of *Atmosfera Zemli s Salyuta-6*. Leningrad: Gidrometeoizdat, 1981.

Link, M.M. *Space medicine in Project Mercury* (NASA SP-4003). Washington, D.C.: U.S. Government Printing Office, 1965.

Yakovleva, I.Ya., Kornilova, L.N., Tarasov, I.K., & Alekseyev, V.N. Results of studies of cosmonauts' vestibular function and spatial perception. *Space Biology and Aerospace Medicine*, 1982, *16*(1): 26–33 (JPRS 80323).

CHAPTER 10
THE CARDIOPULMONARY SYSTEM

The cardiopulmonary system undergoes substantial adaptive changes during spaceflight. The effects of weightlessness are particularly important because many of the adaptive capabilities of the cardiopulmonary system have evolved with specific mechanisms to counter the continuous pull of Earth's gravity. Some of the changes in this system, such as cephalic fluid shifts, are the direct result of weightlessness; others appear to be concomitant adjustments and only indirectly related to the zero-gravity environment. The major alterations in this system involve (1) fluid shifts, (2) orthostatic intolerance, (3) changes in cardiac dynamics and electromechanics, and (4) changes in pulmonary function and exercise capacity.

Fluid Shifts

The most significant alteration that occurs in the cardiopulmonary system is the cephalic shift estimated at 1.5 to 2 liters from lower extremities. Evidence for this shift of fluids comes from a number of observations.

Photographic Evidence

Photographs taken of the Skylab crew inflight show signs of periorbital puffiness, facial edema, and thickening of the eyelids (Thornton, Hoffler, & Rummel, 1977). The jugular veins and the veins in the temple, scalp, and forehead appear full and distended. Figure 1 shows photographs of the Skylab 3 Commander (CDR) taken inflight (a), and postflight (b). Although the inflight uplifting in the facial tissue may have been partly due to the absence of gravity, the fluid shifts probably also played a significant role, particularly through venous engorgement. These photographs were taken near the end of the Skylab 3 mission (59 days), indicating that edema and venous engorgement do not subside even after some months in space. The crews' subjective observations of nasal stuffiness, head "fullness," and puffy faces support the hypothesis of substantial fluid shifts.

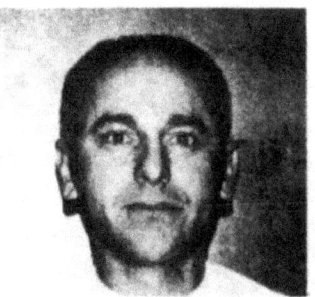

Figure 1. Skylab 3 Commander comparing the puffy face inflight (a) to the normal face preflight (b). (From Thornton, Hoffler, & Rummel, 1977)

Changes in Calf Girth and Leg Volume

Astronauts have typically demonstrated inflight decrements in calf girth of up to 30 percent. For Skylab and Apollo-Soyuz, multiple circumferential measurements of the lower extremities were taken pre-, in- and postflight to obtain volumetric estimates and to ascertain the time course of fluid shifts. Figure 2 shows the measurement technique.

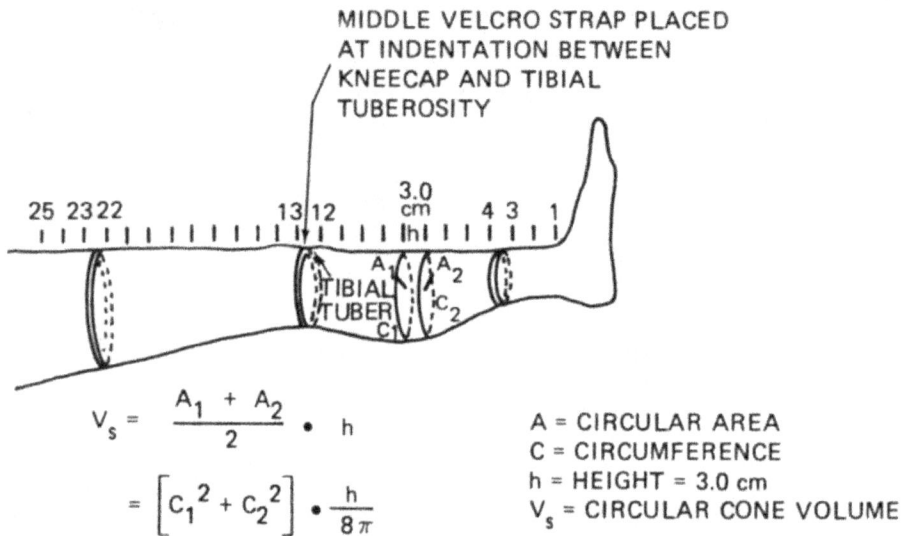

MIDDLE VELCRO STRAP PLACED AT INDENTATION BETWEEN KNEECAP AND TIBIAL TUBEROSITY

$$V_s = \frac{A_1 + A_2}{2} \cdot h$$

$$= \left[c_1^2 + c_2^2 \right] \cdot \frac{h}{8\pi}$$

A = CIRCULAR AREA
C = CIRCUMFERENCE
h = HEIGHT = 3.0 cm
V_s = CIRCULAR CONE VOLUME

Figure 2. The estimation of leg volume by measurement of multiple circumferences. Calculation is based on assumed circular geometry and summation of multiple, truncated, conical volumes.

Figure 3 shows limb volume data for the Skylab 4 CDR, demonstrating that lower, but not upper, limb volumes decrease early inflight, and return rapidly to preflight values upon return to Earth (Thornton, Hoffler, & Rummel, 1977). Figure 4 shows limb volume data from Apollo-Soyuz, in which it was possible to obtain measurements as early as six hours inflight from one crewmember. From these data, it appears that the major shift of fluids does not occur rapidly and upon insertion into orbit. The rate of fluid shifts appears instead to follow an exponential course, attaining a maximum within 24 hours and reaching a plateau or a new steady state within three to five days (Hoffler, Bergman, & Nicogossian, 1977). Calf volume measurements taken during long-term Soviet missions aboard Salyut 6 showed a similar pattern of decrease, with values reaching a plateau on the twelfth inflight day. Over 140-, 175-, and 185-day missions, fluctuations in leg volume followed a wavelike course of loss and recovery (Gazenko, Genin, & Yegorov, 1981a).

The changes in limb volume, particularly the rapid postflight recovery, combined with photographic evidence and the crews' subjective observations, indicate the occurrence of substantial cephalic fluid shifts inflight. On the basis of inferences drawn from the results of ground-based simulation studies (Burkovskaya et al., 1980; Katkov, Chestukhin et al., 1981; Nicogossian et al., 1979; Sandler, 1977) and inflight data (Gazenko, Genin, & Yegorov, 1981b), it appears that these

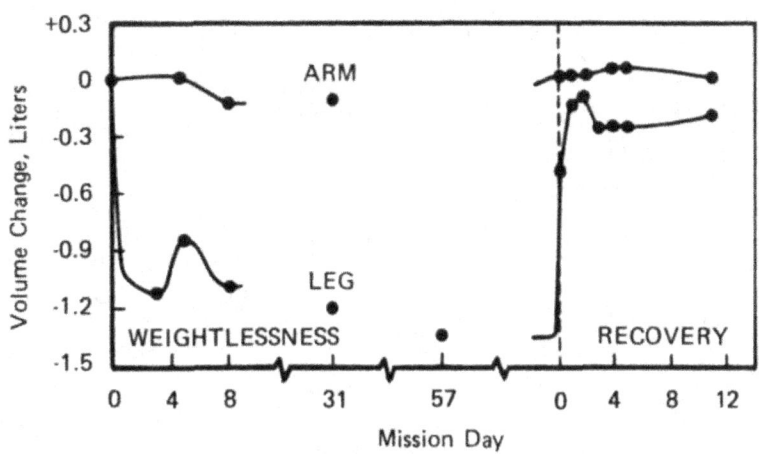

Figure 3. Change in left limb volumes, Skylab 4
CDR. (From Thornton, Hoffler, & Rummel, 1977)

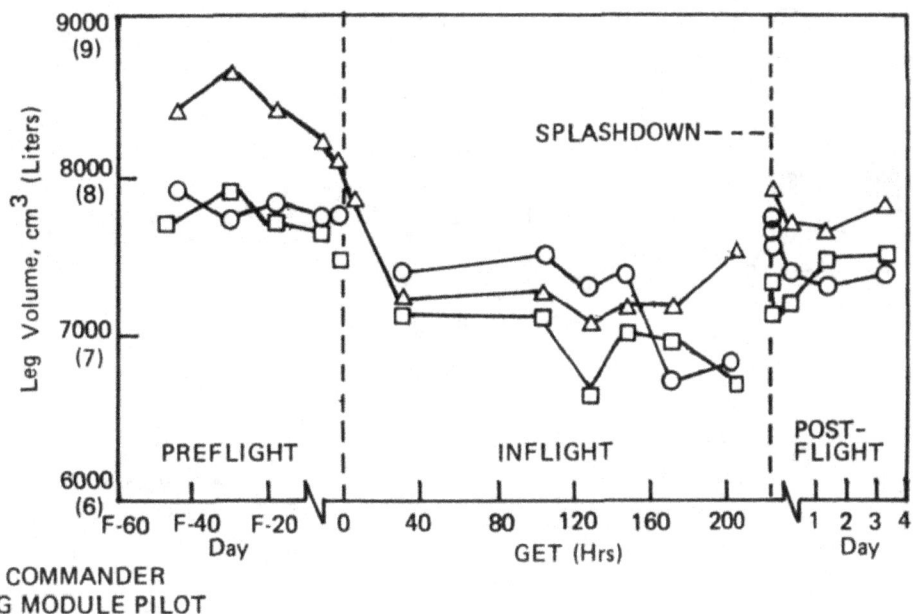

○ APOLLO COMMANDER
□ DOCKING MODULE PILOT
△ CMP

Figure 4. Left leg volume measurements of U.S. crewmen in Apollo-Soyuz
Test Project. (From Hoffler, Bergman, & Nicogossian, 1977)

fluid shifts result in a number of physiological readjustments, as diagrammed in Figure 5. The immediate response is an initial increase in central blood volume, resulting in stimulation of cardiopulmonary mechanoreceptors. The activation of these low-pressure receptors leads to an inhibition of the medullary vasomotor center, decreased sympathetic nervous system tone, and indirect effects on renal function via both neural and hormonal pathways. The complex events triggered by the fluid shifts initiate, at least initially, compensatory diuresis and decreased water intake until the cardiovascular system finally becomes stabilized at a new pressure/volume relationship. The remainder of this chapter discusses inflight and postflight observations of cardiopulmonary changes in more detail. Alterations in renal function and in the neuroendocrine system are examined in other chapters within this section.

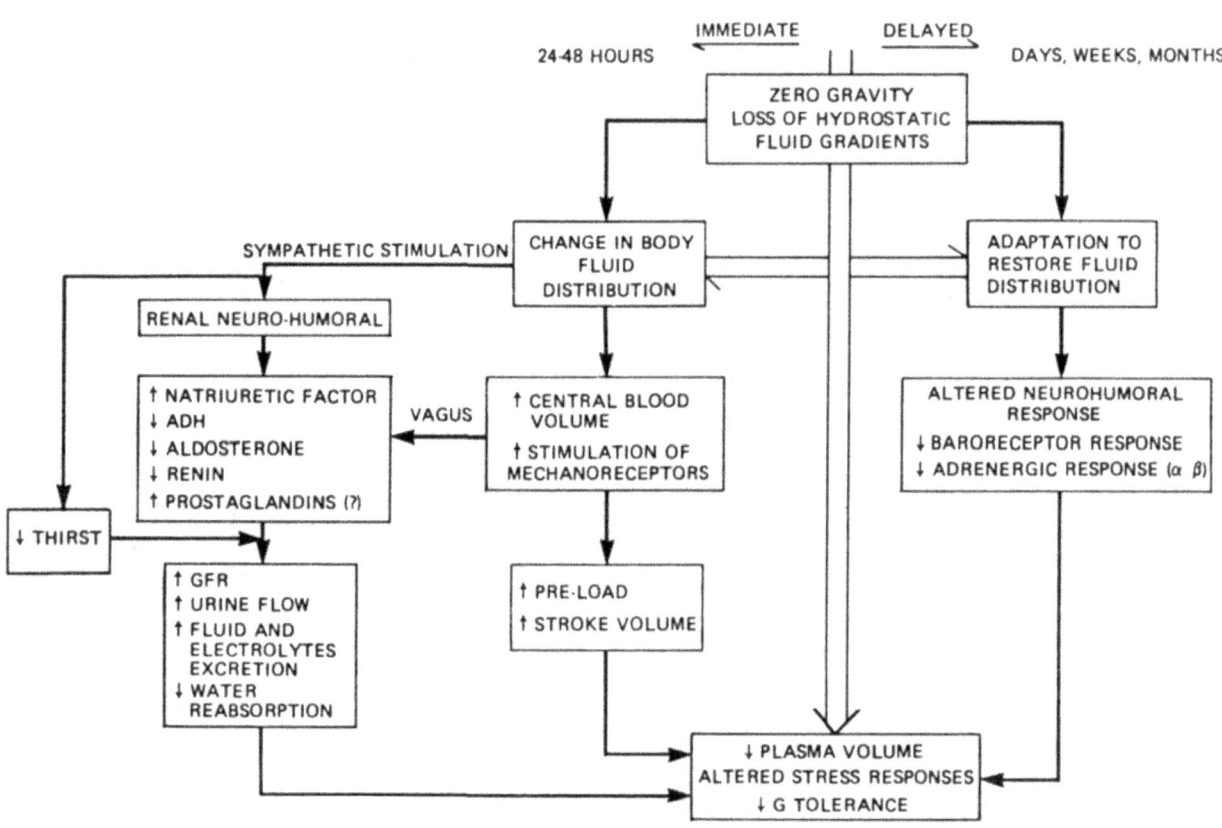

Figure 5. Suggested cardiovascular response to weightlessness.

Decreased Orthostatic Tolerance

Decreased orthostatic tolerance has invariably been observed in both American and Soviet crewmen postflight. Symptoms have ranged from increased heart rate and decreased pulse pressure to a tendency toward spontaneous syncope. The methods most widely used to quantify orthostatic tolerance include the tilt test and the stand test and lower body negative pressure (LBNP). Figure 6 shows the protocols for these tests, and the cardiovascular parameters that have been measured.

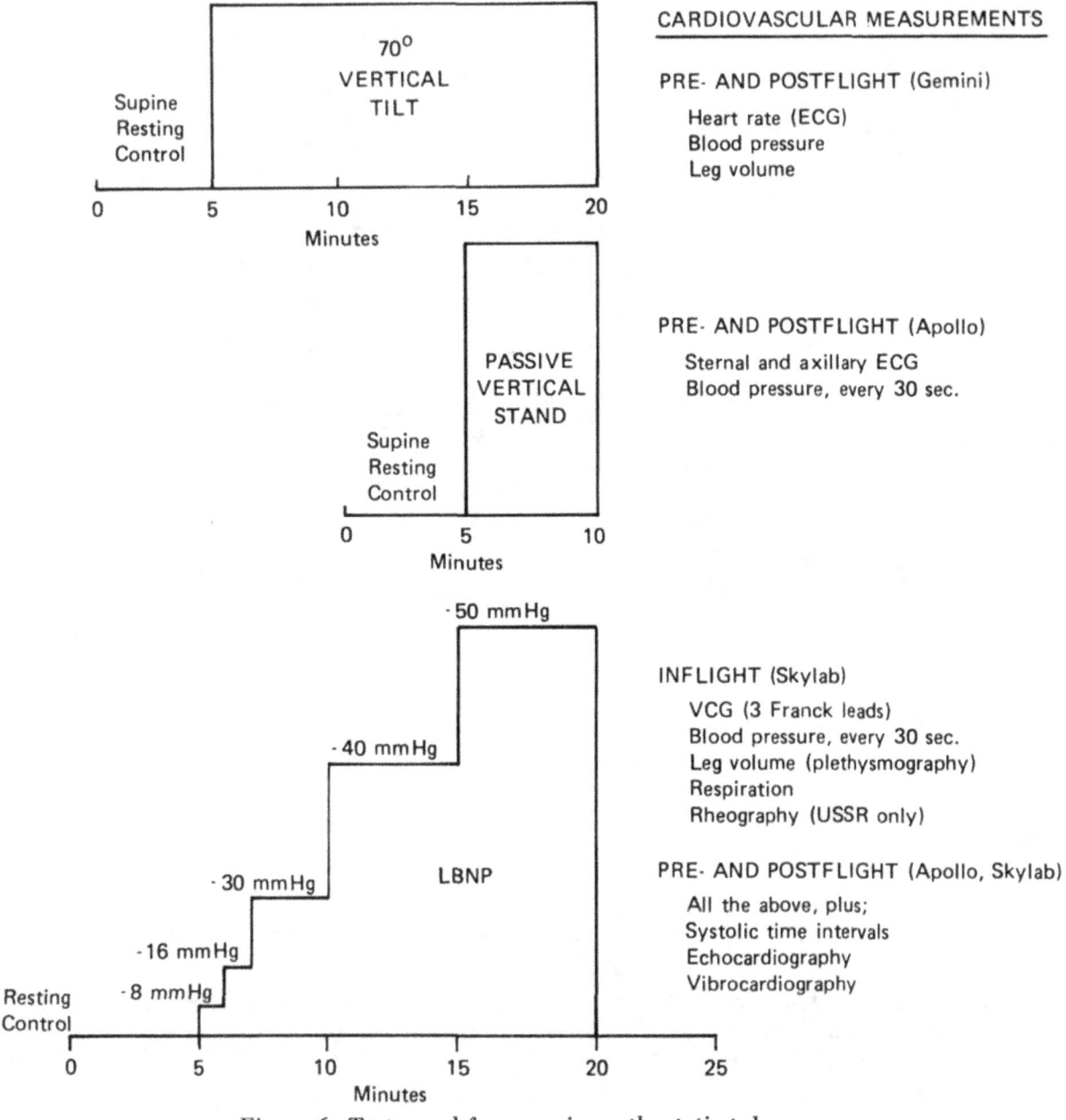

Figure 6. Tests used for assessing orthostatic tolerance.

Passive Testing

The Gemini crewmen were tested for orthostatic tolerance utilizing a tilt table from a supine control position to a 70° head-up tilt. Figure 7 shows a typical response, which was aborted because of presyncopal symptoms. The stand test, in which the subject leans against a wall in a relaxed manner with his heels 15 cm from the wall, was used on some of the Apollo crewmen. Similar exaggerations in postflight cardiovascular responses are observed in this type of test (Hoffler & Johnson, 1975).

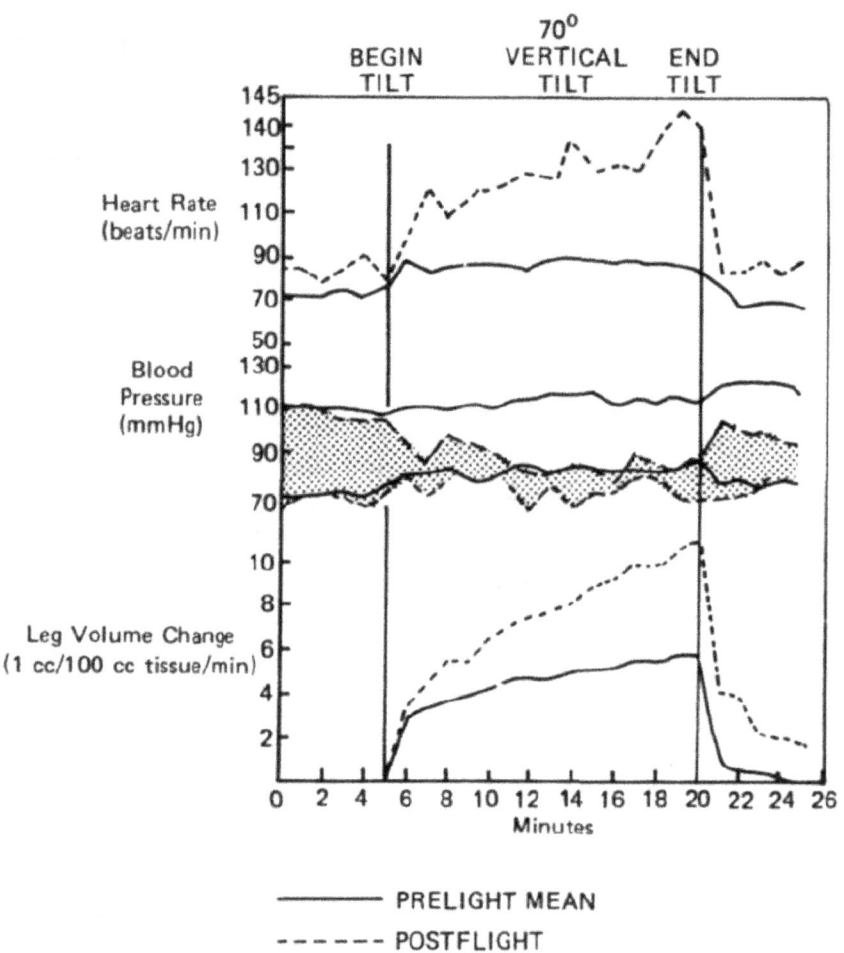

Figure 7. Heart rate (beats per minute), blood pressure (systolic and diastolic, in mmHg), and change in leg volume (in percentage) during a 25-min, 70° tilt test protocol, the first and last 5 min being in the horizontal, supine position. Preflight mean curves are solid lines and the first postflight test values are dashed (or hatched). The crewman was the Command Pilot of the 14-day Gemini VII flight. (After Berry & Catterson, 1967)

170

Lower Body Negative Pressure Test (LBNP)

LBNP, shown in Figure 8, is a useful means of assessing orthostatic tolerance because, unlike the passive methods, it can be used in zero gravity. Different levels of negative pressure can be applied to the lower part of the body, resulting in footward displacement of body fluids (Wolthius et al., 1974). An appropriate protocol can simulate the effects of erect posture on the cardiovascular system in one gravity. Figure 9 shows a comparison of the pre- and postflight responses of the Apollo 8 Commander to the LBNP protocol.

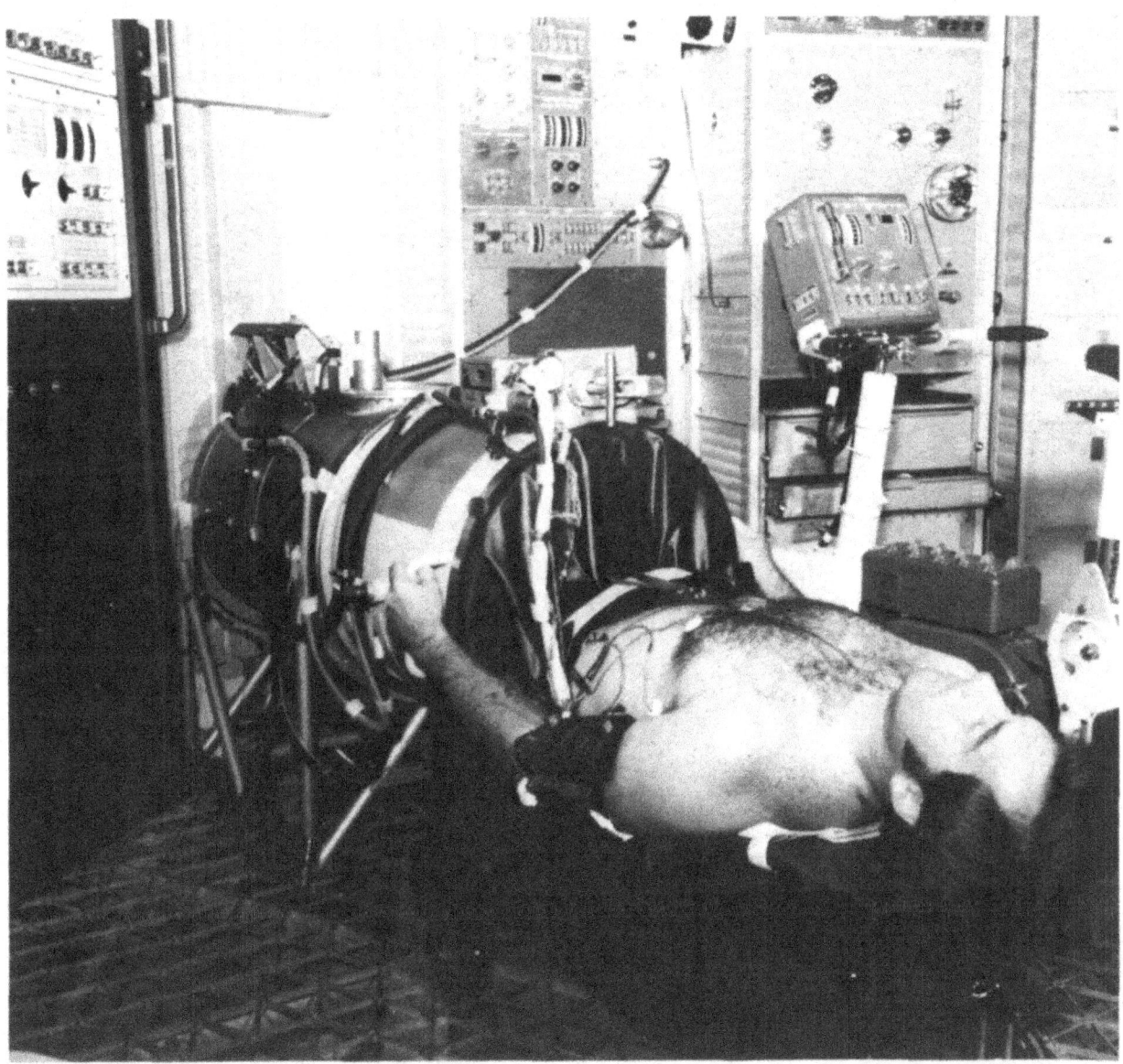

Figure 8. Subject undergoing test in lower body negative pressure device.

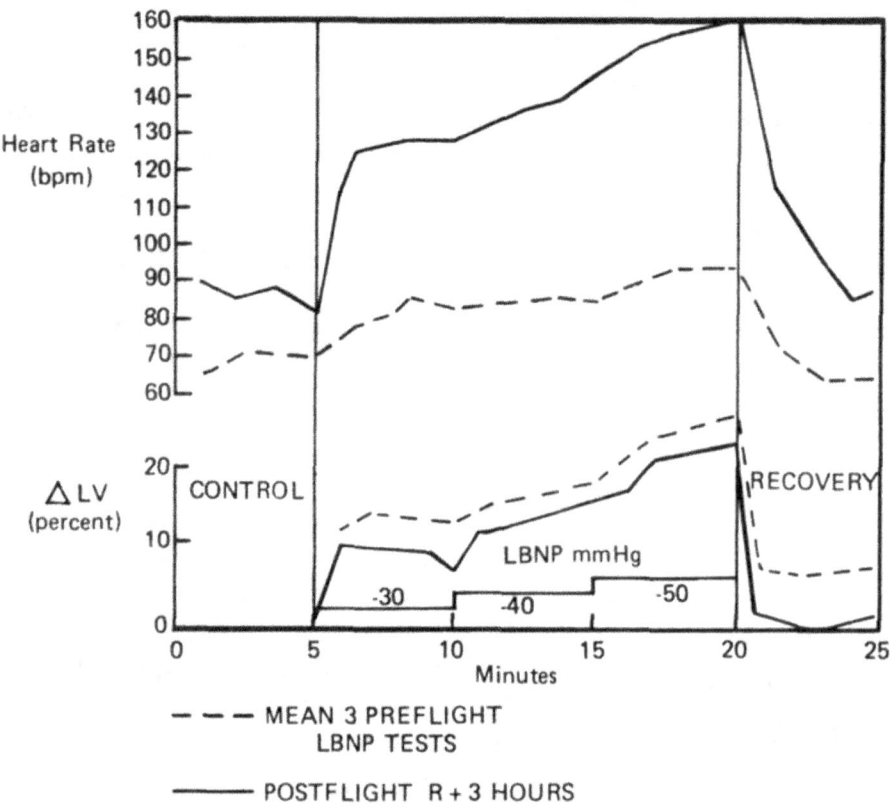

Figure 9. Heart rate (in beats per minute) and change in leg volume (in percentage) during a 25-min LBNP test protocol. Preflight mean curves are dashed and the first postflight test data 3 hr after splashdown are solid lines. The crewman was Commander of the Apollo 8 flight.

LBNP tests were performed inflight at regular intervals of 3 days during Skylab missions, affording an opportunity to observe the time course of changes in physiological responses to this kind of orthostatic stress. Figure 10 shows a comparison of the responses of the Skylab 4 Scientist Pilot (SPT) for tests conducted 21 days prior to flight and for the first inflight test (Mission Day 6) (Johnson et al., 1977). Figure 11 shows resting and orthostatically stressed heart rates for the Skylab 4 Pilot (PLT). The elevations in both parameters were evident in the first test (four to six days inflight), and cardiovascular responses to LBNP continued to show instability, especially during the first three weeks.

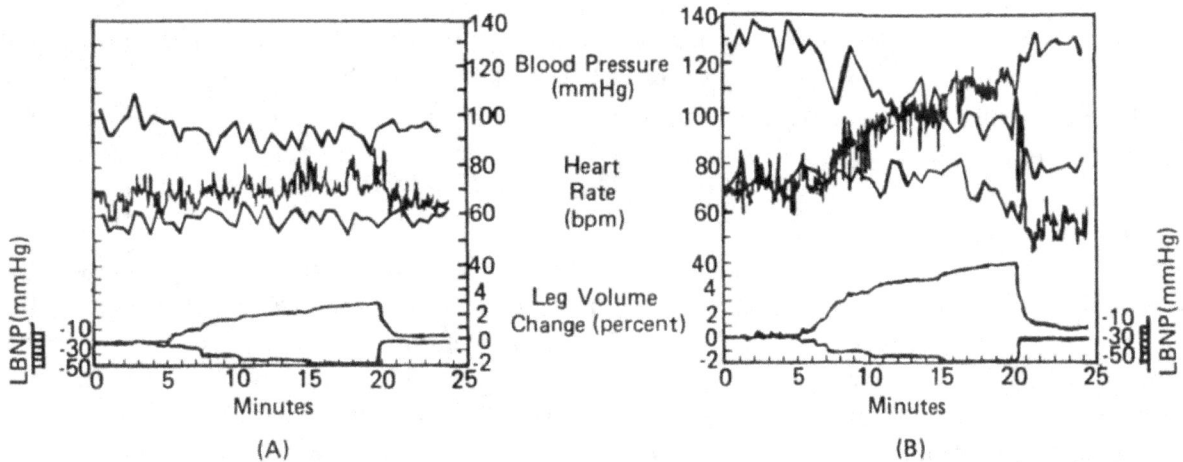

Figure 10. Cardiovascular responses of the Skylab 4 Scientist Pilot during the LBNP test 21 days prior to flight (A), and during the first inflight test on mission day 6 (B). (From Johnson et al., 1977)

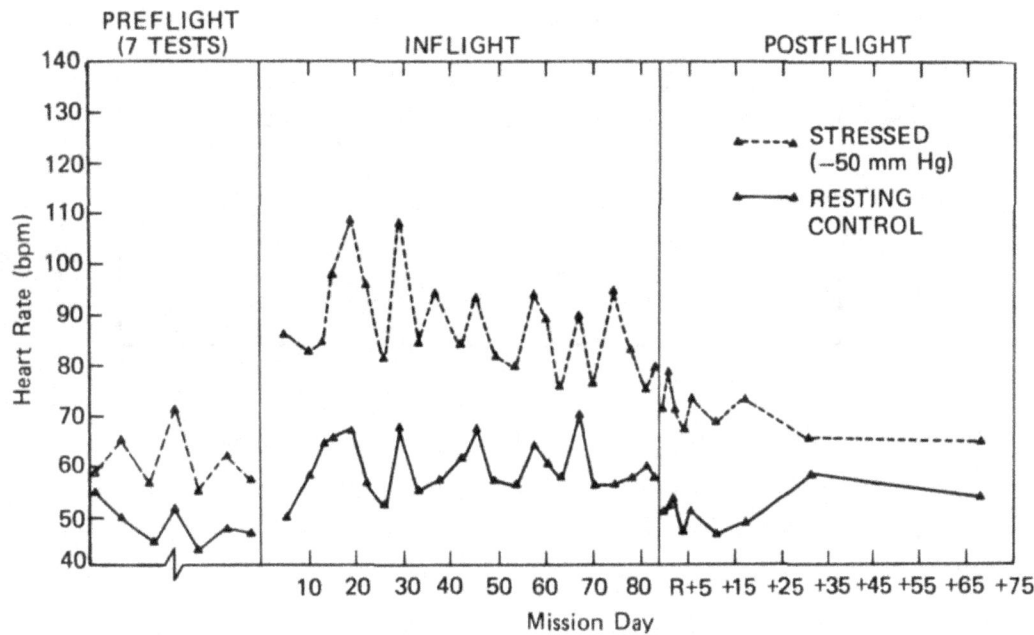

Figure 11. Mean heart rate of the Skylab 4 Pilot during resting and 50 mmHg phases of lower body negative pressure. The high heart rates that appeared periodically declined in magnitude after the first month inflight. A slight downward trend in stressed heart rates was apparent during the latter period. A presyncopal episode on mission day ten may have been associated with the lower mean stressed heart rate on that day. (From Johnson et al., 1977)

These results, along with corresponding Soviet findings (Gazenko et al., 1981b) indicate that inflight LBNP presents a greater stress to the cardiovascular system than the same levels of LBNP preflight, apparently due to alterations attending the large inflight cephalic migration of body fluids. The diminishing total blood volume probably plays a significant role, but the time course of the fluid loss inflight is not yet clearly defined. Nevertheless, there is an inverse relationship between the change in orthostatically stressed heart rates pre- versus postflight, and the corresponding changes in blood volume. This relationship is shown in Figure 12 as percent change from preflight reference values. Recent findings that rigorous inflight exercise, which moderates the fluid loss along with other aspects of deconditioning, also normalizes certain parameters of the response to LBNP inflight, appear to corroborate this relationship (Yegorov et al., 1981). These data indicate that depleted blood volume is at least associated with the exaggerated cardiovascular responses to LBNP observed postflight, and perhaps with those observed inflight as well (Hoffler, 1977).

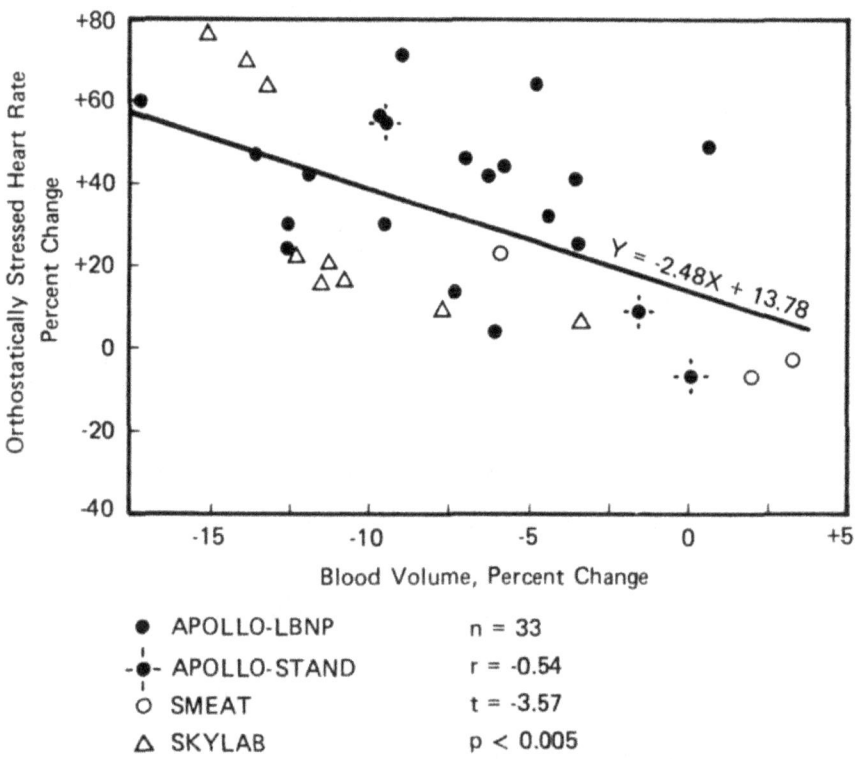

Figure 12. The change in immediate postflight orthostatically stressed heart rates from preflight mean reference values as a function of change in corresponding post- and preflight total blood volumes (measured by radioisotope dilution methods). Both LBNP and passive 90° standing orthostatic stress techniques are included. The equation of the best-fit linear regression through all 33 data points is given, and the r value of -0.54 is statistically significant (p<0.005).

Peripheral Circulation

Blood Flow

Using an arm pressure cuff and a leg band plethysmograph, it was possible to obtain measurements of calf blood flow inflight on Skylab 4 (Thornton & Hoffler, 1977). Figure 13 shows those blood flow measurements during preflight, inflight, and postflight tests for each crewmember. Although there is great individual variability, it appears clear that leg blood flow was significantly increased inflight. Measurements rapidly reverted to preflight levels upon recovery. Transient increases in intrathoracic venous pressure, increased cardiac output, or reduced peripheral resistance may have accounted for these blood flow changes inflight.

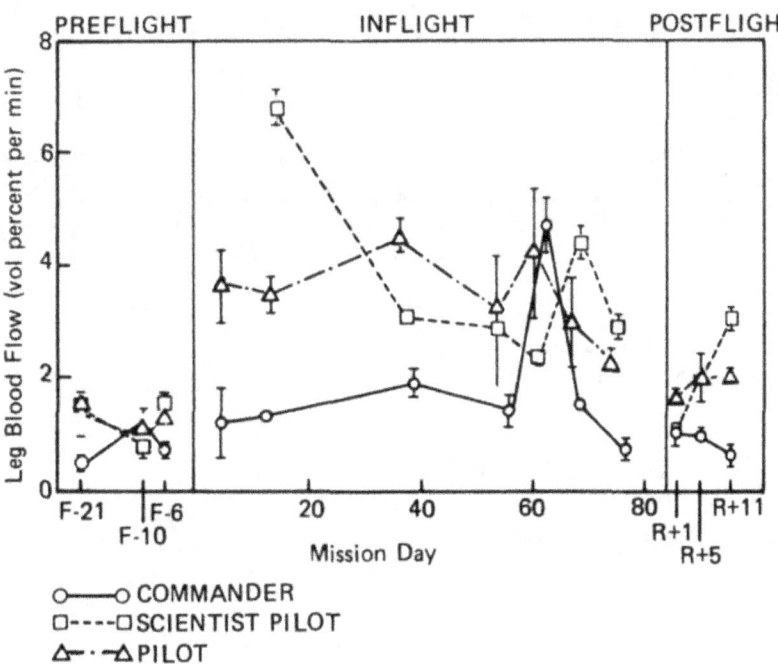

Figure 13. Skylab 4 crew leg blood flow measurements, -50 mmHg. (From Thornton & Hoffler, 1977)

Venous Compliance

Measures of venous compliance in the lower extremities were obtained in conjunction with the measures of blood flow on Skylab 4. Figure 14 shows vascular compliance change (in volume percent) for tests conducted preflight, inflight, and postflight. Vascular compliance increased inflight, and did not reach maximum values for at least ten days. In the later weeks of the flight, compliance appeared to reverse this trend, and rapidly returned to preflight levels after recovery. Changes in venous compliance combined with cephalic fluid shifts and depleted blood volume appear to contribute to the stressful effects of inflight LBNP (Thornton & Hoffler, 1977).

175

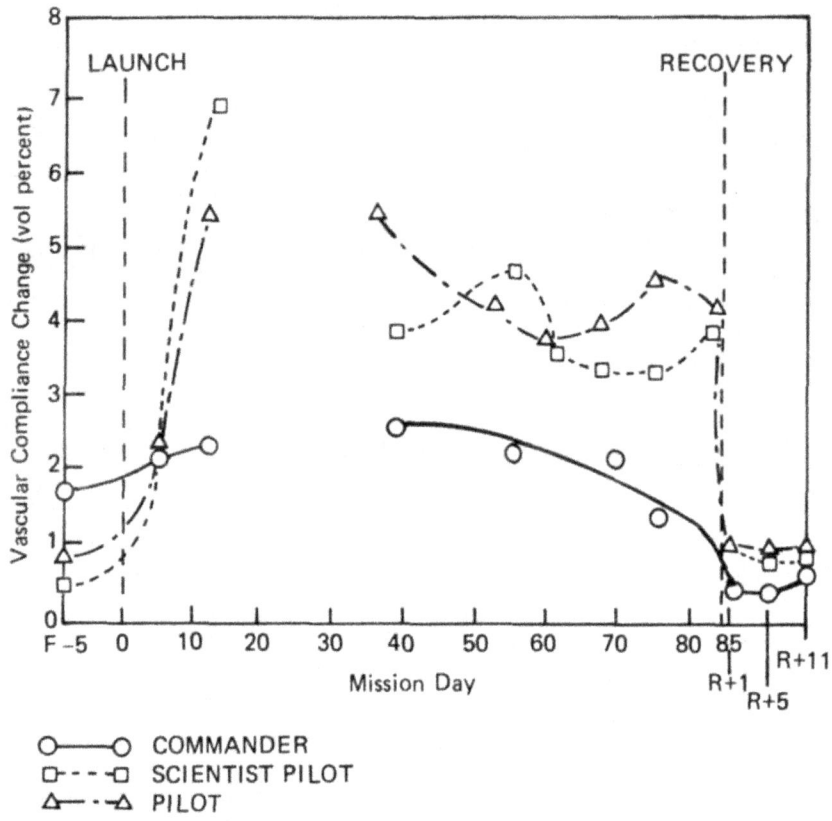

Figure 14. Skylab 4 crew vascular compliance.
(From Thornton & Hoffler, 1977)

Cardiac Dynamics and Electromechanics

A variety of measures has been obtained to ascertain the effects of spaceflight on cardiac dynamics and electromechanics, including studies of cardiac size, systolic time intervals, echocardiography, rheography, and vectorcardiography.

Cardiac Size

Standard posterior-anterior chest X-rays have been taken before and after each U.S. space flight, and these films have provided a base for determining changes in cardiac silhouette size. Each of the nine Skylab crewmen showed modest decreases in cardiothoracic (C/T) ratios (Nicogossian et al., 1976). Combining the data from 4 Mercury, 18 Gemini, 30 Apollo, and 9 Skylab crewmen, postflight decrements in C/T ratio averaged -.018.

Systolic Time Intervals

To assess ventricular function pre- and postflight, systolic time intervals were obtained during rest and LBNP stress. Figure 15 shows the average ratios of pre-ejection period (PEP) over left ventricular ejection time (ET) for the Skylab 4 crewmen. The postflight changes in PEP/ET ratio are probably due to markedly decreased ventricular filling and diminished total blood volume, although decreased contractility could not be excluded (Bergman et al., 1977; Hoffler, 1977).

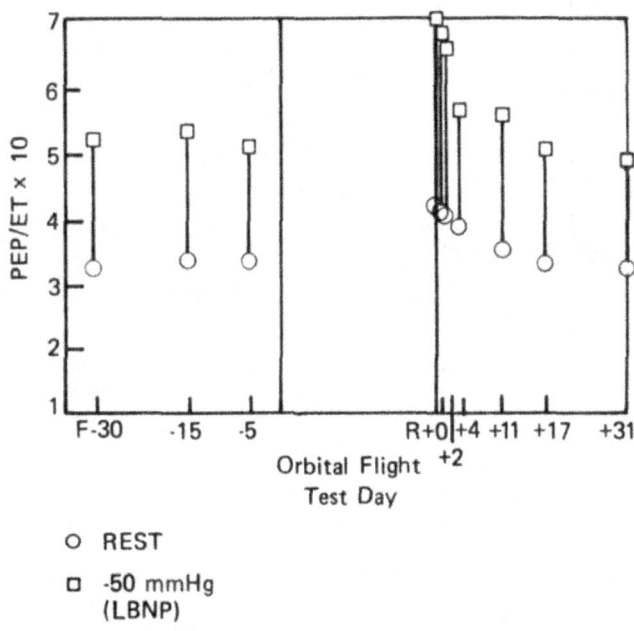

Figure 15. Resting and LBNP-stressed pre-ejection period/ejection time (PEP/ET): mean values for Skylab 4 crewmen. (From Bergman, Johnson, & Hoffler, 1977)

Echocardiography

Echocardiograms (ECGs) were obtained pre- and postflight for the Skylab 4 crewmen during rest and LBNP stress in order to further assess ventricular function. These studies demonstrated postflight decreases in stroke volume, left ventricular end-diastolic volume, and estimated left ventricular mass. Ventricular function curves were constructed by plotting left ventricular end-diastolic volume versus stroke volume. The pre- and postflight curves for the CDR are shown in

177

Figure 16. Since the curves fall on a straight line, it appears that cardiac function and myocardial contractility had not deteriorated despite the decreases in cardiac size and stroke volume (Henry et al., 1977). Additional studies concluded in conjunction with the Salyut 6 missions tend to corroborate the above findings (Gazenko et al., 1981b).

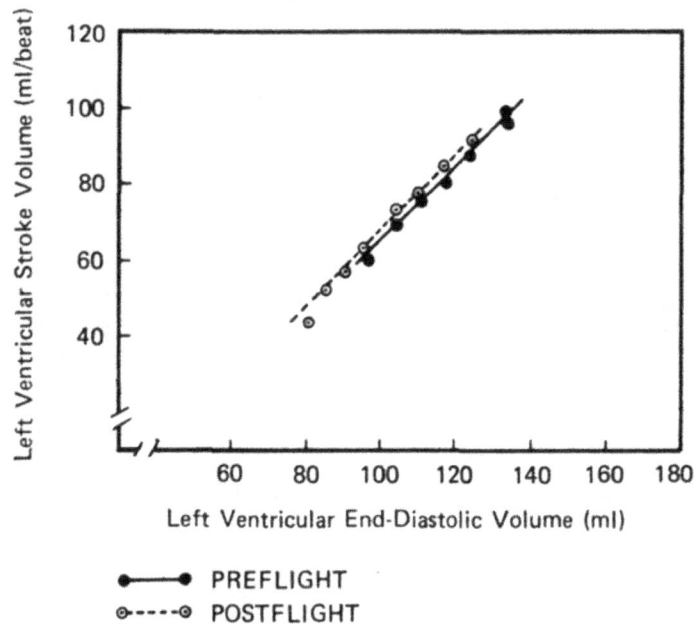

Figure 16. Ventricular function curve, Skylab 4 Commander. (From Henry et al., 1977)

Rheography

Soviet scientists use rheographic measurement of central and peripheral hemodynamic responses at rest to detail compensatory mechanisms in the adaptation to weightlessness. These data have consistently shown decreases in cardiac stroke output, with a shorter isovolumetric contraction (Yegorov et al., 1981). In one study (Turchaninova & Domracheva, 1980), an initial increase in stroke volume was demonstrated, subsiding after the first week. Changes in all cardiovascular functions, as measured by this technique, are seen to vary phasically throughout a mission (Yegorov et al., 1981).

178

Vectorcardiography

Table 1 presents a summary of the changes (as percent change from preflight resting values) in vectorcardiography determinations obtained from Skylab crewmen during rest and LBNP (Hoffler et al., 1977). One common observation was the increase in the QRS maximum vector inflight, a phenomenon that probably also stems from the cephalic-centripetal shift of fluid volume in weightlessness. The mechanism may be related to augmented preload producing the Brody effect. Another observation was an inflight increase in the PR interval duration, a measure that provides an estimate of atrioventricular conduction time. This change may have been due to an increase in vagal tone (Hoffler et al., 1977; Smith et al., 1977).

Table 1
Percent Change After Designated Condition from
Preflight Supine Resting Reference Values

Vectorcardiogram Measurement	Preflight	Condition = LBNP		Condition = Rest	
		Inflight[†]	Postflight[†] R+0	Inflight	Postflight R+0
Heart rate (bpm)	+20*	+54	+57	+ 9	+ 2
PR interval (ms)	- 11	- 6	- 5	+ 3*	+ 2
QRS duration (ms)	- 6*	- 6	- 4	- 3	+ 2
QT interval (ms)	- 6*	- 13	- 14	- 2	- 1
QT_0 interval (ms)	+ 2*	+ 7	+ 7	+ 2*	+ 0
P_{max}MAG (mV)	+27*	+78	+75	+24*	+16
QRS_{max}MAG (mV)	- 6*	+13	+12	+12*	+18*
QRS-E circ (mV)	+ 3	+24	+32	+19*	+21*
ST_{max}MAG (mV)	- 15*	- 32	- 37	- 10	- 6
J-Vector (mV)	+28*	+13	+26	+ 6	+18*
ST Slope (mV/s)	- 1	- 21	- 18	+ 5	+ 7
QRS-T angle (deg)	+61*	+13	+105	- 17	+12

*Statistically significant change from preflight reference values.

[†]Statistical tests were not performed.

Adapted from Hoffler et al., 1977.

Vectorcardiograms (VCGs) and ECGs have been used to detect cardiac arrhythmias. The majority of arrhythmias consist of PVBs and PABs, scattered throughout the pre-, in-, and postflight periods. Astronauts typically exhibit a characteristic sinus arrhythmia with a heart rate of 40 to 50 bpm over a five- to ten-second period after release of LBNP. Soviet cosmonauts have exhibited pronounced postflight decreases in the integral repolarization vector (Degtyarev, Doroshev et al., 1980; Golubchikova et al., 1981). Table 2 lists additional significant inflight arrhythmias.

Table 2
Significant Inflight Cardiac Arrhythmias

Apollo 15

LMP
PVC's; nodal bigeminal rhythm during lunar surface operations, accompanied by extreme fatigue; PAC's observed during 60 hrs after bigeminal rhythm. (This astronaut developed myocardial infarction 2 years after the flight.)

CDR
PVC's during sleep and upon awakening.

Skylab 3

CDR
Occasional PVC's; A-V block at the time of release of LBNP.

SPT
Ectopic beats, probably of supraventricular origin, particularly during LBNP; A-V junctional rhythm after release of LBNP; PVB's and wandering pacemaker during EVA.

Skylab 4

PLT
Occasional junctional rhythm following release of LBNP; nonspecific ST-segment and T-wave changes during maximum LBNP and/or 3rd-level exercise.

Pulmonary Function and Exercise Capacity

Pulmonary function tests postflight have generally revealed no abnormalities. However, inflight decreases in vital capacity approaching 10 percent were observed in the Skylab 3 Pilot and the entire Skylab 4 crew. Figure 17 shows the results of vital capacity tests for the three crewmen of Skylab 4. These inflight decreases may have been due to a combination of factors such as redistribution of body fluids into the thoracic cavity, cephalic shift of the diaphragm, or a reflection of the decrease in cabin ambient pressure to 1/3 of sea-level pressure (Sawin et al., 1976).

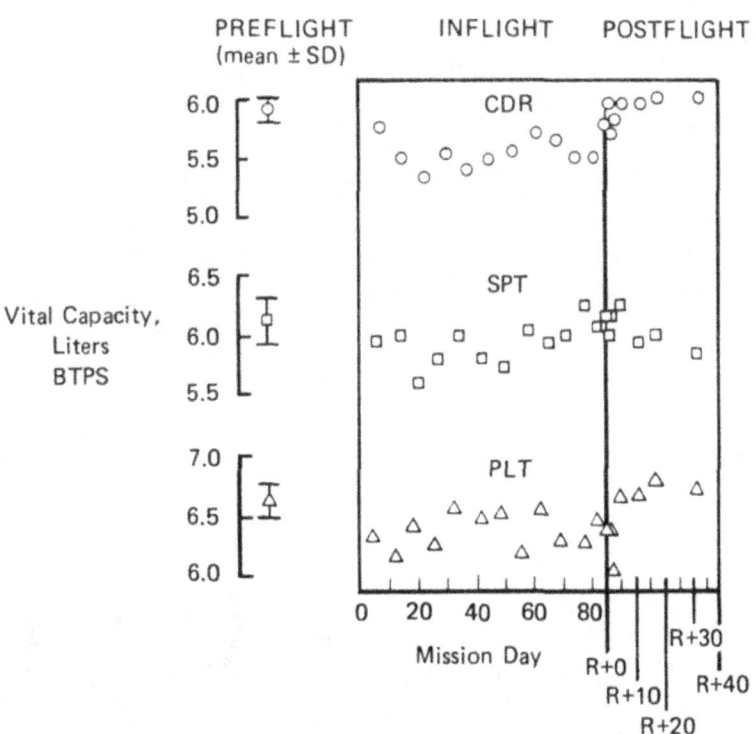

Figure 17. Vital capacities of Skylab 4 Crewmen.
(From Sawin et al., 1976)

Measures of postflight cardiopulmonary responses to exercise have consistently revealed decreases in exercise capacity. For example, measurements were obtained pre- and postflight on Apollo 7 through 11 (Rummel et al., 1973). These studies demonstrated significant decreases immediately postflight in workload, oxygen consumption, systolic blood pressure, and diastolic blood pressure, at a heart rate of 160 beats/minute. Mechanical efficiency (oxygen required to perform a given amount of work) showed no gross changes postflight. Studies of Skylab crewmen revealed similar postflight decrements in exercise capacity, evidenced by decreases in oxygen uptake, pulse, cardiac output, and stroke volume (Michel et al., 1977). Most of the cardiovascular responses returned to normal within three weeks. Similar results were reported by Soviet investigators (Gazenko et al., 1981b; Gazenko, Kakurin, & Kuznetsov, 1976).

Studies aboard both Skylab and Salyut 6, however, demonstrated that exercise capacity is not adversely affected inflight (Gazenko, Genin, & Yegorov, 1981a, 1981b; Michel et al., 1977). For example, Figure 18 shows Skylab preflight, inflight, and postflight heart rates during 75 percent maximum exercise. Most of the Skylab crewmen exhibited increased heart rates postflight relative to preflight baselines, but very little change in inflight heart rate during exercise. Similarly, Salyut cosmonauts have shown no change in inflight performance under physical load, as reflected in measures of oxygen sufficiency. Crewmembers of the longer missions in both programs did not require more time for readaptation postflight. In Soviet missions lasting 96, 140, 175, and 185 days, readaptation time has not varied substantially. All cardiovascular parameters returned to normal in 18 to 21 days for the Skylab 2 crew (28-day mission), 5 days for the Skylab 3 crew (59-day mission), and 4 days for the Skylab 4 crew (84-day mission). Since the crew of Skylab 4 performed the most exercise inflight and the crew of Skylab 2 the least, it appears that the amount of exercise performed inflight is inversely related to the amount of time required for the cardio-vascular system to readapt to the one-gravity environment. Of course, factors other than the amount of inflight exercise may have contributed to this apparent paradox concerning mission length and postflight recovery time. For example, high initial levels of conditioning may be a factor, especially if the vigorous inflight exercise required to maintain this level of conditioning is not performed. Loss of muscle mass might also contribute to this phenomenon, since significant losses and decreased strength will result in early fatigue and inability to complete the stress test protocols.

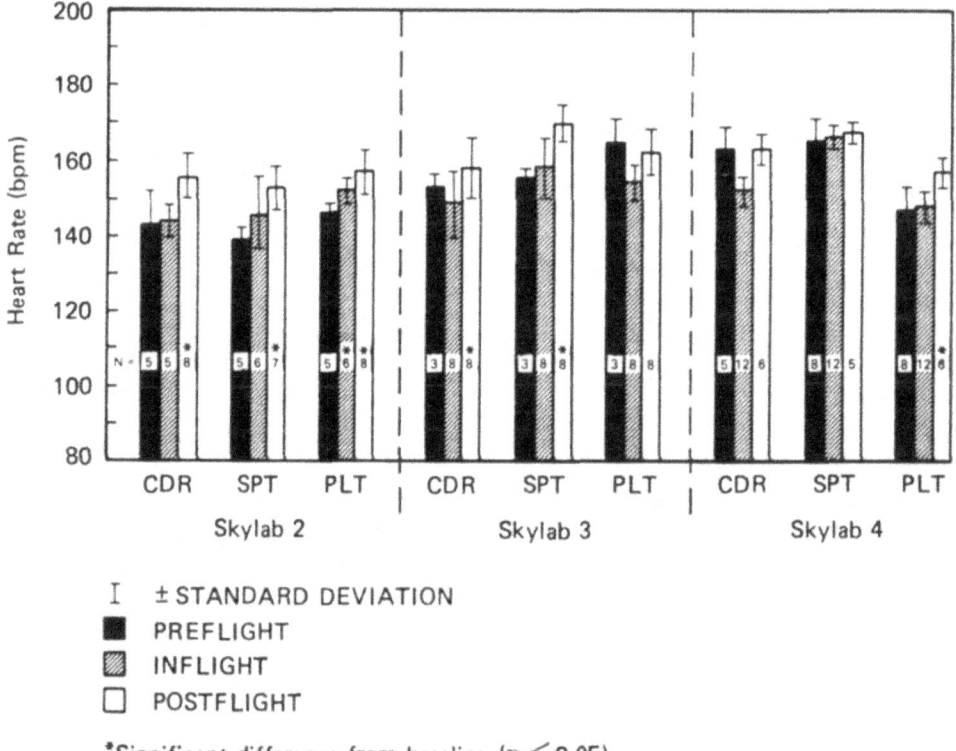

Figure 18. Heart rates pre-, in-, and postflight, at 75 percent maximum exercise. (From Michel et al., 1977)

Summary and Conclusions

From the knowledge gained from previous space missions, it is clear that the cardiopulmonary system acclimates promptly and satisfactorily to the space environment, and that human beings can maintain functional capacity for prolonged periods in space. These inflight adaptations are not appropriate in a one-gravity environment, however, and some cardiopulmonary responses may require up to three weeks postflight to return to preflight baselines.

Although the observations thus far suggest optimism with respect to the ability of the cardiopulmonary system to acclimate to the space environment, several questions remain unanswered. In particular, it is important to elucidate further the effects and time course of the substantial cephalic fluid shifts and the resulting reduction in total blood volume on central venous and cardiopulmonary pressures. Although horizontal and head-down bedrest or water immersion studies can provide partial answers to these and other questions about the functioning of the cardiopulmonary system in the presence of cephalic fluid shifts, complete information can only be provided by critical, long-term observations and experimentation in space.

References

Bergman, S.A., Johnson, R.L., & Hoffler, G.W. Evaluation of the electromechanical properties of the cardiovascular system after prolonged weightlessness. In R.S. Johnston and L.F. Dietlein (Eds.), *Biomedical results from Skylab* (NASA SP-377). Washington, D.C.: U.S. Government Printing Office, 1977.

Berry, C.A., & Catterson, A.D. Pre-Gemini medical predictions vs. Gemini flight results. In Gemini Summary Conference, February 1 and 2, 1967, Manned Spacecraft Center, Houston, Texas (NASA SP-138). Washington, D.C.: U.S. Government Printing Office, 1967.

Burkovskaya, T.Ye., Ilyukhin, A.V., Lobachik, V.I., & Zhidkov, V.V. Erythrocyte balance during 182-day hypokinesia. From JPRS, *USSR Report: Space Biology and Aerospace Medicine*, 30 October 1980, *14*(5):75-80.

Gazenko, O.G., Kakurin, L.I., & Kuznetsov, A.G. (Eds.). *Space flights onboard the spacecraft Soyuz*. Moscow: Nauka, 1976.

Gazenko, O.G., Genin, A.M., & Yegorov, A.D. Major medical results of the Salyut-6/Soyuz 185-day space flight. NASA NDB 2747. *Proceedings of the XXXII Congress of the International Astronautical Federation*, Rome, Italy, 6-12 September 1981a.

Gazenko, O.G., Genin, A.M., & Yegorov, A.D. Summary of medical investigations in the U.S.S.R. manned space missions. *Acta Astronautica*, 1981b, *8*(9-10): 907-917.

Golubchikova, Z.A., Yegorov, A.D., & Kalinichenko, V.V. Results of vectorcardiographic examinations during and after long-term space flights aboard the Salyut 6-Soyuz orbital complex.
Space Biology and Aerospace Medicine, 5 March 1981, *15*(1):31-35.

Henry, W.L., Epstein, S.E., Griffith, J.M., Goldstein, R.E., & Redwood, D.R. Effect of prolonged
space flight on cardiac function and dimensions. In R.S. Johnston and L.F. Dietlein (Eds.),
Biomedical results from Skylab (NASA SP-377). Washington, D.C.: U.S. Government Printing
Office, 1977.

Hoffler, G.W. Cardiovascular studies of U.S. space crews: An overview and perspective. In
N.H.C. Hwang and N.A. Normann (Eds.), *Cardiovascular flow dynamics and measurements*.
Baltimore, Maryland: University Park Press, 1977.

Hoffler, G.W., Bergman, S.A., & Nicogossian, A.E. In-flight lower limb volume measurement. In
A.E. Nicogossian (Ed.), *The Apollo-Soyuz Test Project medical report* (NASA SP-411).
Washington, D.C.: U.S. Government Printing Office, 1977.

Hoffler, G.W., & Johnson, R.L. Apollo flight crew cardiovascular evaluations. In R.S. Johnston,
L.F. Dietlein, and C.A. Berry (Eds.), *Biomedical results of Apollo* (NASA SP-368). Washington,
D.C.: U.S. Government Printing Office, 1975.

Hoffler, G.W., Johnson, R.L., Nicogossian, A.E., Bergman, S.A., & Jackson, M.M. Vectorcardio-
graphic results from Skylab medical experiment M092: Lower body negative pressure. In
R.S. Johnston and L.F. Dietlein (Eds.), *Biomedical results from Skylab* (NASA SP-377).
Washington, D.C.: U.S. Government Printing Office, 1977.

Johnson, R.L., Hoffler, G.W., Nicogossian, A.E., Bergman, S.A., & Jackson, M.M. Lower body
negative pressure: Third manned Skylab mission. In R.S. Johnston and L.F. Dietlein (Eds.),
Biomedical results from Skylab (NASA SP-377). Washington, D.C.: U.S. Government Printing
Office, 1977.

Katkov, V.Ye., Chestukhin, V.V., Rumyantsev, V.V., Troshin, A.Z., & Zybin, O.Kh. Jugular, right
atrial pressure and cerebral hemodynamics of healthy man submitted to postural tests. *Space
Biology and Aerospace Medicine*, 1981, *15*(5):68-73.

Michel, E.L., Rummel, J.A., Sawin, C.F., Buderer, M.C., & Lem, J.D. Results of Skylab medical
experiment M171—metabolic activity. In R.S. Johnston and L.F. Dietlein (Eds.), *Biomedical
results from Skylab* (NASA SP-377). Washington, D.C.: U.S. Government Printing Office,
1977.

Nicogossian, A., Hoffler, G.W., Johnson, R.L., & Gowen, R.J. Determination of cardiac size
following space missions of different durations: The second manned Skylab mission. *Aviation,
Space, and Environmental Medicine*, 1976, *47*(4):362-365.

Nicogossian, A.E., Whyte, A.A., Sandler, H., Leach, C.S., & Rambaut, P.C. (Eds.). *Chronological summaries of United States, European, and Soviet bedrest studies.* Washington, D.C.: NASA, 1979.

Rummel, J.A., Michel, E.L., & Berry, C.A. Physiological responses to exercise after space flight— Apollo 7 to Apollo 11. *Aerospace Medicine*, 1973, *44*:235-238.

Sandler, H. Cardiovascular effects of weightlessness. In P.M. Yu and J.F. Goodwin (Eds.), *Progress in cardiology, Volume 6.* Philadelphia: Lea & Febiger, 1977.

Sawin, C.F., Nicogossian, A.E., Rummel, J.A., & Michel, E.L. Pulmonary function evaluation during the Skylab and Apollo-Soyuz missions. *Aviation, Space, and Environmental Medicine*, 1976, *47*(2):168-172.

Smith, R.F., Stanton, K., Stoop, D., Brown, D., Janusz, W., & King, P. Vectorcardiographic changes during extended space flight (M093). Observations at rest and during exercise. In R.S. Johnston and L.F. Dietlein (Eds.), *Biomedical results from Skylab* (NASA SP-377). Washington, D.C.: U.S. Government Printing Office, 1977.

Thornton, W.E., & Hoffler, G.W. Hemodynamic studies of the legs under weightlessness. In R.S. Johnston and L.F. Dietlein (Eds.), *Biomedical results from Skylab* (NASA SP-377). Washington, D.C.: U.S. Government Printing Office, 1977.

Thornton, W.E., Hoffler, G.W., & Rummel, J.A. Anthropometric changes and fluid shifts. In R.S. Johnston and L.F. Dietlein (Eds.), *Biomedical results from Skylab* (NASA SP-377). Washington, D.C.: U.S. Government Printing Office, 1977.

Turchaninova, V.F., & Domracheva, M.V. Results of studies of pulsed blood flow and regional vascular tonus during flight in the first and second expeditions aboard the Salyut-6-Soyuz orbital complex. *Space Biology and Aerospace Medicine*, 1980, *14*(3):11-14.

Wolthius, R.A., Bergman, S.A., & Nicogossian, A.E. Physiological effects of locally applied reduced pressure in man. *Physiological Reviews*, 1974, *54*(3):566-595.

Yegorov, A.D., Itsekhovskiy, O.G., Kas'yan, I.I., Alferova, I.V., Polyakova, A.P., Turchaninova, V.F., Bernadskiy, V.I., Doroshev, V.G., & Kobzev, Ye.A. Study of hemodynamics and phase structure of cardiac cycle in second crew of the Salyut 6 orbital station at rest. *Space Biology and Aerospace Medicine*, 4 February 1981, *14*(6):11-15.

CHAPTER 11

LEAN BODY MASS AND ENERGY BALANCE

The weightless environment of space has profound effects on lean body mass, which are apparently related to disuse atrophy of the muscle tissue. These effects have been documented in a number of different ways, including measurements of body weight (or mass), anthropometric measurements, biochemical analyses and metabolic balance studies, and evaluation of muscle condition and neuromuscular function postflight. Furthermore, changes in body composition and energy balance during longer term space flights suggest that the deterioration in muscle tissue may be accompanied by decreases in metabolic efficiency.

Changes in Body Weight

One of the most well documented physiological changes associated with space flight is weight loss. Figure 1 shows postflight weights of astronauts (as percent of preflight baselines) for a number of space missions. Postflight weight losses average around three to four percent of preflight weights; however, there does not appear to be a close relationship between mission length and the magnitude of the weight loss (Gazenko, Genin, & Yegorov, 1981; Hawkins & Zieglschmid, 1975; Thornton & Ord, 1977; Nicogossian et al., 1977). In two Soviet missions of six months' duration, three of the four cosmonauts exhibited increases in body mass throughout most of the mission. The body mass tended to grow linearly with flight time, plateauing at around day 140-160 with increases of two to five percent over preflight weight. However, these gains were attributed to an increase in adipose tissue which more than offset losses in lean body mass (Gazenko, et al., 1981). During missions

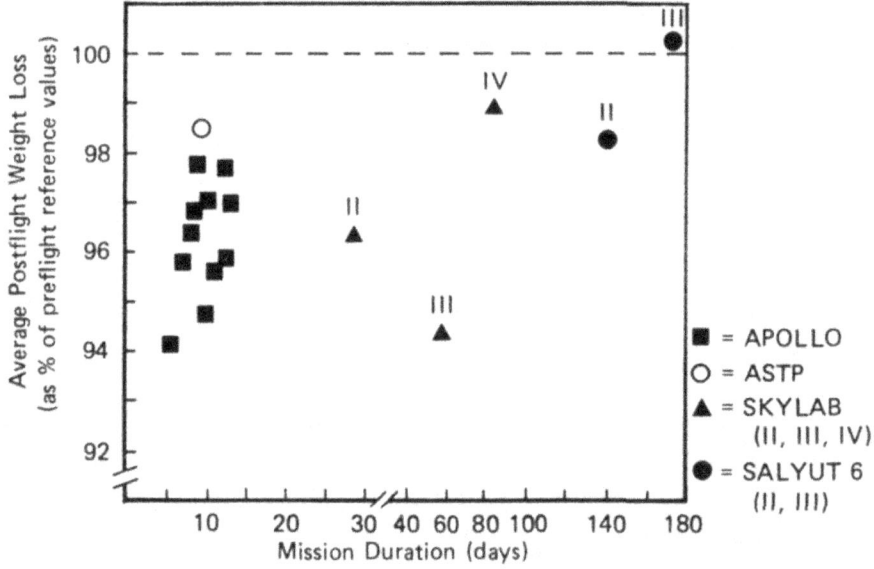

Figure 1. Weight losses found postflight in U.S. space missions. (Calculated and redrawn from data in Gazenko, Genin, & Yegorov, 1981; Hawkins & Zieglschmid, 1975; Thornton & Ord, 1977; and Nicogossian et al., 1977)

186

of less than six months' duration, the loss of about 6–7 kg was independent of mission duration, and was attributed to fluid losses and to a negative nitrogen balance.

During the Skylab missions, a spring-mass oscillator constrained to linear motion was installed onboard to obtain inflight measurements of mass, and to ascertain the time course of the weight loss inflight (Thornton & Ord, 1977). A similar massmeter was employed aboard Salyut 6 (Yegorov, 1980). In general, inflight observations of physiological mass aboard Skylab revealed that most of the weight was lost early in the mission, and that most of the lost weight is regained rapidly postflight. For example, Figure 2 shows the changes in body mass early in flight and during recovery for the crew of Skylab 3. This pattern of initial rapid weight loss and rapid postflight recovery suggests that much of the weight loss associated with space flight is due to loss of fluids. However, in the two Soviet flights where there were weight gains, the single instance of progressive decrease in body mass was attributed to a slight metabolic deficiency and to the effects of exercise (Yegorov, 1980).

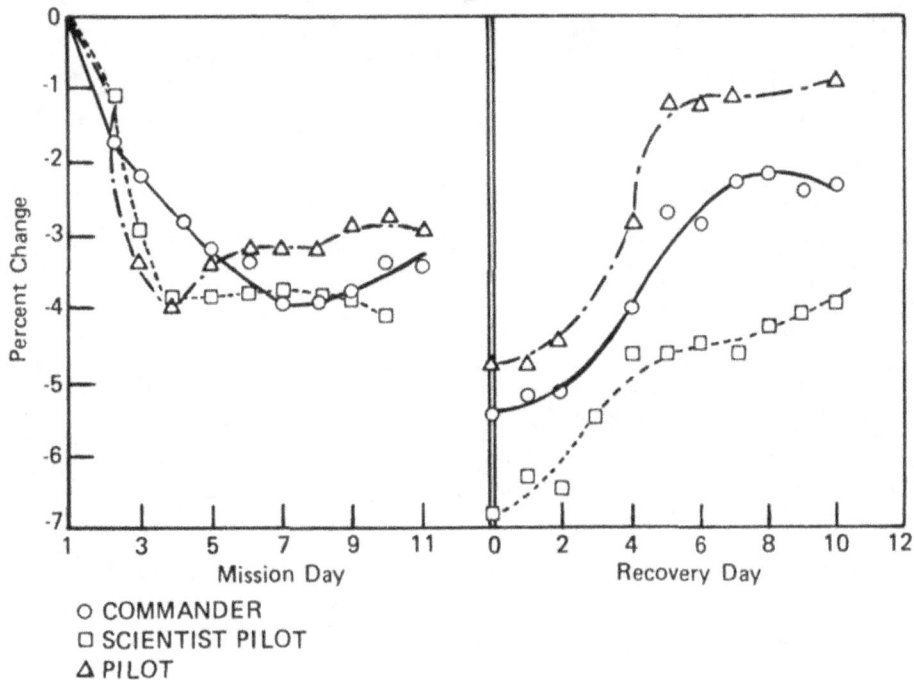

Figure 2. Changes in body mass early inflight on Skylab 3 and during recovery. (From Thornton & Ord, 1977)

Metabolic losses appear to account for part of the weight loss in returned Skylab astronauts as well. After the large, early inflight weight loss, a slower loss was seen that persisted throughout the remainder of the mission. An example of this pattern is illustrated in Figure 3. The slower loss appears to be a metabolic one, a hypothesis that is supported by the observed negative correlation between weight loss and caloric intake, shown in Figure 4. Those astronauts who ate less per kilogram of body weight also tended to lose more weight.

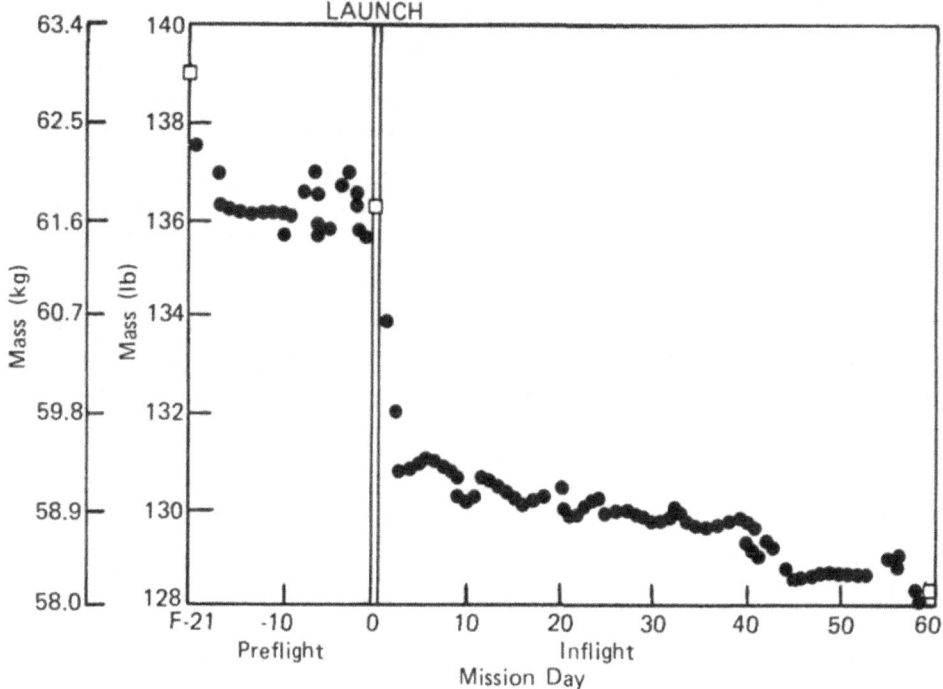

Figure 3. Body mass measurements of the Skylab 3 Scientist Pilot, showing the gradual long-term loss of weight inflight. (From Thornton & Ord, 1977)

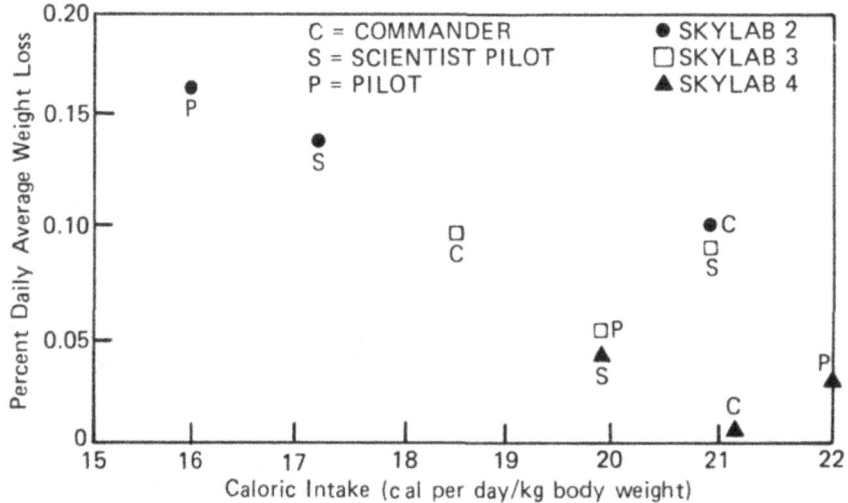

Figure 4. Weight loss vs. caloric intake for the nine Skylab astronauts inflight. (From Thornton & Rummel, 1977)

C-3

Changes in Muscle Tissue

The weight losses observed in astronauts and cosmonauts have been accompanied by a variety of changes in muscle tissue and function, all of which suggest the deterioration of muscle inflight. These muscular changes are demonstrated by biochemical analyses of blood, urine, and fecal samples; assessment of autoimmune processes; metabolic balance studies; evaluations of muscle condition; anthropometric measurements; and studies of neuromuscular function.

Biochemical Analyses

Analyses of blood, urine, and fecal samples from Skylab astronauts have supported the hypothesis that space flight is associated with substantial changes in muscle tissue. Of particular relevance to the state of muscle tissue are the inflight increases in plasma calcium, phosphorus, potassium, and creatinine. Urine analyses have revealed changes in a number of substances that are considered indices of muscle condition, including increases in inflight excretion of calcium, sodium, potassium, creatinine, phosphates, magnesium, total hydroxylysine, N^+-methylhistidine, and almost all amino acids (Gazenko, Grigor'yev, & Natochin, 1980; Leach & Rambaut, 1977; Leach et al., 1976). This biochemical pattern strongly points to muscle breakdown inflight. According to Gazenko et al. (1980), fluid losses on long-term flights are largely attributable to a decrease of intracellular fluid due to muscle atrophy. The ratio of sodium/fluid in the urine is restored to normal more rapidly after short flights than after long flights.

Autoimmune Processes

The serum of Soviet cosmonauts is frequently analyzed postflight for the presence of auto-antibodies, which would indicate muscle atrophy. The method used is indirect immunofluorescence measurement of reactions of the sera with human heart tissue. In one such study (Tashpulatov et al., 1979), 45 percent of the cosmonauts tested exhibited the autoimmune reaction; of three individuals with particularly strong reactions, two had flown twice before. The reaction was not evident after readaptation.

Metabolic Balance Studies

In the Skylab missions, dietary intake of several key elements was carefully monitored to investigate metabolic balances during space flight, and to further assess the condition of the musculoskeletal system. The key elements included calcium, nitrogen, phosphorus, magnesium, potassium, and sodium. Inflight urinary excretion of nitrogen and phosphorus increased, and both substances displayed a negative balance throughout most of the early inflight period. Negative or only slightly positive balances persisted for the latter parts of the Skylab missions, despite high protein and caloric intake.

Figure 5 shows a typical example of nitrogen balance across the pre-, in-, and postflight periods of a crewmember on Skylab 3. Postflight nitrogen and phosphorus balances were markedly positive, indicating a return to retention of these substances (Whedon et al., 1977; Rambaut et al., 1979).

The pattern of changes in nitrogen balance suggests that muscle tissue is most affected by weightlessness early in flight; however, nitrogen continues to be lost throughout the duration of the space mission.

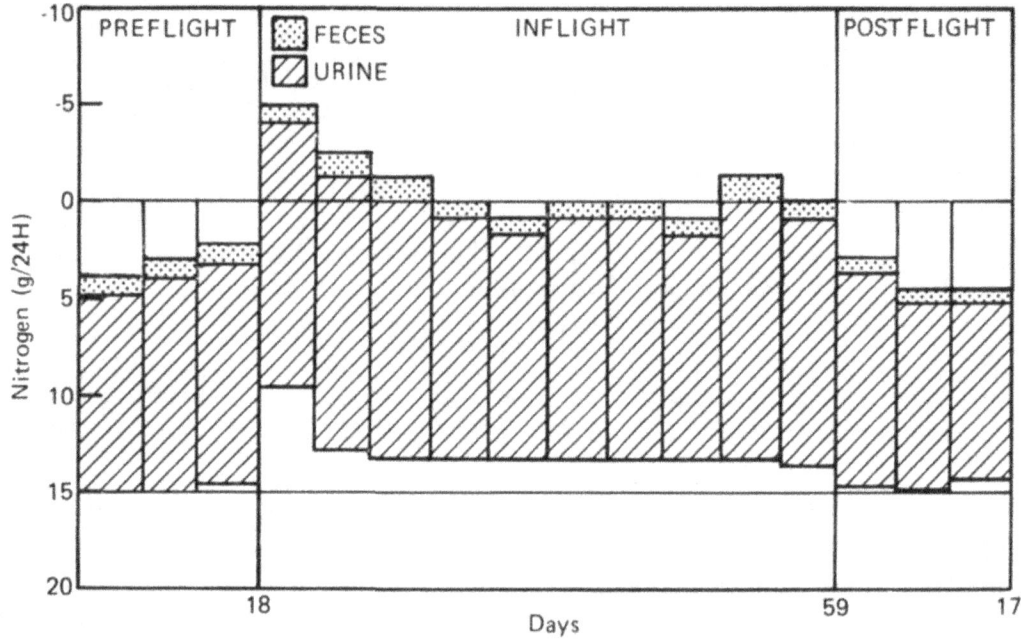

Figure 5. Nitrogen balances pre-, in-, and postflight for the Commander
of Skylab 3. (From Whedon et al., 1977)

Negative potassium balances have been recorded by Soviet researchers even five days postflight (Gazenko, Grigor'yev, & Natochin, 1980). This is probably also a manifestation of muscular atrophy, since reduction of cell mass results in a concomitant loss of potassium from cells.

Evaluation of Muscle Condition

The biochemical evidence indicating breakdown of muscle tissue is supported by evaluations of arm and leg muscles conducted pre- and postflight for all the Skylab missions. Evaluations were conducted using a dynamometer, shown in Figure 6. Figures 7 and 8 show the average loss in strength for flexors and extensors in the arm and leg for each of the three Skylab crews (Thornton & Rummel, 1977). The legs of all three crews showed larger postflight decrements in strength compared to the arms, probably because muscular activity in weightlessness provides comparatively larger loads for the arms. Under weightless conditions, the legs are without the normal stress of load bearing, while the arms have the additional stress of controlling locomotion.

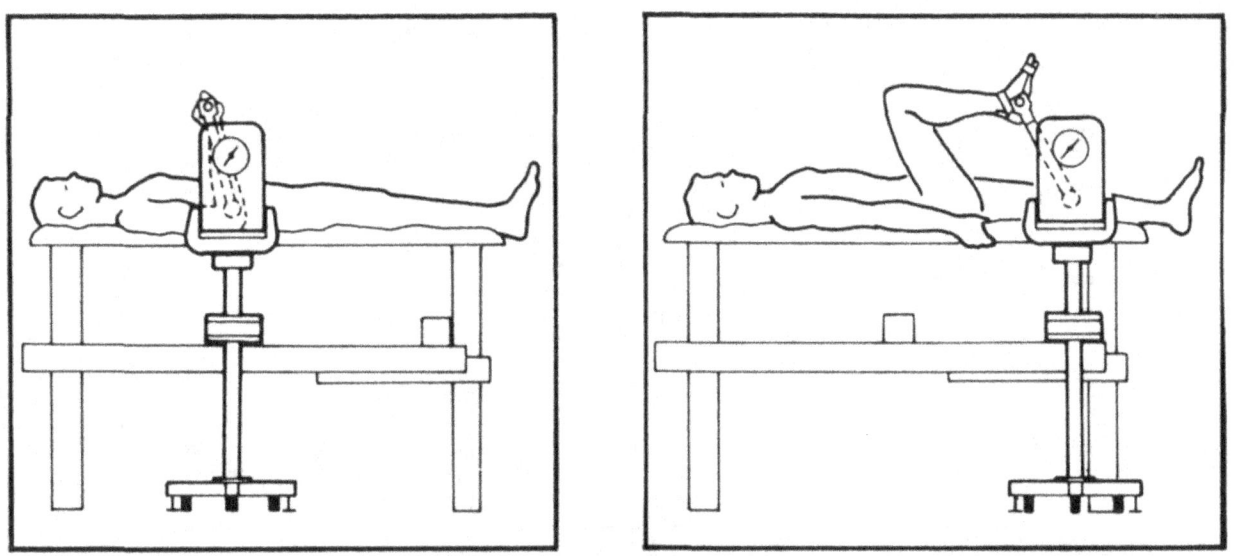

Figure 6. Dynamometer used for evaluating muscular condition pre- and postflight.
(From Thornton & Rummel, 1977)

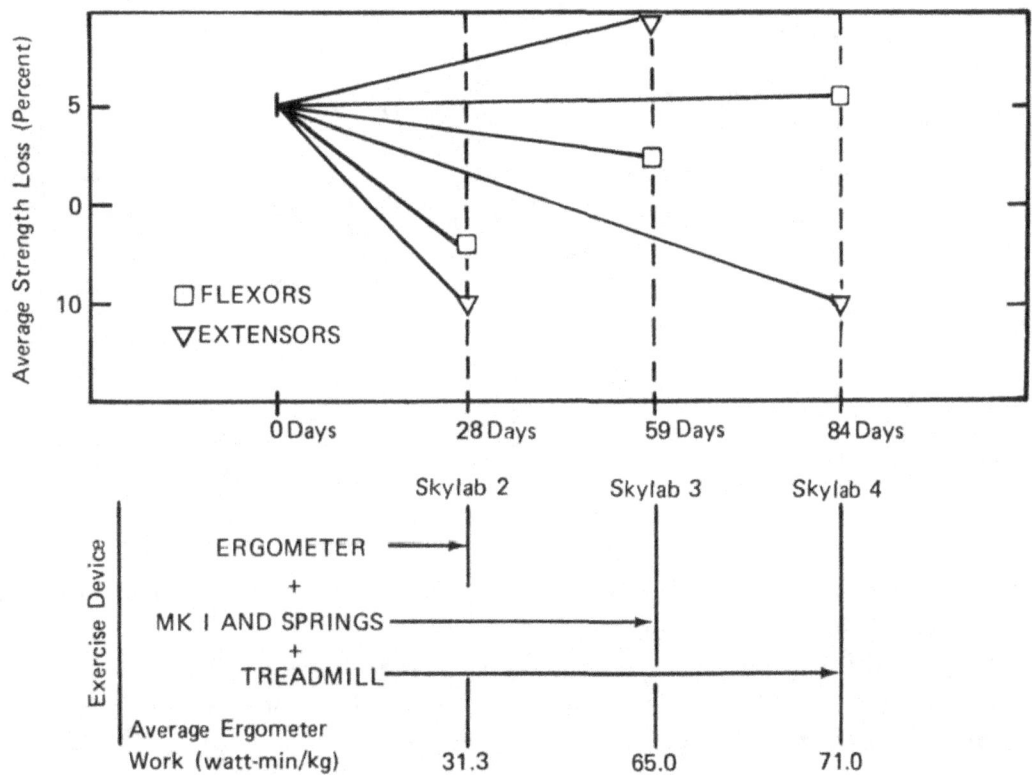

Figure 7. Average postflight strength changes in arm. (From Thornton & Rummel, 1977)

191

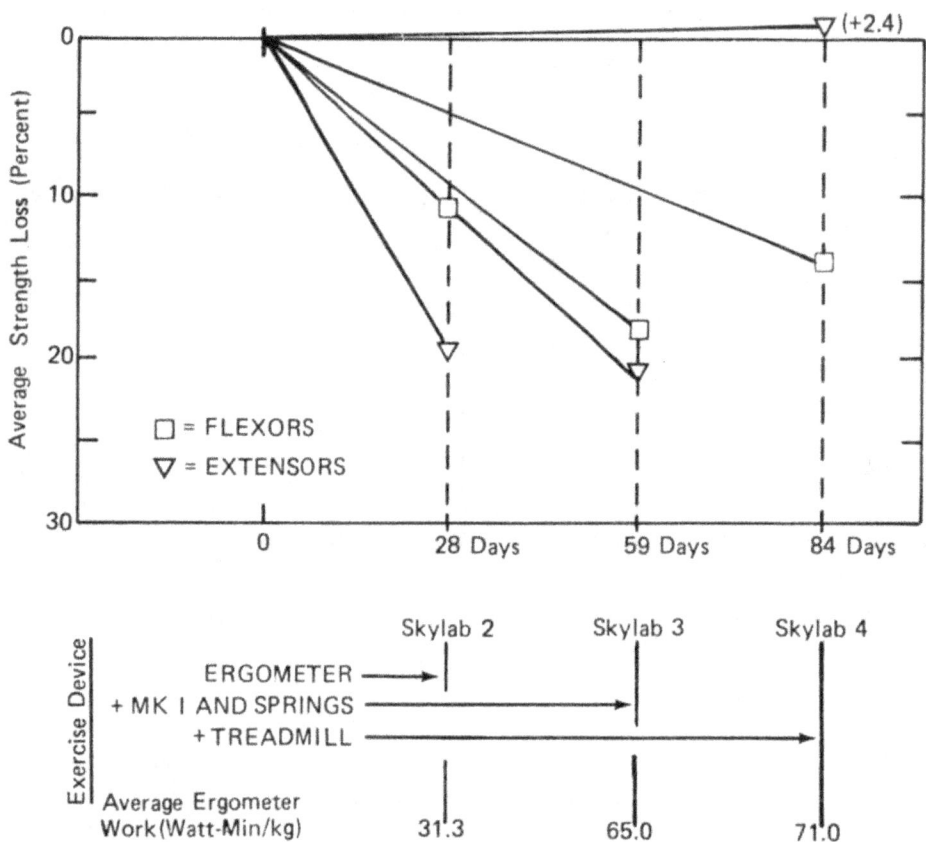

Figure 8. Average postflight strength changes in leg. (From Thornton & Rummel, 1977)

Figures 7 and 8 also show the kinds of exercise devices available to each of the Skylab crews, and the average amount of inflight ergometer work. The variety of different devices available to the crews was increased for each mission, and the total amount of exercise in which the crews engaged was also increased. The crew of Skylab 4 showed the smallest decrements in strength in both arms and legs, even though they participated in the longest mission (84 days). These findings suggest that inflight exercise is capable of diminishing or possibly preventing the deterioration in muscular strength associated with disuse. However, the extremely rigorous exercise regimens on Skylab 3 and 4 did not prevent the loss of nitrogen and phosphorus (Thornton, 1981).

Comparable studies of pre- and postflight muscle condition have been carried out in the Soviet space program. After long-term missions aboard Salyut 6, the force and speed of muscle groups was evaluated with an isokinetic dynamometer known as the "Cybex" (Gazenko et al., 1981; Yegorov, 1980). Extensive exercise eliminated substantial changes in peripheral muscles, usually with the exception of reversible atony of posterior calf muscles and sub-atrophy of latissimus and trapezius muscles. However, reductions in the force of the gastrocnemius and frontal tibialis muscles, as measured by dynamometry, were persistent, with only a gradual increase in maximum effort.

Anthropometric Measurements

Measurements of leg volume and body form were conducted postflight on Skylab and both in- and postflight on Salyut 6. Data from these studies provide additional information about the effects of space flight on lean body mass. Figure 9 shows average changes in leg volume during the post-Skylab recovery period. The increases in leg volume during the first two recovery days are primarily due to gains in fluid. The slower gains during the remainder of the 11 days postflight shown in this figure are probably related to the recovery of muscle mass lost inflight (Thornton & Rummel, 1977). Salyut 6 data have generally shown pronounced decreases in leg volume in the first three weeks of flight, followed by a wavelike pattern of large decreases, periods of stability, and smaller recoveries. In the longest of these flights, leg volume decreased during the first half of the 185-day mission, and then tended to increase (Gazenko et al., 1981).

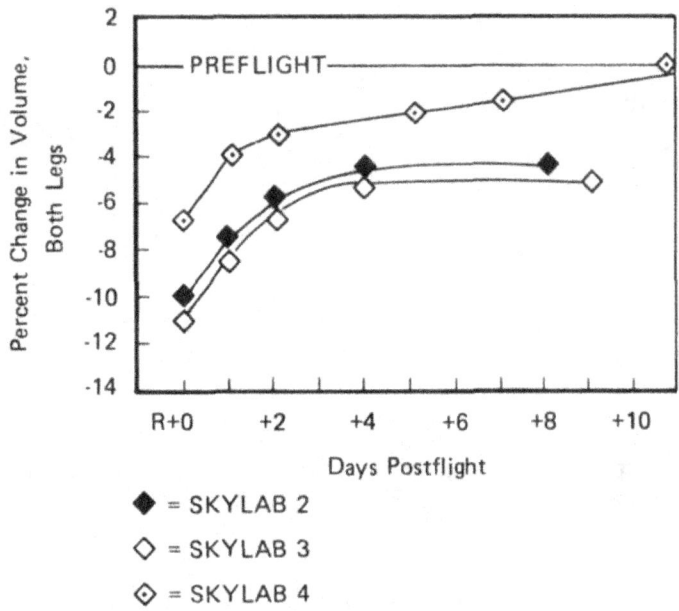

Figure 9. Average leg volume changes postflight.
(From Thornton & Rummel, 1977)

Biostereometric measurements of the Skylab astronauts were also made postflight in order to estimate changes in the distribution of muscle and fat tissue (Whittle et al., 1977). Figure 10 shows a typical example of preflight and postflight volume distribution. In general, these analyses of body form revealed striking losses of volume in the abdomen, buttocks, and calves, and less striking losses in the thighs. The authors concluded that the losses observed in the abdomen and buttocks are probably due mainly to loss of fat, and those observed in the legs, particularly the calves, are due partly to fluid losses and partly to the reduction in muscle mass associated with space flight.

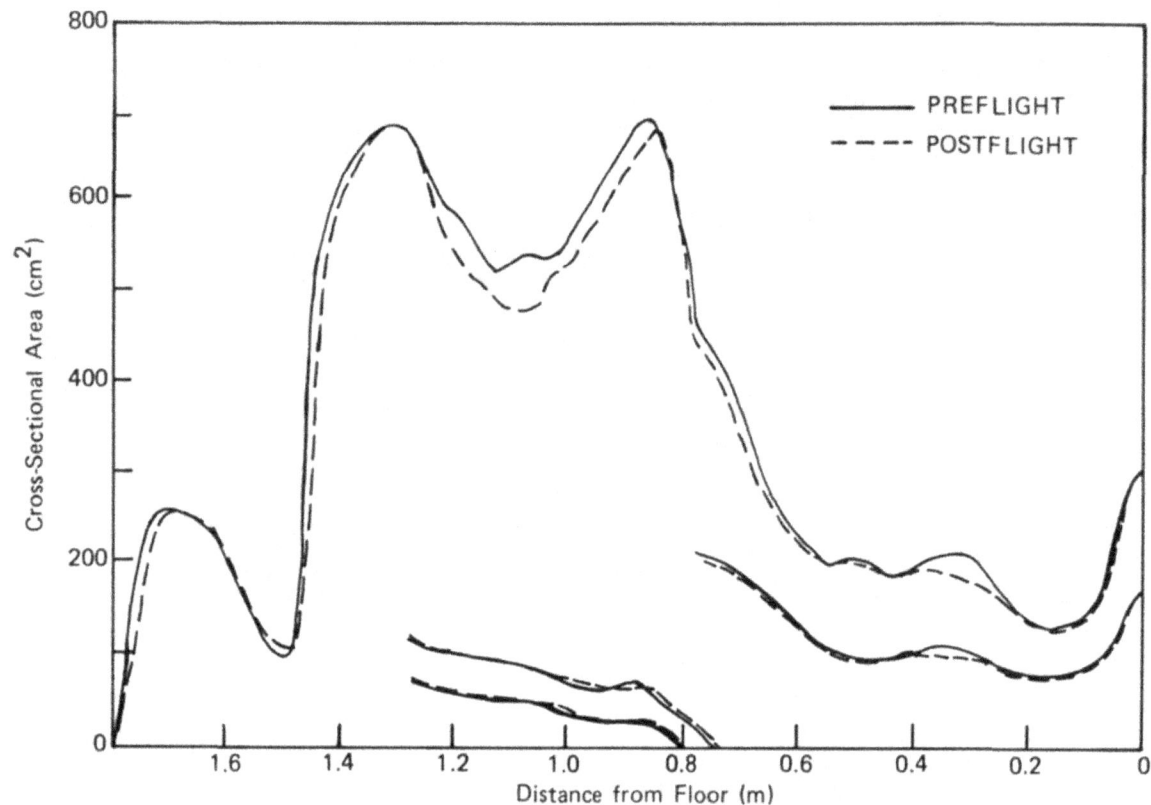

Figure 10. Preflight and postflight bodily volume distribution,
SPT, Skylab 3. (From Whittle, Herron, & Cuzzi, 1977)

Neuromuscular Function

Studies of neuromuscular function have been conducted on astronauts and cosmonauts by means of electromyographic (EMG) spectral power analyses and by measurements of reflex durations. Both of these measurements have revealed alterations in neuromuscular function postflight.

Electromyographic Analyses

Comparisons of pre- and postflight integrated electromyograms (EMGs) from the m. gastro-cnemius of the crew of Skylab 3 demonstrate changes that suggest the muscle had atrophied during space flight. Figure 11 shows the average spectral power for successive bandwidths of 10 Hz for the three crewmen. On recovery day, the average predominant frequency had shifted to the higher bands, compared to preflight baselines. This pattern is similar to that observed in neuropathologic muscle (LaFevers et al., 1975).

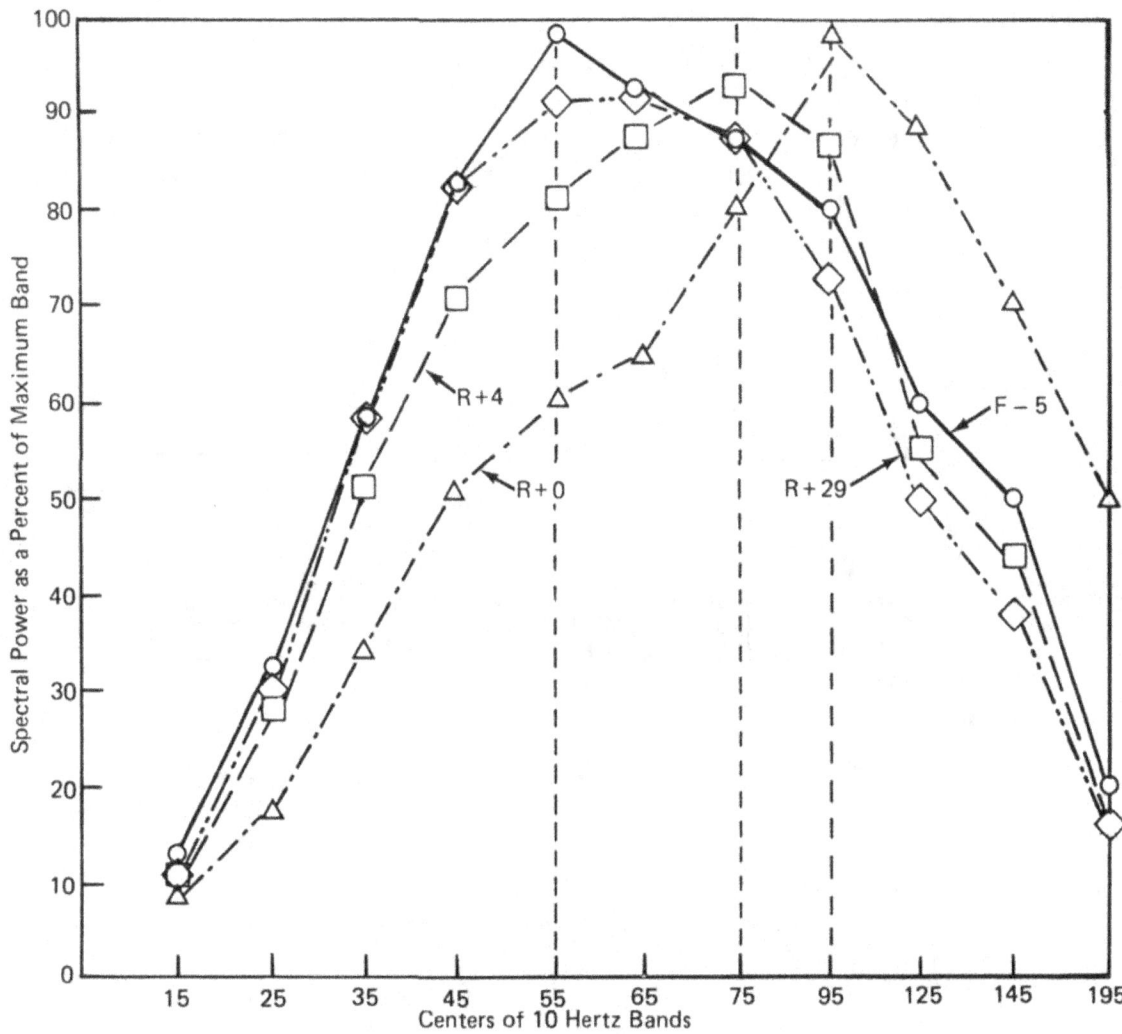

Figure 11. Plot of EMG data averaged across 3 crewmen for preflight and postflight tests. (From LaFevers et al., 1975)

EMGs from the gastrocnemius of one member of the Apollo-Soyuz crew showed the same trend, and also exhibited a pattern after muscular contractions that suggested an increased susceptibility to fatigue and reduced muscular efficiency. EMGs from the arm muscles, however, did not show patterns suggestive of muscular deterioration (LaFevers et al., 1976).

Electromyographic analyses of Salyut 6 cosmonauts were carried out by means of the "Miotest" apparatus. Static and dynamic evaluations of the condition of calf and thigh muscles postflight generally indicated a reduction in the functional capability of the posterior group of calf muscles. According to this test, the electromyographic value of a standard muscular effort typically increased more than twofold following a six-month flight (Yegorov, 1980).

195

Changes in Reflexes

Alterations in neuromuscular function have also been suggested by measurements of pre- and postflight Achilles tendon reflex duration (Baker et al., 1976; Burchard & Nicogossian, 1977). Sharp reductions have been seen for more than a month postflight in the thresholds of tendon reflex in crewmembers of long-term Soviet flights (Yegorov, 1980). A generalized hyperreflexia was observed in the crew of Skylab 2, and the changes in the Achilles reflex for the Skylab 3 crew and for Apollo-Soyuz were quantified. Figure 12 shows both the recording technique and the method of measuring reflex duration and muscle potential interval.

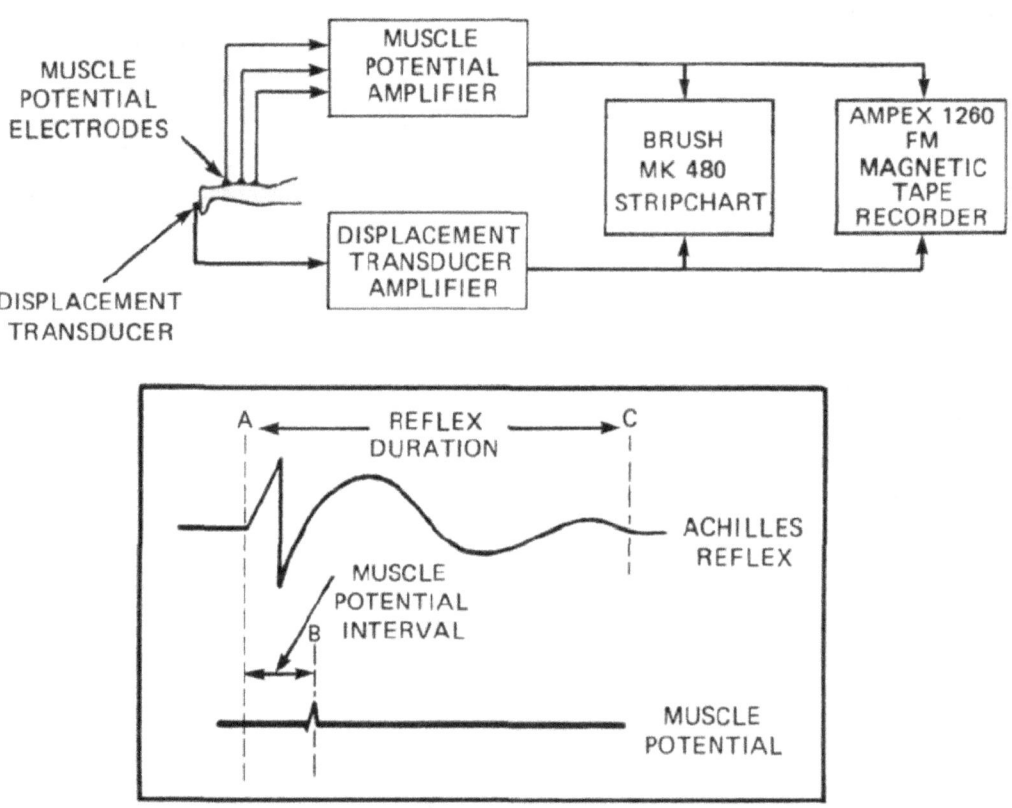

Figure 12. Method of measuring reflex duration and muscle potential interval. (From Baker et al., 1976)

Postflight changes in reflex durations for the crews of Apollo-Soyuz and Skylab 3 are shown in Figure 13. Although the sample size is small, the data suggest the magnitude of the immediate postflight depression in reflex duration is related to mission length. Furthermore, the data from the

196

Skylab crew suggest the duration of the Achilles tendon reflex may become slightly longer than normal about 2 weeks postflight, and only return to preflight baselines after a month or more.

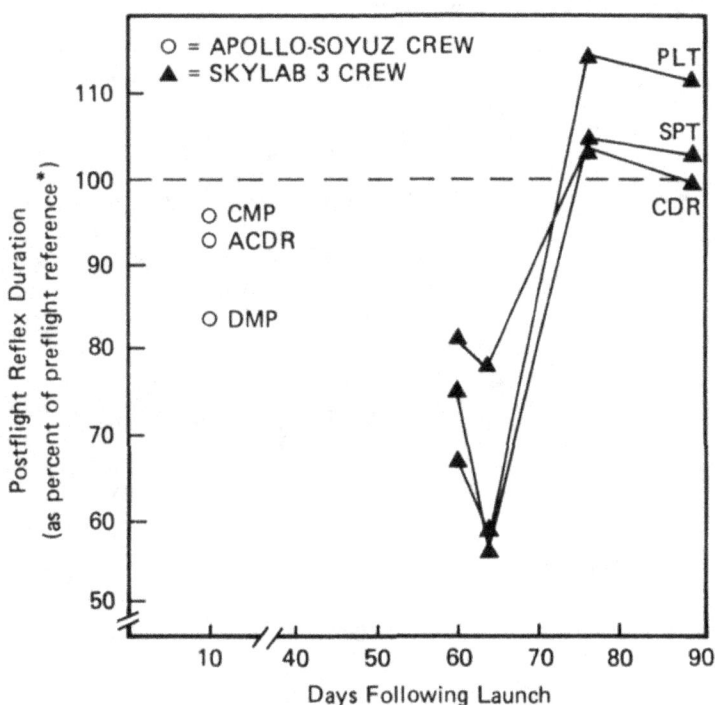

*Preflight references are either single measures, or means of all measurement taken preflight if more than one was obtained.

Figure 13. Changes in reflex duration postflight, Apollo-Soyuz and Skylab 3. (Calculated and redrawn from data in Baker et al., 1976, and Burchard & Nicogossian, 1977)

Body Composition and Energy Balance

Early speculation on the amount of food required to maintain body weight in weightlessness generally hypothesized that space crews would require fewer calories than normal, because of the reduced muscular load. These speculations were not correct, partly because the Gemini missions on which they were based involved little movement and energy expenditure. The information obtained from later space missions indicated that movement in a weightless environment entails a higher metabolic cost than predicted. Although locomotion demands less energy than in one gravity, those tasks that ordinarily depend on friction for their reactive force require muscular work to supply that force. Furthermore, only a small amount of the basal energy expenditure is attributable to direct gravity effects.

197

As a result of observed weight losses, which were partly due to insufficient caloric intake, food supply was increased for astronauts on subsequent space missions. On Skylab, for example, 3100 calories/day were provided. However, inflight weight losses cannot be entirely countered by increased food intake because of muscle breakdown and the likelihood of changes in body composition. Table 1 presents the overall changes in body composition of the Skylab astronauts for the preflight period, and for three consecutive inflight periods of 28 days each. Protein changes were estimated from nitrogen losses, and changes in water content were estimated by isotopic dilution (Rambaut et al., 1979). The preflight period, during which the astronauts engaged in intense physical training, was characterized by protein accretion and loss of fat. The first 28 days of flight were characterized by losses of water, protein, and fat. Later inflight periods showed more losses of fat, but slight gains in protein, water, and, eventually, mass.

Table 1

Average Daily Changes in Body Composition

	Preflight (n=9)	Inflight I (n=9)	Inflight II (n=6)	Inflight III (n=3)
Mass, g	-0.3	-79.4	-2.4	15.5
Water, g	8.0	-48.6	16.7	15.5
Protein, g	19.0	-10.9	1.3	1.1
Fat, g	-27.3	-19.9	-20.4	-1.1

From Rambaut, Leach, & Leonard, 1977

Changes in energy balance accompanied these alterations in body composition during the Skylab missions. Figure 14 shows the components of average daily energy balance for the preflight period, and for the three inflight periods. Energy input includes calories from food intake and from utilization of endogenous fat and protein stores. Energy output includes loss of calories through feces and urine, measured by bomb calorimetry; protein accretion through the addition of muscle mass; and fat deposition. This figure demonstrates that fat was utilized in the preflight period, and also during each of the inflight periods. Protein accretion occurred during the preflight period and later inflight as well, but endogenous protein was broken down during the first 28 days inflight. Food intake was markedly reduced during the first inflight period, but increased thereafter.

198

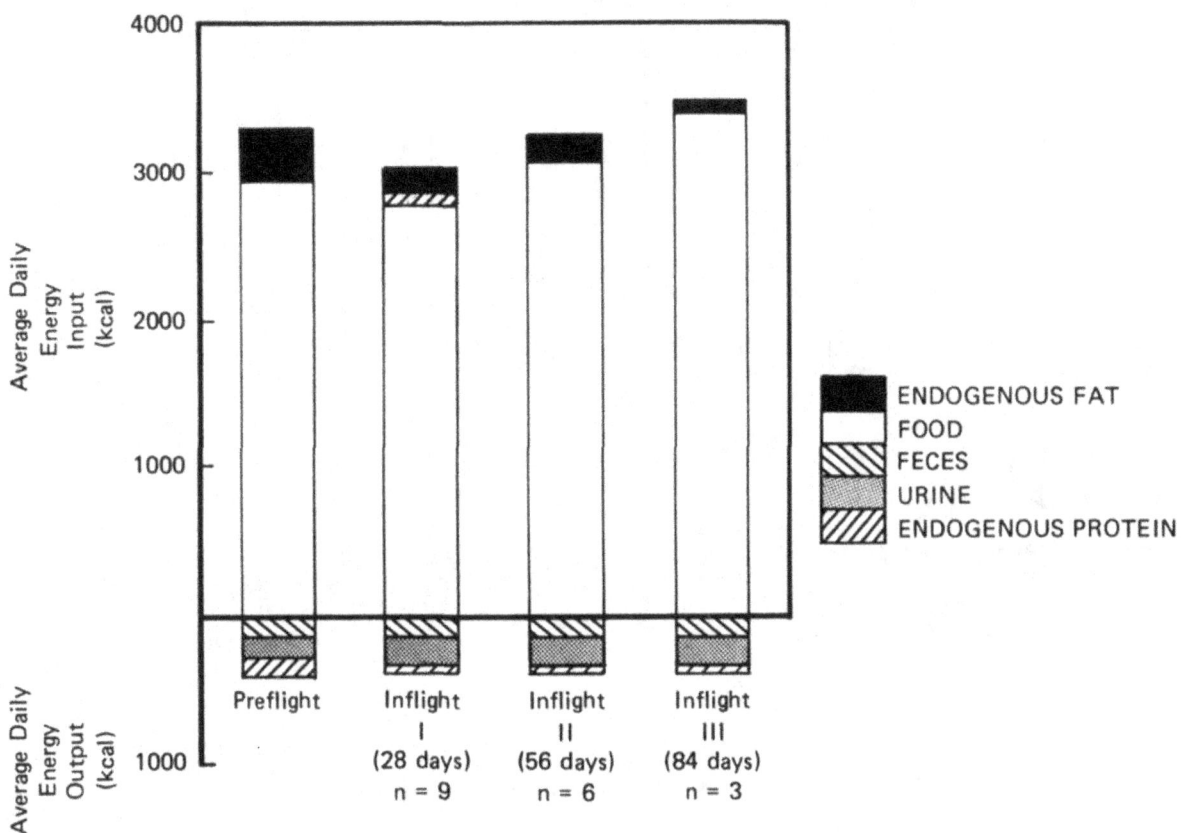

Figure 14. Components of average daily energy balance, for preflight period and 3 inflight periods (Skylab).

Figure 15A shows the mean net energy input for the same time period. These data represent the algebraic sum of energy gained and energy lost. The mean net energy input for the first inflight period is significantly lower than the preflight mean, an event due partly to decreased food intake and partly to protein catabolism. Figure 15B shows the same data, corrected for changes in the amount of total body potassium. This normalizing calculation takes into account the progressive loss of active metabolic mass, a quantity that is proportional to the body's total potassium content.

Assuming that the amount of work performed and the amount of heat output were constant across these four time periods (an assumption that may not be accurate), it is possible to estimate changes in metabolic efficiency by comparing net energy utilization or input. Figure 16A shows shifts from preflight references in net energy utilization for each inflight period. Figure 16B shows the same values corrected for changes in total body potassium. Both of these graphs indicate that net energy utilization declines during the early part of a space mission, but increases beyond preflight levels during the later phases of the mission.

199

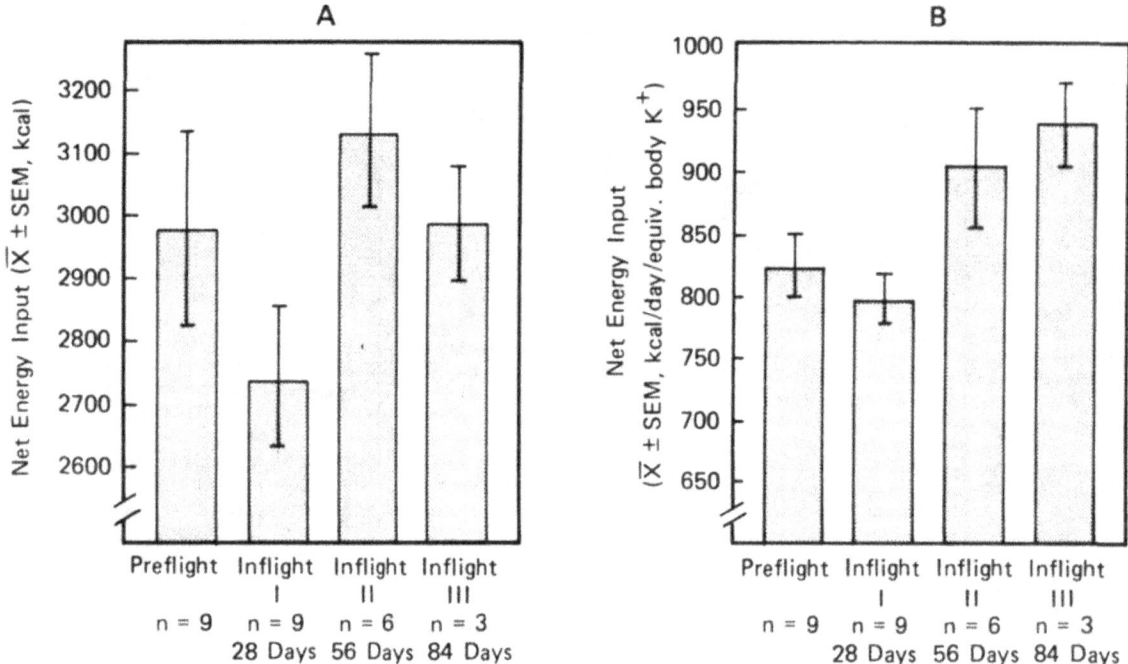

Figure 15. A. Net energy input for preflight period and three inflight periods.
　　　　　　B. Net energy input normalized for changes in active metabolic mass.

(Calculated and redrawn from data in Rambaut, Leach, & Leonard, 1979)

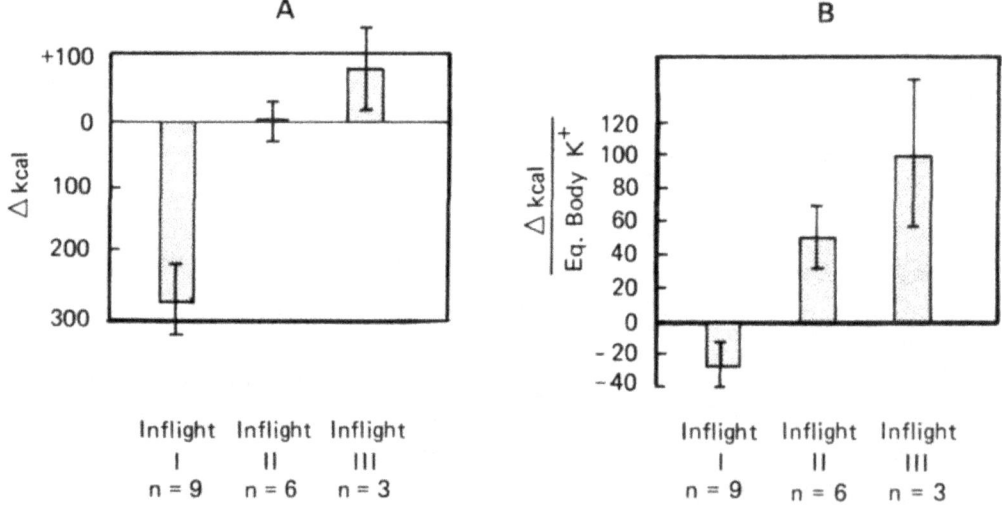

Figure 16. A. Shift from preflight control in net energy utilization for each inflight period.
　　　　　　B. Shift from preflight control in net energy utilization (normalized by changes in total body potassium) for each inflight period.
　　　　　　(From Rambaut, Leach, & Leonard, 1977)

The progressive inflight increases in net energy utilization suggest that either energy expenditure is increasing or metabolic efficiency is decreasing. As lean body mass declines, proportionately more work is required of the muscle tissue. Nevertheless, these calculations raise the possibility that the weightless environment has a deleterious effect on metabolic efficiency.

Gastroenterology

In a recent review of digestive system function under spaceflight conditions, Smirnov and Ugolev (1981) report measurable changes, principally alterations in pancreatic enzyme secretions and, in particular, lipolytic activity. Postflight recovery appears to be related to some extent to mission duration. For long flights, alteration of trypsin activity also was found.

Smirnov and Ugolev conclude that the altered gastrointestinal activity during early flight phases involves an increase in the hydrolysis of food. Postflight findings include hypersecretion of gastric juices, especially an increase in hydrochloric acid and pepsinogen; a decrease in gastric motor activity; an increase in gastric proteolytic activity as well as that of the pancreas and jejunum; an increase in lipolytic ferment secretions; and a decrease in carbohydrate ferments. The significance of these changes is not well understood as yet. However, it is assumed that under conditions of weightlessness certain vitamins and/or amino acids might not be readily absorbed via the digestive tract. The continuous loss of calcium through the feces, or the decreases noted in blood cholesterol and triglyceride levels, might be a function of selective absorption through the gastrointestinal tract.

Summary and Conclusions

The evidence indicates that space flight is associated with a loss in weight that is partly attributable to a breakdown of muscle tissue, particularly in the antigravity muscles. With respect to muscle metabolism, space flight appears to be similar to the fasting state, in which muscle protein is broken down to amino acids. Data related to energy input and output suggest that space flight might also be associated with a decrease in metabolic efficiency of the diminishing muscle mass. The effects of space flight on muscle tissue are reversible, and most parameters return to normal within one month postflight.

Increased food intake and more rigorous exercise regimens appear to at least partially counter the deterioration of muscle tissues inflight. However, the crew of Skylab 4 engaged in substantial amounts of exercise and, while they showed less deterioration in muscle function postflight, they lost nitrogen and phosphorus inflight. Likewise, in crews of long-term Salyut 6 missions, extensive exercise did not prevent chemical imbalances and a degree of muscle function degradation. On shorter missions, the changes in muscle tissue and metabolic efficiency are not likely to present significant health hazards. Long-term space flight will require more effective countermeasures to prevent or ameliorate the loss of lean body mass and alterations in energy balance.

The importance of changes observed in gastrointestinal absorption and secretory rates under spaceflight conditions to the general physiology of body metabolism requires further evaluation.

References

Baker, J.T., Nicogossian, A.E., Hoffler, G.W., & Johnson, R.L. Measurement of a single tendon reflex in conjunction with a myogram: The second manned Skylab mission. *Aviation, Space, and Environmental Medicine*, 1976, *47*(4):400-402.

Burchard, E.C., & Nicogossian, A.E. Achilles tendon reflex. In A.E. Nicogossian (Ed.), *The Apollo-Soyuz Test Project: Medical Report* (NASA SP-411). Washington, D.C.: U.S. Government Printing Office, 1977.

Gazenko, O.G., Genin, A.M., & Yegorov, A.D. Major medical results of the Salyut-6/Soyuz 185-day space flight. NASA NDB 2747. *Proceedings of the XXXII Congress of the International Astronautical Federation*, Rome, Italy, 6-12 September 1981.

Gazenko, O.G., Grigor'yev, A.I., & Natochin, Yu.V. Fluid-electrolyte homeostasis and weightlessness. From JPRS, *USSR Report: Space Biology and Aerospace Medicine*, 30 October 1980, *14*(5):1-11.

Hawkins, W.R., & Zieglschmid, J.F. Clinical aspects of crew health. In R.S. Johnston, L.F. Dietlein, and C.A. Berry (Eds.), *Biomedical results of Apollo* (NASA SP-368). Washington, D.C.: U.S. Government Printing Office, 1975.

LaFevers, E.V., Nicogossian, A.E., Hoffler, G.W., Hursta, W., & Baker, J. Spectral analysis of skeletal muscle changes resulting from 59 days of weightlessness in Skylab II (Report No. JSC 09996). NASA TM X-58171. Houston, Texas: Lyndon B. Johnson Space Center, November 1975.

LaFevers, E.V., Nicogossian, A.E., & Hoffler, G.W. Electromyographic analysis of the skeletal muscle changes arising from long and short periods of spaceflight weightlessness. Preprints of the 47th Aerospace Medical Association Meeting, Bal Harbour, Florida, May 1976.

Leach, C.S., & Rambaut, P.C. Biochemical responses of the Skylab crewmen: An overview. In R.F. Johnston and L.F. Dietlein (Eds.), *Biomedical results from Skylab* (NASA SP-377). Washington, D.C.: U.S. Government Printing Office, 1977.

Leach, C.S., Rambaut, P.C., & DiFerrante, N. Amino aciduria in weightlessness. *Acta Astronautica*, 1976, *6*:1323-1333.

Nicogossian, A.E., Burchard, E.C., & Hordinsky, J.R. General biomedical evaluation. In A.E. Nicogossian (Ed.), *The Apollo-Soyuz Test Project: Medical Report* (NASA SP-411). Washington, D.C.: U.S. Government Printing Office, 1977.

Rambaut, P.C., Leach, C.S., & Leonard, J.I. Observations in energy balance in man during spaceflight. *American Journal of Physiology*, 1977, *233*: R208-R212.

Rambaut, P.C., Leach, C.S., & Whedon, G.D. A study of metabolic balance in crewmembers of Skylab IV. *Acta Astronautica*, 1979, *6*(10).

Smirnov, K.M., / Ugolev, A.M. *Space gastroenterology*. Moscow: Nauka Press, 1981.

Tashpulatov, R.Yu., Danilova, T.A., Lesnyak, A.T., Legen'kov, V.I., Znamenskiy, V.S., & Dedyuyeva, Ye.Yu. Antibodies to myofibril antigens in astronauts after spaceflight. NASA TM-76300. Translated into English from *Zh. Mikrobiologii, Epidemiologii i Immunobiologii*, 1979, *12*:36-39.

Thornton, W.E. Rationale for exercise in spaceflight. In J.F. Parker, Jr., C.S. Lewis, and D.G. Christensen (Eds.), *Conference Proceedings: Spaceflight deconditioning and physical fitness*. Prepared by BioTechnology, Inc., under Contract No. NASW-3469 for National Aeronautics and Space Administration, Washington, D.C., August 1981.

Thornton, W.E., & Ord, J. Physiological mass measurement in Skylab. In R.F. Johnson and L.F. Dietlein (Eds.), *Biomedical results from Skylab* (NASA SP-377). Washington, D.C.: U.S. Government Printing Office, 1977.

Thornton, W.E., & Rummel, J.A. Muscular deconditioning and its prevention in space flight. In R.F. Johnson and L.F. Dietlein (Eds.), *Biomedical results from Skylab* (NASA SP-377). Washington, D.C.: U.S. Government Printing Office, 1977.

Whedon, G.D., Lutwak, L., Ramsey, P.C., Whittle, M.W., Smith, M.C., Reid, J., Leach, C., Stadler, C.R., & Stanford, D.D. Mineral and nitrogen metabolic studies, experiment M071. In R.F. Johnston and L.F. Dietlein (Eds.), *Biomedical results from Skylab* (NASA SP-377). Washington, D.C.: U.S. Government Printing Office, 1977.

Whittle, M.W., Herron, R., & Cuzzi, J. Biostereometric analysis of body form. In R.S. Johnston and L.F. Dietlein (Eds.), *Biomedical results from Skylab* (NASA SP-377). Washington, D.C.: U.S. Government Printing Office, 1977.

Yegorov, A.D. Results of medical research during the 175-day flight of the third prime crew on the Salyut-6–Soyuz orbital complex (NASA TM-76450). Moscow: Academy of Sciences USSR, 1980.

CHAPTER 12
BONE AND MINERAL METABOLISM

Among the most striking biomedical findings from space missions are the continuous progressive loss of calcium and the skeletal changes observed postflight. These alterations in bone and mineral metabolism appear to be among the most serious biomedical hazards associated with long-term space flight. Information on skeletal changes has been obtained from a variety of studies conducted in both simulated and actual spaceflight.

Bone Density Studies

Early studies of bone mineral changes using X-ray densitometry suggested that large amounts of bone may be lost during relatively brief periods of space flight. For 12 of the crewmembers who participated in Gemini 4, 5, and 7 and Apollo 7 and 8, the average postflight loss from the os calcis was 3.2 percent, compared to preflight baselines (Mack et al., 1967; Mack & Vogt, 1971; Vose, 1974). Large losses were also observed from the radius and ulna after these early flights.

For later Apollo missions and Skylab, the more precise method of photon absorptiometry was used to assess pre- and postflight bone mineral mass. Figure 1 shows the results of measurements taken on the os calcis in terms of percent change from preflight values (Rambaut & Johnston, 1979). The largest losses were observed in the crew of Skylab 4. In contrast to the data obtained from earlier missions, large bone mineral losses were not evident in the radius. This supports the hypothesis that mineral loss occurs only in weight-bearing bones during space flight. Further substantiation of that hypothesis is given by findings from Soviet space flights. Mineral loss from the os calcis had been seen to increase in rough proportion to the increase in mission length, until extensive exercise countermeasures began to be employed. Percentage reductions seen after six-month flights then decreased to approximately the level (3.2-8.3 percent) seen after three-month flights (Gazenko, Genin, & Yegorov, 1981). This affects postflight readaptation as well, since recovery of the skeletal mass is gradual and appears to take about the same length of time as the loss (Vogel & Whittle, 1976).

Calcium Balance Studies

Studies of metabolic balance conducted on a few of the crewmembers participating in the Gemini and Apollo missions suggested that space flight is accompanied by an increased excretion of calcium and phosphorus. These studies were expanded for the Skylab missions and dietary intake was carefully monitored, permitting more accurate balance determinations.

Figure 2 shows the changes in urine and fecal calcium content inflight (Rambaut & Johnston, 1979). The urine calcium content increased rapidly but reached a plateau after 30 days inflight. In contrast, fecal calcium content continued to increase for the duration of the flights.

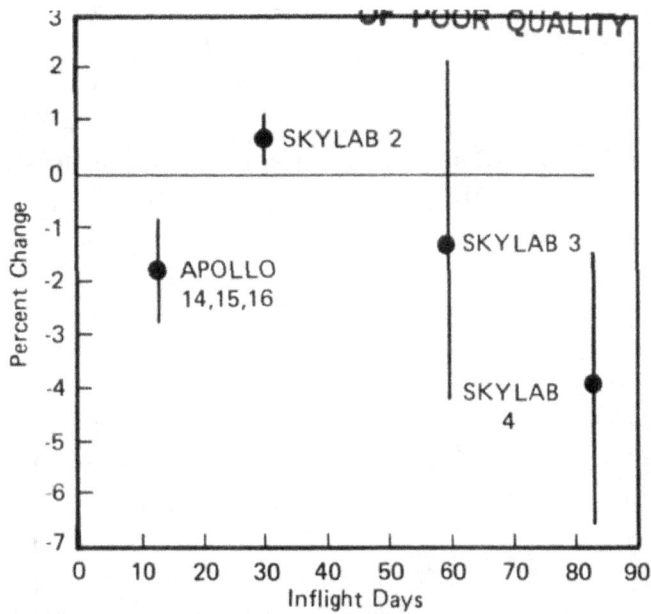

Figure 1. Postflight bone mineral changes in spaceflight crews (mean ± SE). (From Rambaut & Johnston, 1979)

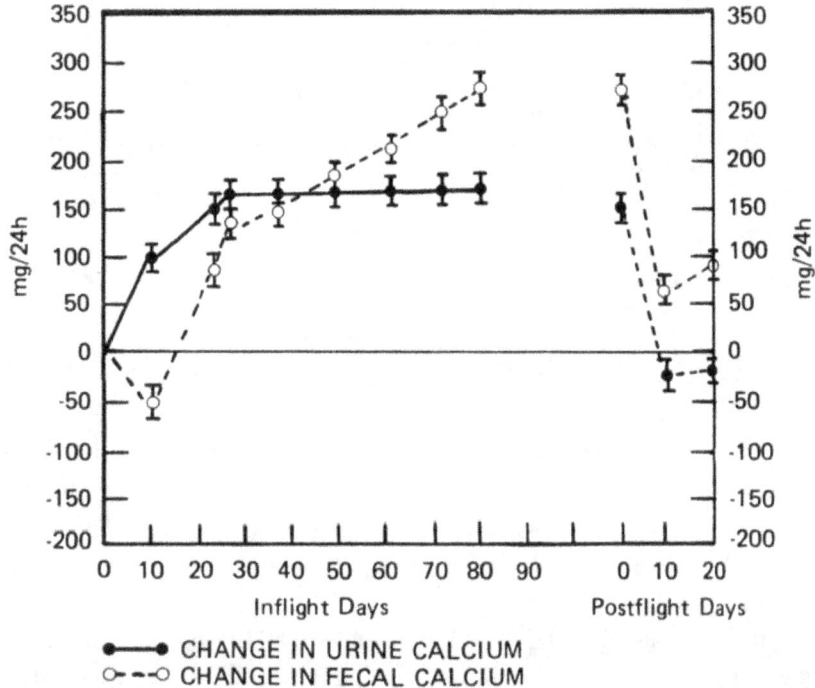

Figure 2. Change in urine and fecal calcium as a function of Skylab flight duration (Mean ± SE). (From Rambaut & Johnston, 1979)

Figure 3 shows the daily intake of calcium and its output in urine and feces. Within ten days inflight, the preflight positive calcium balances were abolished, and the body as a whole began to lose calcium. The rate of loss was slow at first, but increased to almost 300 mg per day by the 84th day of flight. For the three Skylab 4 crewmen, the average loss was 25 g of calcium from the overall body pool (about 1,250 g). Based on the trends in calcium loss during the first 30 days inflight, Rambaut and Johnston (1979) calculated that one year inflight might result in the loss of 300 g, or 25 percent, of the initial body pool. This is much larger than the predicted loss of calcium from bedrest studies, and suggests that calcium losses are more severe in space than in bedrest. Similar conclusions can be drawn from Soviet research (Gazenko, Grigor'yev et al., 1980), in which an increased calcium excretion is attributed to weightlessness.

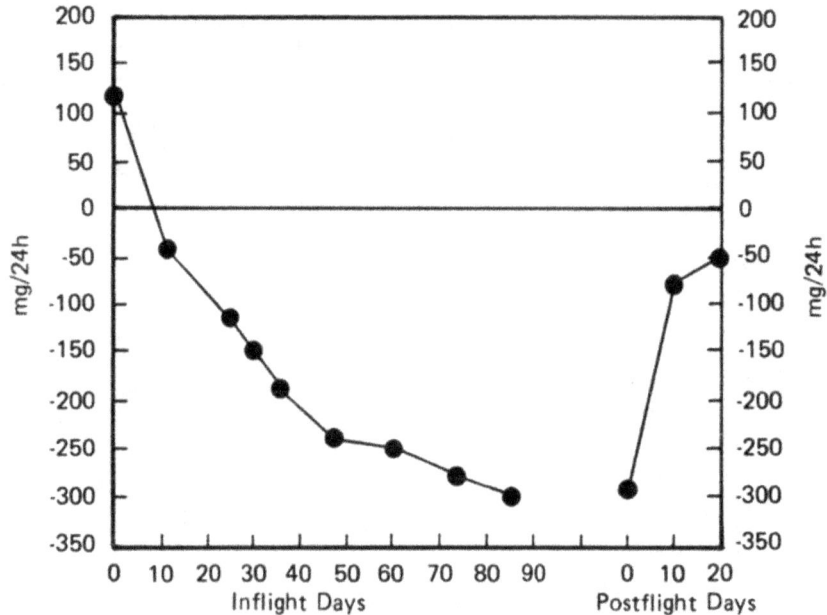

Figure 3. Calcium balance as a function of Skylab flight duration (mean ± SE). (From Rambaut & Johnston, 1979)

The Skylab calcium balance studies suggest that the losses in bone mineral from the os calcis contribute relatively little to the overall calcium loss. The 4 percent loss observed in the os calcis after the 84-day mission would represent a loss of only about 100 mg of calcium, while overall calcium losses for this mission averaged 25 g. In one Soviet mission in which extensive exercise was used, serum calcium content upon recovery was actually lower than preflight, even though there was a significant loss of os calcis mass (Gazenko, Genin, & Yegorov, 1981). Thus it is clear that other weight-bearing skeletal sites account for the major portion of the depleted mineral.

Recovery of the lost calcium begins soon after return to one gravity, as shown in Figures 2 and 3. Urine calcium content dropped below preflight baselines by postflight day 10, but fecal calcium content had not dropped to preflight levels by 20 days postflight. Figure 3 shows that the markedly negative calcium balance also had not returned to zero by day 20. Evidence from the studies on recovery of the os calcis mineral content after space flight, and evidence from bedrest studies, suggests that after a period of some weeks or months the astronaut would return to a condition of positive balance. Nevertheless, Rambaut and Johnston (1979) point out the possibility that the calcium balance might return to zero long before the loss from space flight had been made up, and irreversible damage to the skeleton might result.

Biochemical Analyses

Analyses of inflight urine, fecal, and plasma samples from Skylab missions revealed changes in a number of biochemicals. Urinary output of hydroxyproline gradually increased, indicating the deterioration of the collagenous substance of weight-bearing bones. Output of nitrogen also increased. The proportion of stearic acid in the total fecal fat increased throughout the flight as more and more calcium was available to form nonabsorbable salts. Analyses of plasma revealed inflight increases in calcium and phosphate; PTH levels decreased later in the flight (Whedon et al., 1977; Leach & Rambaut, 1977; Rambaut & Johnston, 1979).

Ground-Based Simulation Models

Bedrest provides a reasonably accurate model for the effects of weightlessness on bone and mineral metabolism, although it now appears that more calcium is lost in space than in bedrest studies. These studies have added pertinent information that bears on the mechanisms underlying bone loss during hypokinesic states.

Bedrest studies have suggested a means to predict the amount of mineral that will be lost from the os calcis during bedrest or in space (Donaldson et al., 1970; Lockwood et al., 1973; Smith et al., 1977; Volozhin et al., 1981). The wide variability in the amount of lost mineral in bedrested subjects can partly be accounted for by two other variables: (1) the initial os calcis mineral content and (2) the urinary hydroxyproline excretion rate (corrected for creatinine excretion). Figure 4 shows the regression of the prediction term (initial mineral divided by urinary hydroxyproline excretion rate) on the amount of mineral loss in subjects bedrested for 59 days (Vogel & Whittle, 1976). The data from two of the Skylab 3 astronauts fit well (open circles), suggesting that these variables also can be used to predict the effects of spaceflight on os calcis mineral content.

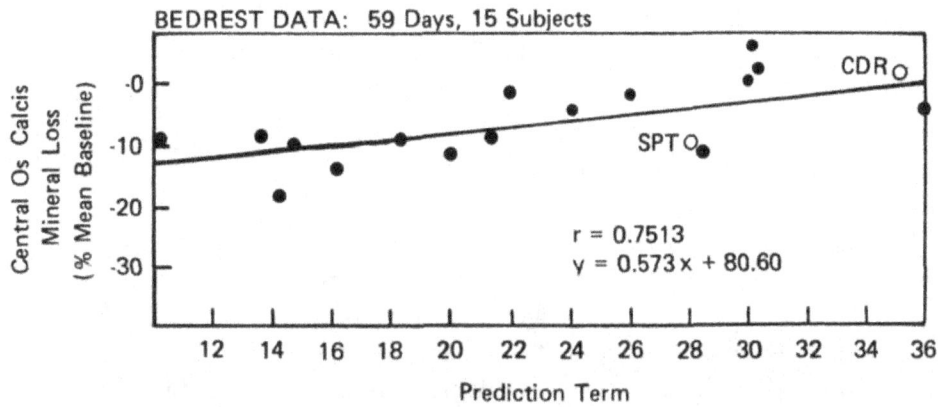

Figure 4. Os calcis mineral loss for varying prediction terms during 59 days of bedrest (solid circles) or space flight (open circles). (From Vogel & Whittle, 1976)

Studies of animals with immobilized limbs have suggested that disuse produces changes in both bone formation and bone resorption, depending upon the length of immobilization. For example, Landry and Fleisch (1964) used osseous tetracycline incorporation corrected by changes in bone weight as a direct index of bone formation, and as an indirect index of bone resorption. They found a short initial phase during which formation decreased, and a second phase in which formation increased but bone weight decreased, indicating an even greater increase in resorption. After 49 days of immobilization, formation again decreased below normal levels.

Didenko and Volozhin (1981) exposed rabbits to 30 days of confinement in order to study changes in bone mineral composition. Levels of calcium in bone did not change, although calcium excretion decreased. This was attributed to an inhibition of bone reorganization, in which bone mass was reduced without a corresponding alteration of crystalline structure.

The most pronounced changes are seen to occur in weight-bearing bones. Mechanical stimulation apparently has a critical effect on bone structure and metabolism, as numerous studies involving bone strain measurement have shown (Hinsenkamp et al., 1981). There also appears to be an age-dependent variation in the relative rates of bone formation and resorption (Novikov & Il'yin, 1981). Older animals show the highest net rate of bone loss during immobilization.

These and other results indicate that immobilization produces a number of time-dependent changes in bone accretion and resorption, and suggest that proportionately larger increases in resorption may be a key factor in the loss of bone mineral mass. It is not yet known whether the skeletal losses in space are due to relatively larger increases in bone resorption compared to formation, but autopsies of the three Soviet cosmonauts who died after a 21-day flight revealed "a good number of unusually wide osteocytic lacunae," which may have been due to increased bone resorption.

Inflight Animal Experiments

Studies of animals flown aboard satellites have also revealed changes in bone mineral content. Monkeys experiencing 8.8 days of weightlessness showed larger losses in bone mineral than ground controls (Mack, 1971). Studies of rats flown onboard the Cosmos biosatellites also demonstrated skeletal changes. For example, skeletal changes in rats exposed to 19 days of space flight included osteoporosis, decreased density and mineralization, and a 30 percent decrease in mechanical bending strength (Gazenko, Il'in et al., 1980). In addition to these changes, other functional rearrangements such as inhibition of periosteal neoformation and tubular bone growth have been demonstrated (Kaplanskiy et al., 1980). These and other studies have suggested that the loss of bone mineral might be primarily due to inhibited bone formation rather than increased bone resorption (Morey & Baylink, 1978; Yagodovsky et al., 1976). Rats on the 22-day Cosmos 605 flight showed decreased metaphyseal bone in the vicinity of the epiphyseal cartilagenous plate, suggesting an inhibition of bone growth during flight. It is not yet possible to integrate these findings with the findings from hypokinesia studies on humans and animals in one gravity because of the complicating factors of time-dependent changes, species differences, and potential differences in the mechanisms by which bone is lost in space and in bedrest or immobilization.

Countermeasures

The major countermeasures being explored to reduce the effects of space flight on the skeleton are the use of various weight-loading exercises or artificial gravity regimens that counteract the loss of gravitational and muscular stress, and nutritional and pharmacological manipulations. The crews of Skylab 3 and 4 exercised heavily inflight. Three of these six people showed substantial mineral losses, which casts doubt on the effectiveness of exercise as a countermeasure. However, as was mentioned earlier, later Soviet findings regarding the effect of inflight exercise have been somewhat more positive (Gazenko, Genin, & Yegorov, 1981). Nutritional supplements of calcium and phosphorus, and drugs such as EHDP, show some promise as countermeasures for the effects of bedrest on the skeleton and may be effective for space flight. Due to technical and hardware constraints, artificial gravity has so far been employed only in animal studies, but results have been quite promising. Centrifugation has been shown to prevent changes in calcium and phosphorus content of rat long bones (Gazenko, Il'in et al., 1980) and to prevent osteoporosis (Stupakov, 1981).

Summary and Conclusions

Based on the information obtained from space missions, particularly Skylab, it is clear that bone and mineral metabolism is substantially altered during space flight. Calcium balance becomes increasingly more negative throughout the flight, and the bone mineral content of the os calcis declines. The major health hazards associated with skeletal changes include the lengthy recovery of lost bone mass postflight, the possibility of irreversible bone loss (particularly the trabecular bone), the possible toxic effects of the increased release of calcium and phosphate on soft tissue such as kidney, and the potentially increased possibility of fracture.

For these reasons, major research efforts are currently being directed toward elucidating the fundamental mechanisms by which bone is lost in space and developing more effective countermeasures to prevent these losses.

References

Didenko, I.Ye., & Volozhin, A.I. Chemical composition of mineral component of rabbit bones as related to 30-day hypokinesia. *Space Biology and Aerospace Medicine*, 1981, *15*(1): 123-127.

Donaldson, C.L., Hulley, S.B., Vogel, J.M., Haltner, R.S., Bayers, J.H., & McMillan, D.E. Effect of prolonged bed rest on bone mineral. *Metabolism*, 1970, *19*:1071-1084.

Gazenko, O.G., Genin, A.M., & Yegorov, A.D. Major medical results of the Salyut-6/Soyuz 185-day space flight. NASA NDB 2747. *Proceedings of the XXXII Congress of the International Astronautical Federation*, Rome, Italy, 6-12 September 1981.

Gazenko, O.G., Grigor'yev, A.I., & Natochin, Yu.V. Fluid-electrolyte homeostasis and weightlessness. *Space Biology and Aerospace Medicine*, 1980, *14*(5):1-11.

Gazenko, O.G., Il'in, Ye.A., Genin, A.M., Kotovskaya, A.R., Korol'kov, V.I., Tigranyan, R.A., & Portugalov, V.V. Principal results of physiological experiments with mammals aboard the Cosmos-936 biosatellite. *Space Biology and Aerospace Medicine*, 1980, *14*(2):33-37.

Hinsenkamp, M., Burny, F., Bourgois, R., & Donkerwolcke, M. In vivo bone strain measurements: Clinical results, animal experiments, and a proposal for a study of bone demineralization in weightlessness. *Aviation, Space, and Environmental Medicine*, 1981, *52*(2):95-103.

Kaplanskiy, A.S., Savina, Ye.A., Portugalov, V.V., Il'ina-Kakuyeva, Ye.I., Alexeyev, Ye.I., Durnova, G.N., Pankova, A.S., Plakhuta-Plakutina, G.I., Shvets, V.N., & Yakovleva, V.I. Results of morphological investigations aboard biosatellites Cosmos. *The Physiologist*, Supplement, December 1980, *23*(6):S51-S54.

Landry, M., & Fleisch, H. The influence of immobilization on bone formation as evaluated by osseous incorporation of tetracycline. *Journal of Bone and Joint Surgery—British Volume*, 1964, *46B*(4):764-771.

Leach, C.S., & Rambaut, P.C. Biochemical responses of the Skylab crewmen: An overview. In R.S. Johnston and L.F. Dietlein (Eds.), *Biomedical results from Skylab* (NASA SP-377). Washington, D.C.: U.S. Government Printing Office, 1977.

Lockwood, D.R., Lammert, J.E., Vogel, J.M., & Hulley, S.B. Bone mineral loss during bed rest. In *Clinical Aspects of Metabolic Bone Disease. Excerpta Medica*, 1973, 261-265.

Mack, P.B. Bone density changes in a *Macaca nemestrina* monkey during the biosatellite II project. *Aerospace Medicine*, 1971, *42*:828-833.

Mack, P.B., LaChance, P.A., Vose, G.P., & Vogt, F.B. Bone demineralization of foot and hand on Gemini-Titan IV, V, and VII astronauts during orbital flight. *Journal of Roentgenology, Radium Therapy and Nuclear Medicine*, 1967, *3*:503-511.

Mack, P.B., & Vogt, F.B. Roentgenographic bone density changes during representative Apollo space flight. *American Journal of Roentgenology*, 1971, *113*:621-623.

Morey, E.R., & Baylink, D.J. Inhibition of bone formation during spaceflight. *Science*, 1978, *201*:1138.

Novikov, V.E., & Il'in, E.A. Age-related reactions of rat bones to their unloading. *Aviation, Space, and Environmental Medicine*, 1981, *52*(9):551-553.

Rambaut, P.C., & Johnston, R.S. Prolonged weightlessness and calcium loss in man. *Acta Astronautica*, 1979, *6*:1113-1122.

Smith, M.C., Rambaut, P.C., Vogel, J.M., & Whittle, M.W. Bone mineral measurement—Experiment M078. In R.S. Johnston and L.F. Dietlein (Eds.), *Biomedical results from Skylab* (NASA SP-377). Washington, D.C.: U.S. Government Printing Office, 1977.

Stupakov, G.P. Artificial gravity as a means of preventing atrophic skeletal changes. *Space Biology and Aerospace Medicine*, 1981, *15*(4):88-90.

Vogel, J.M., & Whittle, M.W. Bone mineral changes: The second manned Skylab mission. *Aviation, Space, and Environmental Medicine*, 1976, *47*:396-400.

Volozhin, A.I., Didenko, I.Ye., & Stupakov, G.P. Chemical composition of mineral component of human vertebrae and calcaneus after hypokinesia. *Space Biology and Aerospace Medicine*, 1981, *15*(1):60-63.

Vose, G.P. Review of roentgenographic bone demineralization studies of the Gemini space flight. *American Journal of Roentgenology*, 1974, *121*:1-4.

Whedon, G.D., Lutwak, L., Rambaut, P.C., Whittle, M.W., Smith, M.C., Reid, J., Leach, C.S., Stadler, C.R., & Sanford, D.D. Mineral and nitrogen metabolic studies—Experiment M071. In R.S. Johnston and L.F. Dietlein (Eds.), *Biomedical results from Skylab* (NASA SP-377). Washington, D.C.: U.S. Government Printing Office, 1977.

Yagodovsky, V.S., Trifranidi, L.A., & Goroklova, G.P. Space flight effects on skeletal bones of rats. *Aviation, Space, and Environmental Medicine*, 1976, *47*:734-738.

CHAPTER 13

BLOOD, FLUID, AND ELECTROLYTES

Exposure to the space environment produces a wide range of effects on body fluids. Many of the changes observed in returning astronauts are thought to be directly or indirectly due to the cephalad shift of fluids accompanying weightlessness. However, some of these changes, particularly hormonal ones, may be related to stress or other variables associated with space flight.

The most significant hematological changes include a reduction in plasma volume, alterations in red blood cell (RBC) mass, changes in the distribution of RBC shapes, and changes in the number and function of T-cells. With respect to fluid and electrolytes, the most significant changes accompanying space flight involve a reduction in total body fluid and a gradual and progressive loss of electrolytes.

Hematology

Plasma Volume and Red Blood Cell Mass

Plasma volume decreases as a function of exposure to the weightlessness of space flight. With one exception, a decrease has been recorded for every U.S. mission, with a rapid postflight recovery. Table 1 shows the decreases found in plasma volume and for red blood cell mass as a function of time spent in space. Also shown is the percent of loss in hemoglobin mass recorded following Soviet missions. In the Soviet program, hemoglobin mass, rather than RBC, was measured by spectroscopic methods, thereby avoiding the injection of radioisotopes (Cogoli, 1981). The results

Table 1

Hematological Changes Found in
U.S. and Soviet Manned Missions
(Mean Percent Change from Preflight Baseline)

Days in Space	U.S. Data		Soviet Data
	Plasma Volume	RBC Mass	Hb Mass
1 to 5	- 9		
6 to 10	- 5.6	- 9.4	
11 to 20	- 0.4	- 6.8	-12.7
21 to 50	- 9	-14	-33
51 to 100	-14.5	- 9	-20
Over 100			-16

for both the U.S. and Soviet programs show considerable variability in the hematological response of individual crewmen. However, it does appear that there may be a leveling of the decrease after about 60 days of exposure and a gradual recovery from that point. Figure 1 shows the recovery curves, based on both inflight and postflight measurements, for the crewmen of the three Skylab missions (Kimzey, 1977). The Skylab data also indicate that the loss in RBC mass, once triggered, may continue for some time independent of the effects of the space environment. For example, the RBC mass of the crew of Skylab 2 did not begin to recover until two weeks postflight.

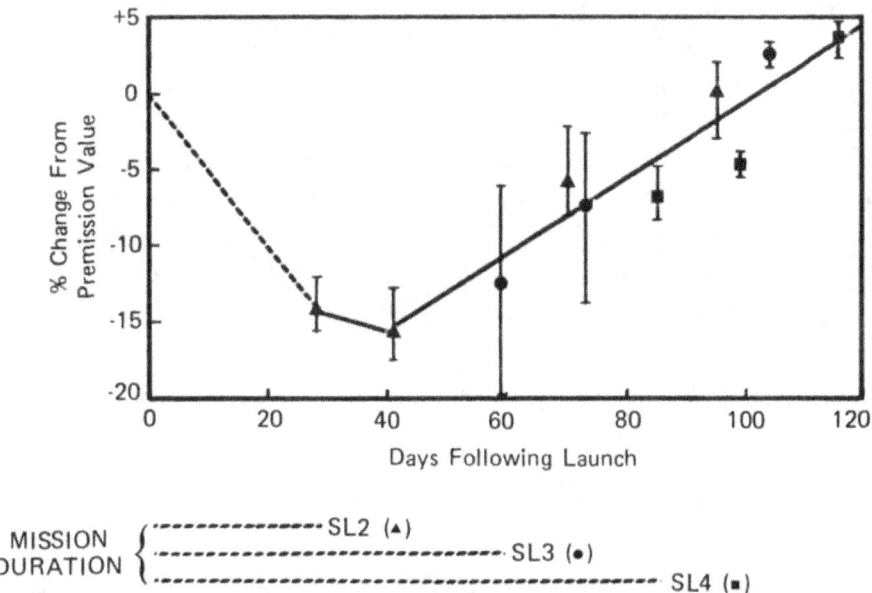

Figure 1. Red cell mass determinations in Skylab crewmen. Vertical bars represent the range (Kimzey, 1977).

Soviet scientists (Gazenko, Kakurin, & Kuznetsov, 1976; Gazenko, Genin, & Yegorov, 1981) have reported that erythrocyte counts show no change for flights of three days or less. For flights up to 18 days, erythrocyte counts have been variable. Cosmonauts participating in missions of 18 days to six months in length have shown postflight decreases in erythrocyte counts which returned to baseline values within six weeks.

Red Blood Cell Metabolism

A variety of cellular enzymes and metabolites involved in RBC metabolism were assessed postflight on Skylab in order to evaluate the maintenance of erythrocyte integrity (Mengel, 1977). The studies carried out included an analysis of red cell components involved with peroxidation of red cell lipids, enzymes of red cell metabolism, and levels of 2,3-diphosphoglyceric acid and adenosine triphosphate (ATP). Although the enzymes and metabolites showed changes, there was no consistent pattern and no evidence of lipid peroxidation.

Soviet investigators found that changes in the ATP content and the activity of glycolytic enzymes in erythrocytes were a function of mission duration (Ushakov et al., 1981). A one-week flight produced no alterations. During 30- and 63-day flights, glycolytic activity declined while ATP content remained stable. Cosmonauts on 96- and 175-day flights displayed a strong decline in ATP content with more intensive glycolysis and elevated RBC metabolic rate (Yegorov, 1980). These changes are described as being adaptive, and not a result of the basal metabolic processes in red blood cells. Thus there is no support for the hypothesis that space flight is associated with irreversible RBC damage.

Red Blood Cell Shapes

Samples of pre-, in-, and postflight blood of Skylab crewmen were examined with light and scanning electron microscopy for alterations in the shape of red blood cells. Table 2 lists the categories used for classifying RBC shapes. Discocytes accounted for between 80 and 90 percent of the red blood cells in preflight samples. In inflight samples, the proportion of discocytes declined and the proportion of other cell types, particularly echinocytes, increased (Kimzey, 1977). The cosmonauts of Salyut 6 missions (96, 140, 175, and 185 days) also showed decreases in discoid-shaped red blood cells and increases in elliptical and spherical shapes. Although such changes are considered to be of potential clinical significance, the alteration in RBC shape distribution readily reversed postflight, suggesting that no permanent cellular damage or alteration of bone marrow function had occurred (Gazenko, 1979; Yegorov, 1980).

Table 2
Distribution of Red Blood Cell Shapes

Designation	Characteristic	Description	Gradual Distribution Changes
Discocyte	Disc	Normal biconcave erythrocyte	Preflight, this shape constituted 80-90 percent of red cells. The proportion decreased inflight.
Leptocyte	Thin, flat	Flattened cell	No change or slightly increased.
Codocyte	Bell	Bell-shaped erythocyte (appearance depends on side of cell uppermost)	
Stomatocyte	Single concavity	Various stages of shapes	Maximum increase prior to MD* 27, gradual reduction with continued time inflight.
Knizocytes	Pinched	Triconcave erythrocyte	Maximum increase prior to MD 27, gradual reduction with continued time inflight.
Echinocyte	Spiny	Various stages of crenation	Increased inflight, rapidly decreased postflight.

*MD = Mission day.

Adapted from Kimzey, 1977

Possible Mechanisms Underlying Loss of Red Blood Cell Mass

The decrease in RBC mass associated with space flight has been observed since the early Gemini missions. Initially it was proposed that these losses were related to the 100 percent oxygen atmosphere in the spacecraft, which might have produced mild oxygen toxicity resulting in alterations in RBC integrity and premature intravascular hemolysis. Subsequent hematological data from Skylab and from the ground-based 59-day Skylab Medical Experiments Altitude Test (SMEAT) (Johnson, 1973) did not support this hypothesis. Both Skylab and the SMEAT simulation chamber contained the same environmental conditions (70 percent oxygen and 30 percent nitrogen, at 258 torr), but only the Skylab astronauts showed decreases in RBC mass. Furthermore, studies of RBC metabolism in astronauts have not suggested changes in RBC functional integrity, and have not supported the hypothesis of increased intravascular hemolysis.

Two alternative hypotheses to explain the loss of red blood cells and the transient low postflight reticulocyte counts are (1) increased destruction of red blood cells by selective sequestration and removal in the reticuloendothelial system (RES) and (2) a transient suppression of red blood cell production. The changes in the distribution of RBC shapes may contribute to premature sequestration of the cells by the spleen. However, measurement of spleen-liver ratios on the Apollo-Soyuz Test Project mission and in control subjects showed increases in the controls, but not in the astronauts. These findings dispute an inflight increase in reticuloendothelial trapping (Kimzey & Johnson, 1977).

The weight of the evidence suggests instead that the decrease in RBC mass is due to a transient suppression of RBC production, resulting from a temporarily inhibited bone marrow. As plasma volume decreases early inflight, it appears that RBC and hemoglobin mass losses are induced via a feedback mechanism regulated by the production of plasma volume (Cogoli, 1981). The simultaneous reduction of plasma volume and erythrocyte level results in a constant RBC and Hb concentration. This feedback mechanism might operate on the blood system either by inhibiting the production of reticulocytes and/or of erythropoietin, or by increasing the hemolytic activity of the spleen. As noted, however, hemolysis is considered unlikely. It appears, then, that a crewman adapts to the zero gravity environment with an initial and significant reduction of reticulocytes. As adaptation occurs, reticulocyte production increases dramatically and persists during the postflight recovery.

Immunology

Humoral Immune System

Certain alterations in serum proteins were consistently observed postflight during the Apollo program, suggesting that space flight is associated with changes in the humoral immune system (Kimzey et al., 1975). These alterations included a significant rise and subsequent decrease of $\alpha2$-macroglobulin, and a rise in haptoglobin levels postflight. In order to evaluate the humoral component of the immune system, a number of plasma proteins were assayed during the Skylab missions. No significant changes were found in any of the plasma proteins (Kimzey, 1977).

The effect of space flight on plasma proteins was also examined by Soviet investigators (Guseva & Tashpulatov, 1980). The blood of cosmonauts after space flights lasting 2, 16, or 49 days was analyzed for protein composition. The two-day flight was characterized by a decrease in gamma-globulin (especially IgG and IgA) and β2-glycoprotein fractions. Increases in albumin and most globulins were seen after the 16- and 18-day space flights. Upon completion of the 49-day flight, changes were observed only in globulin fractions: C_{3c} and C_4 complement factors were monitored at higher-than-baseline levels, as were IgG, IgA, and IgM. By contrast, after the 175-day flight a significant reduction in serum levels of the immunoglobulin IgC was observed (Yegorov, 1980).

Cellular Immune System

This component of the immune system was evaluated by measuring RNA and DNA synthesis in purified lymphocyte cultures in response to an in vitro mitogenic challenge using phytohemagglutinin (PHA). The PHA responsiveness of lymphocytes was markedly decreased on the day of recovery for Skylab astronauts, but not for the Apollo crews (Kimzey, 1977; Kimzey et al., 1975; Taylor, 1981). For the Skylab crews, PHA responsiveness returned to preflight levels by three to seven days postflight. Skylab astronauts also had reduced numbers of T-cells postflight. The lymphocytes of Salyut 6 cosmonauts who participated in missions of 96 days and 140 days also showed decreased nucleic acid synthesis in response to PHA challenge. These findings suggest that long-duration space flight can be accompanied by a transient impairment in the function of T-cells and the cellular immune system, but it is not yet clear what the clinical significance of the observed changes might be (Gazenko, 1979; Gazenko et al., 1981).

Hormonal and Associated Fluid and Electrolyte Changes

Many of the endocrine and biochemical changes observed in conjunction with space missions fit a consistent picture of homeostatic adjustments of circulatory dynamics, renal function, and endocrine response, probably initiated by fluid shifts and associated environmental stresses. A model describing these homeostatic adjustments was presented in the cardiovascular section of this report. Apparently, the cephalad shift of fluids produces at least a transient increase in circulating central blood volume. This redistribution of fluid is interpreted as an increase in total blood volume by stretch receptors in the left atrium, thus causing a compensatory loss of water, sodium, and potassium from the renal tubules through a series of neural, humoral, and direct hydraulic mechanisms (Leach, 1979, 1981).

The key events initiated by the redistribution of fluids, and which eventually promote the establishment of a new steady state of reduced extracellular fluid volume, appear to be the following (Alvioli et al., 1979):

- Reflex peripheral vasodilatation
- Suppression of the renin-angiotensin-aldosterone system
- Suppression of antidiuretic hormone (ADH)
- Increased secretion of humoral natriuretic substances
- Reduced thirst.

216

Biochemical and fluid measurements that have been made inflight or postflight appear to fit the general model. For example, physiological mass measurements on Skylab documented the rapid loss of weight within the first few days inflight. Most of this weight loss is rapidly regained postflight, indicating that it is primarily due to fluid loss (Thornton & Ord, 1977). The observation of elevated ADH in blood and urine postflight appears to corroborate this explanation (Gazenko, Grigor'yev, & Natochin, 1980).

The early diuresis predicted by the model has been observed in bedrest studies but not in space flight, probably for a number of reasons. First, operational constraints have made it difficult to accurately document urine volumes early inflight. And, second, water intake is usually markedly reduced during the first few days inflight, perhaps because of motion sickness and drugs. Nevertheless, data obtained from Skylab indicate that the nine crewmen decreased their water intake by approximately 700 ml/day during the first six days inflight (Leach & Rambaut, 1977). Their urine volume decreased an average of only 400 ml/day during the same period, indicating a net loss of water as predicted by the model. Additional observations which lend support to this general hypothesis include the observed inflight increases in urinary output of sodium, potassium, and chloride, an inflight decrease in ADH, and a reduced postflight excretion of sodium (Yegorov, 1980).

Although the events depicted in this model appear to reflect the physiological adjustments to space flight reasonably well, some biochemical observations are puzzling. In particular, although the model predicts that urinary levels of aldosterone would decrease in weightlessness, the opposite has been observed in Skylab astronauts. The inflight increase in urinary aldosterone is consistent with the observed losses of potassium, but not with the losses of sodium. Since space flight is also associated with events such as emotional stresses, space motion sickness, increased drug usage, G-loads, and altered work/rest cycles, it is likely that many biochemical adjustments occur simultaneously, which may mask or negate the predicted homeostatic responses to fluid shifts.

Summary and Conclusions

The cephalad shift in fluids that accompanies weightlessness produces a wide range of changes in blood and in fluid and electrolyte balance. The fluid shift results in a reduction of plasma volume and in an initial decrease in RBC mass, probably due to a temporary inhibition of bone marrow activity. After a 30- to 60-day delay, the RBC mass begins to recover, even if weightlessness is still present. The cephalad shift of fluids and the resulting transient increase in central blood volume initiate a series of other compensatory mechanisms, including a loss of fluid and electrolytes. The losses of electrolytes appear to continue throughout the flight.

Countermeasures that may prove useful against the effects of space flight on blood, fluid, and electrolytes include water and electrolyte replenishment and vigorous isotonic and isometric exercise regimens. Although these countermeasures are unlikely to prevent the loss of RBC mass, they may partly alleviate the problem of reduced plasma volume and decreased orthostatic tolerance postflight. Exercise appears to have multiple benefits as a countermeasure, and may diminish the loss of electrolytes and metabolites associated with changes in muscle and bone and in mineral metabolism.

References

Alvioli, L., Biglieri, E.G., Daughaday, W., Raisz, L.G., Reichlin, S., & Vogel, J. Endocrinology. In *Life beyond the Earth's environment*. Space Science Board, National Research Council, Washington, D.C.: National Academy of Sciences, 1979.

Cogoli, A. Hematological and immunological changes during spaceflight. *Acta Astronautica*, 1981, *8*(9-10): 995-1002.

Gazenko, O.G. (Ed.). Summaries of reports of the 6th All Soviet Union Conference on Space Biology and Medicine. Vol. I and II. Kaluga, USSR, 5-7 June 1979.

Gazenko, O.G., Genin, A.M., & Yegorov, A.D. Major medical results of the Salyut-6/Soyuz 185-day space flight. NASA NDB 2747. *Proceedings of the XXXII Congress of the International Astronautical Federation*, Rome, Italy, 6-12 September 1981.

Gazenko, O.G., Grigor'yev, A.I., & Natochin, Yu.V. Fluid-electrolyte homeostasis and weightlessness. From JPRS, *USSR Report: Space Biology and Aerospace Medicine*, 30 October 1980, *14*(5):1-11.

Gazenko, O.G., Kakurin, L.I., & Kuznetsov, A.G. (Eds.). Spaceflights aboard the "Soyuz" spacecraft (and ASTP). Biomedical investigations. Library of Congress, Science and Technology Division abstract. S&T Alert, Item No. 4457, 21 February 1978. Translated into English from *Biomeditsinskiye issledovaniya*, Moscow, "Nauka" Press, 1976.

Guseva, V.V., & Tashpulatov, R.Yu. Effects of flights differing in duration on protein composition of cosmonauts' blood. *Space Biology and Aerospace Medicine*, 1980, *14*(1):15-20.

Johnson, P.C. Hematology/immunology (M110 series). Part C: Blood volume and red cell life span (M113). In BioTechnology, Inc. (Managing Eds.), *Skylab medical experiments altitude test (SMEAT)* (NASA TM X-58115). Prepared for the Lyndon B. Johnson Space Center, Houston TX, under contract NASW-2518. Springfield, VA: Clearinghouse for Federal Scientific and Technical Information, October 1973.

Kimzey, S.L. Hematology and immunology studies. In R.S. Johnston and L.F. Dietlein (Eds.), *Biomedical results from Skylab* (NASA SP-377). Washington, D.C.: U.S. Government Printing Office, 1977.

Kimzey, S.L., Fischer, C.L., Johnson, P.C., Ritzmann, S.E., & Mengel, C.E. Hematology and immunology studies. In R.S. Johnston, L.F. Dietlein, and C.A. Berry (Eds.), *Biomedical results of Apollo* (NASA SP-368). Washington, D.C.: U.S. Government Printing Office, 1975.

218

Kimzey, S.L., & Johnson, P.C. Hematological and immunological studies. In A.E. Nicogossian (Ed.), *The Apollo-Soyuz Test Project: Medical Report* (NASA SP-411). Springfield, VA: National Technical Information Service, 1977.

Leach, C.S. An overview of the endocrine and metabolic changes in manned space flight. *Acta Astronautica*, 1981, 8(9-10): 977-968.

Leach, C.S. A review of the consequences of fluid and electrolyte shifts in weightlessness. *Acta Astronautica*, 1979, 6:1123-1135.

Leach, C.S., & Rambaut, P.C. Biochemical responses of the Skylab crewmen: An overview. In R.S. Johnston and L.F. Dietlein (Eds.), *Biomedical results from Skylab* (NASA SP-377). Washington, D.C.: U.S. Government Printing Office, 1977.

Mengel, C.E. Red cell metabolism studies on Skylab. In R.S. Johnston and L.F. Dietlein (Eds.), *Biomedical results from Skylab* (NASA SP-377). Washington, D.C.: U.S. Government Printing Office, 1977.

Taylor, G.R. Hematological and immunological analyses. In S.L. Pool, P.C. Johnson, Jr., & J.A. Mason (Eds.), *STS-1 medical report* (NASA TM-58240). Washington, D.C.: NASA Scientific and Technical Information Branch, December 1981.

Thornton, W.E., & Ord, J. Physiological mass measurement in Skylab. In R.S. Johnston and L.F. Dietlein (Eds.), *Biomedical results from Skylab* (NASA SP-377). Washington, D.C.: U.S. Government Printing Office, 1977.

Ushakov, A.S. Nutrition during long flight. Translated into English from *Zdorov'ye*, 1980, 4(304): 4-5. NASA TM-76436, 1981.

Yegorov, A.D. Results of medical research during the 175-day flight of the third prime crew on the Salyut-6—Soyuz orbital complex. Translated into English from *Resultaty meditsinskikh issledovaniy vo vremya 175 sutochnogo poleta tretyego osnovnog ekipazha na orbitalnam komplekse Salyut-6—Soyuz*. Academy of Sciences USSR, Ministry of Health, Moscow, 1980. NASA TM-76450, January 1981.

CHAPTER 14

SIMULATIONS AND ANALOGS OF WEIGHTLESSNESS

Although it is not possible to produce zero gravity in ground-based studies (except for 30 or 40 seconds during Keplerian flight maneuvers), one can obtain a great deal of information about the acclimation of humans to weightlessness by using a variety of simulations or analogs. The most common analog is hypokinesia, produced by water immersion, prolonged bed or chair rest, immobilization, or other methods. Other ground-based techniques designed to investigate the physiological alterations associated with weightlessness include dehydration and in vitro studies of rotated cells. This section describes several such techniques, and summarizes the advantages and disadvantages of each.

Bedrest

The most widely used analog of weightlessness is bedrest. The technique has been in use for decades, and hundreds of studies have been performed to determine the physiological consequences of prolonged horizontal posture (Nicogossian et al., 1979). Although bedrest cannot strictly be called a simulation of zero gravity since gravity is obviously still present (only parabolic flight is a true simulation), it does represent an analog of weightlessness because many of the physiological changes that accompany prolonged bedrest are remarkably similar to those observed in space. For example, bedrest is known to result in muscle atrophy, bone demineralization, redistribution of fluids and body mass, and decreases in plasma volume and red blood cell (RBC) mass (Genin, 1977; Sandler, 1976). Cardiovascular changes include decreases in cardiac output, stroke volume, heart size, and end diastolic volume. After bedrest, subjects show a decline in orthostatic tolerance similar to that typically demonstrated by returning astronauts. Although the physiological consequences of bedrest are in most respects quite similar to those of weightlessness, there are a few notable exceptions. For example, calcium losses may be more severe in space than in bedrest (Rambaut & Johnston, 1979), and while diuresis is common during the early days of bedrest, it has not yet been clearly demonstrated in space (Greenleaf, Shvartz, & Keil, 1981). In addition, although postural disturbances are seen that might be attributable to vestibular deconditioning, bedrest does not produce the full range of vestibular disorders characteristic of space travel (Kotovskaya, Gavrilova, & Galle, 1981).

Bedrest With Head-Down Tilt

Soviet scientists have proposed that bedrest with head-down tilt (antiorthostatic bedrest) reproduces the early physiological effects of weightlessness more closely than the horizontal posture (Kakurin et al., 1976). In particular, head-down tilt produces more rapid and pronounced fluid shifts, along with related symptoms of facial puffiness, nasal congestion, and a feeling of fullness in the head, and results in subjective reports of sensory realignments similar to those observed by cosmonauts. Table 1 shows a comparison of physiological and sensory alterations during the recumbent posture and various degrees of head-down tilt. Although the head-down position results in more pronounced fluid shifts, selective catheterization studies indicate that the shifts are not accompanied by a persistent increase in intravascular pressure (Katkov et al., 1979).

Table 1

Clinical Symptoms During Horizontal
Posture and Head-Down Tilt

Symptoms	Bedrest Position			
	0°	-4°	-8°	-12°
Lowered taste and olfactory sensitivity threshold	−	+	+	++
Sensation of blood rushing to, and heaviness in, the head	−	+	+	++
Nasal congestion	−	+	++	+++
Uncomfortable feeling in the nose and throat, hoarse voice	−	+	+	++
Increase in intranasal resistance	+	+	++	+++
Vertigo and nausea	−	−	+	+
Spatial illusions	−	+	+	++
Nystagmus of eyes	−	+	+	++
Facial puffiness and overfilling of sclera and conjunctiva vessels	−	+	++	++
Sensation of fullness in the eyes, fatigue in the eyes in reading, drop in central sight acuity	−	+	++	++

Kakurin et al., 1976.

American and Soviet scientists performed a joint study to compare the effects of seven days of horizontal bedrest and seven days of -6° head-down tilt (Gazenko & Grigor'yev, 1980; Joint U.S./USSR Hypokinesia Program, 1979). The results of this study indicate that the two treatments have similar physiological consequences, but that head-down tilt results in more pronounced cardiac deconditioning as measured by post-bedrest LBNP and exercise tests.

Shuttle Flight Simulations

In preparation for the first Space Shuttle flights, a number of studies were carried out that attempted to simulate Shuttle missions by exposing subjects to 10 to 15 days of bedrest followed by centrifuge runs of up to $+3$ G_z. These simulations provided valuable data on the physiological changes in men and women of varying ages and their tolerance to post-bedrest acceleration (Goldwater et al., 1978; Jacobson et al., 1974; Leverett et al., 1971; Miller & Leverett, 1965; Montgomery et al., 1979; Newsom et al., 1977; Sandler, 1980; Sandler et al., 1978, 1979).

The general conclusion to be gained from these studies is that tolerance to $+G_z$ acceleration degrades after bedrest, primarily because of the reduction in plasma volume. It also appears that male subjects, especially in the older groups, tolerate post-bedrest centrifugation runs up to $+3\ G_z$ better than females or younger subjects. Studies utilizing Shuttle flight simulations are of enormous value in identifying population risk factors in men and women of various age groups, and are also valuable in testing the efficacy of such countermeasures as salt and fluid replenishments, anti-gravity suits, exercise, and periodic application of acceleration and lower body negative pressure (LBNP) during bedrest. Such simulations can also be used to investigate the cumulative effects of repeated short-term space flights.

Chair Rest

Although few studies have utilized prolonged chair rest as an analog for weightlessness, it appears that this treatment results in many of the same physiological consequences as bedrest, particularly loss of plasma volume, decreases in RBC mass, and cardiovascular deconditioning (Lamb et al., 1964; Lamb, 1965). Although the changes are smaller they are nonetheless puzzling, since headward fluid shifts are not associated with this treatment. It is possible that alterations in basal metabolism attendant upon inactivity may be as important as fluid shifts in the etiology of cardiovascular deconditioning in bedrest and in space.

Water Immersion

Water immersion has been used as an analog for weightlessness by a number of investigators, primarily studying renal and circulatory events (Gauer, 1975; Epstein, 1978). This treatment produces rapid fluid shifts by changes in hydrostatic forces and negative pressure breathing. Immersion is rapidly followed by a pronounced involuntary diuresis, with loss of electrolytes and decrease in plasma volume. These events may be triggered by activation of cardiac mechanoreceptors and suppression of antidiuretic hormone (the Gauer-Henry reflex); there is also evidence that early hemodilution and suppression of vasopressin play a role (Greenleaf, Shvartz, & Keil, 1981). However, studies on immersed animals suggest that other pathways may be involved (Gilmore & Zucker, 1978).

Characteristics of the diuresis response are influenced by the state of hydration of the subject (Gauer, 1975; Leach, 1981). Normally, hydrated subjects exhibit a rise in free water clearance, while dehydrated subjects show increased osmolar clearance.

Although a decline in orthostatic tolerance after water immersion is typical, the magnitude of the change depends upon the physical training of the subject. Non-athletes demonstrate less degradation in orthostatic tolerance compared to athletes, despite the fact that diuresis occurs earlier and is more pronounced in non-athletes (Boening et al., 1972; Stegemann et al., 1975). Stegemann et al. (1975) suggest that this discrepancy can be explained by a reduction in the effectiveness of the blood pressure control system in endurance-trained subjects.

Although fluid shifts and other physiological responses occur rapidly during water immersion, other methods of simulating weightlessness are usually preferred due to the problem of maintaining hygiene and the need for precise thermal control (Sandler, 1979). Immersion treatments longer than 24 hours are impractical because of the problem of skin maceration, making longer-term observations difficult. However, Soviet scientists employ a technique (shown in Figure 1), termed "dry" immersion, in which subjects are protected from water contact by a thin plastic sheet, thus making long-term immersion more feasible. Indeed, this may represent the ideal analog of weightlessness, since it has been demonstrated that the physiological changes related to fluid redistribution are more long-lasting when induced by immersion (Gogolev et al., 1980).

Figure 1. Soviet dry immersion technique for simulation of zero-gravity effects.

Immobilization

Animal studies constitute a significant portion of the research on hypokinesia. A number of investigators employ part or whole body casts or confinement in small cages to produce hypokinesia, particularly to study alterations in circulatory dynamics and in the musculoskeletal system. For example, Dickey et al. (1979) found that primates maintained in horizontal body casts for two to four weeks exhibit many changes similar to those observed in bedrested humans and in individuals exposed to space flight. These monkeys demonstrated slight increases in heart rate, decreased plasma volume, slightly decreased RBC mass, decreased orthostatic tolerance, and decreased responsiveness to vasoactive drugs. In a later study (Dickey et al., 1982), it was shown that the cardiovascular deconditioning thus produced results from an inability to maintain central blood volume during orthostatic stress.

Some investigators, particularly in the USSR, confine animals in small cages for prolonged periods to simulate weightlessness. Although confinement does not result in fluid shifts, it does produce a number of physiological consequences similar to those observed in space—particularly in the musculoskeletal system. Animal studies offer the advantage of permitting these changes to be observed directly. Rats confined for up to two months demonstrate weight loss; decreases in the size of the gastrocnemius muscle; and a decrease in protein content in muscle, cardiac, renal, and liver tissue (Fedorov & Shurova, 1973). Degenerative and atrophic changes are seen in mixed (soleus) and red muscle fibers (Kurash et al., 1981). Bone demineralization is also observed in confined animals, and studies of rats suggest that excessive amounts of phosphorus in the diet aggravate demineralization during hypokinesia (Ushakov et al., 1980). Dogs confined for six months demonstrate a number of circulatory and skeletal abnormalities, particularly in the rear extremities. The animals exhibit morphological alterations in the vascular network of the fascia, an increase in blood stasis in the bone marrow vessels, and changes in the femur and tibia indicative of dystrophic processes (Novikov & Vlasov, 1976). In rabbits, three to four weeks of confinement produces extensive enlargement and deformation of intra-organ veins (Muratikova, 1980).

Partial Body Support Systems

In order to reduce some of the problems associated with immobilization, and to more accurately reproduce the events occurring in space, some investigators have turned to a system of partial body support. In a typical system, the animal is attached to a harness and suspended in a head-down position with complete unloading of the rear limbs. The front limbs do not bear weight, but can be used to pull the animal around in a 360° arc.

Studies on rats suspended in harnesses suggest that the concomitant physiological changes are similar to those in space. Compared to controls, suspended rats show less weight gain per gram of food consumed (Morey, 1979); a decrease in metabolic rate (Jordan et al., 1980); inhibition of periosteal bone formation, particularly in older rats (Morey, 1979; Morey et al., 1979; Novikov & Il'in, 1981); and decreased size of thymus without any decrease in immunocompetence (Caren et al., 1980). With the exception of calcium balance, most of the changes in fluid and electrolyte balance during the first few days of suspension appear to be similar to those observed in rats and humans in space (Meininger et al., 1978). Preliminary studies on the cardiovascular system suggest increased right atrial pressure, an absence of diurnal variation in heart rate and mean arterial blood pressure, and increased cardiac output, indicating the occurrence of cephalad fluid shifts (Popovic, cited by Morey, 1979).

Partial body support systems that align subjects at a 9.5° angle from horizontal have been developed both by NASA (Hewes et al., 1966) and by Soviet scientists. This system is an attempt to simulate the physiological changes experienced under 1/6 g on the lunar surface. Tests with

224

monkeys reveal alterations in motor activity, early diuresis, decreased erythropoiesis, EEG abnormalities, weight loss, and post-treatment orthostatic intolerance. These simulations have provided valuable planning information that may be useful for longer-duration space flights in which reduced artificial gravity is available, and for long-term lunar operations.

Dehydration

One model primarily designed to investigate the erythropoietic effects of space flight involves dehydration (Dunn, 1978; Dunn & Lange, 1979; Dunn, Leonard, & Kimzey, 1981). Mice completely deprived of water for days demonstrated weight loss, reduced plasma volume and elevated hematocrit, and decreased RBC production as estimated by ^{59}Fe incorporation into the erythron. Although the suppression of erythropoiesis in hydrated animals with absolute increases in RBC volume is related to decreased production of a humoral regulator, serum titers of this substance do not change after dehydration. This indicates that the mechanisms underlying suppression of erythropoiesis may differ depending upon whether the elevated hematocrit results from an absolute or a relative increase in RBC volume. Dunn and Lange (1979) suggest that changes in RBC formation associated with dehydration, and possibly space flight, may also be related to factors such as negative energy balance. Using computer simulations, Dunn et al. (1981) conclude that the primary cause of erythropoiesis suppression in dehydrated mice is reduced food intake.

Although the mechanisms that produce dehydration in humans are clearly different from those operant in laboratory mice, the factors mediating suppression of erythropoiesis, once dehydration occurs and plasma volume declines, may be similar. For this reason, studies of dehydrated mice may yield important information about hematological changes in space.

In Vitro Simulation

Several researchers simulate low gravity for in vitro studies of cells by means of rotation. When the axis of rotation is perpendicular to the force of gravity and intersects the cell at its center, the viscosity of the cytoplasm prevents the immediate adjustment of particle distribution to account for the changing direction of gravity. At an appropriate rotation rate, a stationary distribution similar to that of weightlessness is thus attained (Schatz & Teuchert, 1972).

Cogoli et al. (1980) use a clinostat to produce this low-gravity environment and to study its effects on lymphocytes. They have found that low-gravity simulation depresses lymphocyte activation by the mitogen concanavalin A, as measured by DNA synthesis and ultrastructure changes. Compared to controls, lymphocyte activation was reduced by 50 percent in cells incubated for three days with the mitogen. By contrast, lymphocytes exposed to 4 g and 16 g demonstrated accelerated activation and accelerated aging, indicated by ultrastructural alterations. Since transient depressions of lymphocyte activation have been observed in returning astronauts and cosmonauts, this model promises to shed light on the mechanisms involved.

Summary and Conclusions

While no ground-based simulation technique can duplicate exposure to space flight completely, studies utilizing these techniques provide important information about the acclimation process. They verify and extend inflight biomedical observations, develop new hypotheses for further inflight testing, and test the efficacy of countermeasures. Considering that the Shuttle era will expose a more heterogeneous population of individuals to space flight, simulation studies make it possible to describe individual differences in response characteristics and to predict potential problems in subgroups.

References

Boening, D., Ulmer, H.V., Meier, U., Skipka, W., & Stegemann, J. Effects of a multi-hour immersion on trained and untrained subjects: I. Renal function and plasma volume. *Aerospace Medicine*, 1972, *43*:300-305.

Caren, L.D., Mandel, A.D., & Nunes, J.A. Effect of simulated weightlessness on the immune system in rats. *Aviation, Space, and Environmental Medicine*, 1980, *51*(3):251-255.

Chernov, I.P. The stress reaction of hypokinesia and its effect on general resistance. *Space Biology and Aerospace Medicine*, 1980, *14*(3):86-90.

Cogoli, A., Valluchi-Morf, M., Mueller, M., & Briegler, W. Effect of hypogravity on human lymphocyte activation. *Aviation, Space, and Environmental Medicine*, 1980, *51*(1):29-34.

Dickey, D.T., Billman, G.E., Teoh, K., Sandler, H., & Stone, H.L. The effects of horizontal body casting on blood volume, drug responsiveness, and $+G_z$ tolerance in the rhesus monkey. *Aviation, Space, and Environmental Medicine*, 1982, *53*(2):142-146.

Dickey, D.T., Teoh, K.K., Sandler, H., & Stone, H.L. Changes in blood volume and response to vaso-active drugs in horizontally casted primates. *The Physiologist*, 1979, *22*(6)(supplement): S27-S28.

Dunn, C.D.R. Effect of dehydration on erythropoiesis in mice: Relevance to the "anemia" of spaceflight. *Aviation, Space, and Environmental Medicine*, 1978, *49*:990-993.

Dunn, C.D.R., & Lange, R.D. Erythropoietic effect of space flight. *Acta Astronautica*, 1979, *6*:725-732.

Dunn, C.D.R., Leonard, J.I., & Kimzey, S.L. Interactions of animal and computer models in investigations of the "anemia" of space flight. *Aviation, Space, and Environmental Medicine*, 1981, *52*(11):683-690.

Epstein, M. Renal effects on head-out water immersion in man; implications for understanding of volume homeostasis. *Physiology Review*, 1978, *58*:529-581.

Fedorov, I.V., & Shurova, I.F. Content of protein and nucleic acids in the tissues of animals during hypokinesia. *Space Biology and Medicine*, 1973, *2*:22-28.

Gauer, O.H. Recent advances in the study of whole body immersion. *Acta Astronautica*, 1975, *2*:39-39.

Gazenko, O.G., & Grigor'yev, A.I. Modeling the physiological effects of weightlessness: Soviet-American Experiment. NDB 92 (NASA TM-76317). Translated into English from *Vestnik Akademii nauk SSSR*, 1980, *2*:71-75.

Genin, A.M. Laboratory simulation of the action of weightlessness on the human organism. NASA TM-75072. Translated into English from *Laboratornoye modelirovaniye deystviya nevesomosti na organism cheloveka*. Interkosmos Council, Academy of Sciences USSR Report, 1977: 1-17.

Gilmore, J.P., & Zucker, I.L.H. Contribution of vagal pathways to the renal responses to head-out immersion in the nonhuman primate. *Circulation Research*, 1978, *42*(2):263-267.

Gogolev, K.I., Aleksandrova, Ye.A., & Shul'zhenko, Ye.B. Comparative evaluation of changes in the human body during orthostatic (headdown) hypokinesia and immersion. NDB 311. Translated into English from *Fiziologiya Cheloveka*, 1980, *6*(6): 978-983.

Goldwater, D., Sandler, H., Popp, R., Danellis, J., & Montgomery, L. Exercise capacity, body composition, and hemoglobin levels of females during bedrest Shuttle flight simulation. Preprints of the Annual Scientific Meeting, Aerospace Medical Association, New Orleans, Louisiana, 1978.

Greenleaf, J.E., Shvartz, E., & Keil, L.C. Hemodilution, vasopressin suppression, and diuresis during water immersion in man. *Aviation, Space, and Environmental Medicine*, 1981, *52*(6):329-336.

Hewes, D.E., Spady, A.A., & Harris, R.L. Comparative measurements of man's walking and running gaits in Earth and simulated lunar gravity (NASA TN D-3363). Washington, D.C.: National Aeronautics and Space Administration, 1966.

Jacobson, L.H., Hyatt, K.H., & Sandler, H. Effects of simulated weightlessness on responses of untrained men to G_x and G_z acceleration. *Journal of Applied Physiology*, 1974, *36*:745-752.

Joint U.S./USSR Hypokinesia Program (NASA TM-76013). Washington, D.C.: National Aeronautics and Space Administration, 1979.

Jordan, J.P., Sykes, H.A., Crownover, J.C., Schatte, C.L., Simmons, J.B., & Jordan, D.P. Simulated weightlessness: Effects of bioenergetic balance. *Aviation, Space, and Environmental Medicine*, 1980, *51*(5):132-136.

Kakurin, L.I., Lobachik, V.I., Mikhailov, V.M., & Senkevich, Yu.A. Antiorthostatic hypokinesia as a method of weightlessness simulation. *Aviation, Space, and Environmental Medicine*, 1976, 47:1083-1086.

Kaplanskiy, A.S., & Durnova, G.N. Role of dynamic space flight factors in the pathogenesis of involution of lymphatic organs (experimental morphological study). *Space Biology and Aerospace Medicine*, 1980, *14*(2):45-52.

Katkov, V.Y., Chestukhin, V.V., Zybin, O.Kh., Mikhaylov, V.M., Troshin, A.Z., & Utkin, V.N. Effect of brief head-down hypokinesia on pressure in various parts of the healthy man's cardiovascular system. *Space Biology and Medicine*, 1979, *13*(3):86-93.

Kotovskaya, A.R., Gavrilova, L.N., & Galle, R.R. Effect of hypokinesia in head-down position on man's equilibrium function. *Space Biology and Aerospace Medicine*, 1981, *15*(4):34-38.

Kovalev, O.A., Lysak, V.F., Severovostokova, V.I., & Sheremetevskaya, S.K. Local redistribution of blood under the effect of fixation stress against a background of hypokinesia. NDB 57 (NASA TM-76322). Translated into English from *Fiziologii Zhurnal*, 1980, *26*(1):120-124.

Kurash, S., Andzheyevska, A., & Gurski, Ya. Morphological changes in different types of rat muscle fibers during longterm hypokinesia. *Space Biology and Aerospace Medicine*, 4 February 1981, *14*(6):45-52.

Lamb, L.E. Hypoxia—An anti-deconditioning factor for manned space flight. *Aerospace Medicine*, 1965, *36*(2):97-100.

Lamb, L.E., Johnson, R.L., & Stevens, P.M. Cardiovascular deconditioning during chair rest. *Aerospace Medicine*, 1964, *35*:646.

Leach, C.A. A review of the consequences of fluid and electrolyte shifts in weightlessness. *Acta Astronautica*, 1981, *6*:1123-1135.

Leverett, S.D., Shubrooks, S.J., & Shumate, W. Some effects of Space Shuttle reentry profiles on human subjects. Preprints of the Annual Scientific Meeting, Aerospace Medical Association, Washington, D.C., 1971.

Meininger, G.A., Deavers, D.R., & Musacchia, X.J. Electrolyte and metabolic imbalances induced by hypokinesia in the rat. *Federal Proceedings*, 1978, *37*:663.

Miller, P.B., & Leverett, S.D. Tolerance to transverse (+G_x) and headward (+G_z) acceleration after prolonged bedrest. *Aerospace Medicine*, 1965, *36*:13-15.

Montgomery, L.D., Goldwater, D., & Sandler, H. Hemodynamic response of men 45-55 years to +G_z acceleration before and after bedrest. Preprints of the Annual Scientific Meeting, Aerospace Medical Association, Washington, D.C., 1979.

Morey, E.R. Spaceflight and bone turnover: Correlation with a new rat model of weightlessness. *BioScience*, 1979, *29*:168-172.

Morey, E.R., Sabelman, E.E., Turner, R.T., & Baylink, D.J. A new rat model simulating some aspects of spaceflight. *The Physiologist*, 1979, *22*(6)(supplement):S23-S24.

Muratikova, V.A. Effect of hypokinesia on blood vessels of the rabbit sympathetic trunk. NDB 60 (NASA TM-76328). Translated into English from *Arkhiv Anatomii, Gistologii i Embriologii*, 1980, *78*(5):40-45.

Newsom, B.D., Goldenrath, W.L., Winter, W.L., & Sandler, H. Tolerance of females to +G_z centrifugation before and after bedrest. *Aviation, Space, and Environmental Medicine*, 1977, *48*:327-331.

Nicogossian, A.E., Whyte, A.A., Sandler, H., Leach, C.S., & Rambaut, P.C. (Eds.). Chronological summaries of United States, European, and Soviet bedrest studies. Washington, D.C.: NASA, 1979.

Novikov, I.I., & Vlasov, V.B. Morphology of the circulatory bed of elements of the soft stroma and bones of the rear extremities of the dog upon extended hypodynamia. *Biomedical and Behavioral Science*, 1976, *59*:9.

Novikov, V.E., & Il'in, E.A. Age-related reactions of rat bones to their unloading. *Aviation, Space, and Environmental Medicine*, 1981, *52*(9): 551-553.

Rambaut, P.C., & Johnston, R.S. Prolonged weightlessness and calcium loss in man. *Acta Astronautica*, 1979, *6*:1113-1122.

Sandler, H. Cardiovascular effects of weightlessness. In P.N. Yu and J.R. Goodwin (Eds.), *Progress in cardiology*. Volume 6. Philadelphia: Lea & Febiger, 1976.

Sandler, H. Low-g simulation in mammalian research. *The Physiologist*, 1979, *22*(6)(supplement): S19-S22.

Sandler, H. Effects of bedrest and weightlessness on the heart. In G.H. Bourne (Ed.), *Hearts and heart-like organs. Vol. II—Physiology*. New York: Academic Press, 1980.

Sandler, H., Goldwater, D.J., & Rositano, S.A. Physiologic responses of female subjects during bedrest Shuttle flight simulation. Preprints of the Annual Scientific Meeting, Aerospace Medical Association, New Orleans, Louisiana, 1978.

Sandler, H., Goldwater, D., Rositano, S.A., Sawin, C.F., & Booher, C.R. Physiologic response of male subjects ages 46-55 years to Shuttle flight simulation. Preprints of the Annual Scientific Meeting, Aerospace Medical Association, Washington, D.C., 1979.

Schatz, A., & Teuchert, G. Effects of combined 0 g simulation and hypergravity on eggs of the nematode, *Ascaris suum*. *Aerospace Medicine*, 1972, *43*(6):614-619.

Stegemann, J., Meier, U., & Skipka, W., et al. Effects of a multi-hour immersion with intermittent exercise on urinary excretion and tilt table tolerance in athletes and nonathletes. *Aviation, Space, and Environmental Medicine*, 1975, *46*(1): 26-29.

Ushakov, A.S., Smirnova, T.A., Pitts, G.C., Pace, N., Smith, A.H., & Rahlmann, D.F. Body composition of rats flown aboard Cosmos 1129. *The Physiologist*, Supplement, December 1980, *23*(6): S41-S44.

Zagorskaya, Ye.A. Corticosteroid content of rat adrenals in the presence of hypokinesia combined with graded physical exercise. *Space Biology and Aerospace Medicine*, 4 February 1981, *14*(6): 53-56.

SECTION V

Health Maintenance
of Space Crewmembers

CHAPTER 15

SELECTION OF ASTRONAUTS AND SPACE PERSONNEL

Candidates for space flight are screened, using principles and techniques of preventive medicine, to identify individuals with maximum career potential. Employing modern diagnostic and evaluation procedures, the process selects applicants whose long-term health prospects are excellent. Since NASA and the nation make a large investment in the training of an astronaut, the selection process is designed to ensure that this investment is not lost during a normal career because of identifiable medical problems.

History of Astronaut Selection

Development of the Selection Process: 1958-1969

In November 1957, President Dwight D. Eisenhower established the President's Scientific Advisory Committee. One of the first recommendations of this committee was that the nation establish a civilian agency to pursue an aggressive program of space exploration. The White House released a paper in March 1958 listing key reasons for supporting the exploration of space. These included the compelling urge of man to explore and discover, defense considerations, and national prestige. However, the primary focus was on the new opportunities for scientific observation and experimentation that would broaden man's knowledge and understanding of the Earth, the solar system, and, ultimately, the universe.

After lengthy deliberations on Capitol Hill, the National Aeronautics and Space Act of 1958 was enacted by Congress in July of that year. Based on the work of a Space Task Group, which had been established at Langley Field, Virginia, the highest national priority for personnel selection was granted to the first manned spacecraft project, Project Mercury. To meet this requirement, considerable effort was expended by members of the Space Task Group to determine the type of individual who would function most effectively as an astronaut. Since no data were available at that time on physiological changes occurring in space, most of the qualifications for such individuals had to be based on a combination of conjecture and experience with high-performance aircraft operations.

The Space Task Group, together with a Special Committee on the Life Sciences, evolved a selection procedure based on what they felt the duties of an astronaut would be (Link, 1965). Simply stated, these were:

1. To survive; that is, to demonstrate the ability of man to fly in space and to return safely.

2. To perform; that is, to demonstrate man's capacity to act usefully under conditions of space flight.

3. To serve as a backup for the automatic controls and instrumentation; that is, to increase the reliability of space systems.

4. To serve as a scientific observer; that is, to go beyond what the instruments and satellites can observe and report.

PRECEDING PAGE BLANK NOT FILMED

5. To serve as an engineering observer and, acting as a true test pilot, to improve the flight system and its components.

The Space Task Group explored several categories of professionals to determine which group might furnish individuals who could best qualify as astronauts. The categories considered were aircraft pilots, balloonists, submariners, deep-sea divers, mountain climbers, explorers, flight surgeons, and scientists.

By December of 1958 a set of proposed civil service standards had been drafted. It was expected that representatives from the Department of Defense and from industry would nominate approximately 150 men by the end of January 1959. Thirty-nine would be selected for further testing. The Civil Service Commission approved the request and on December 9, 1958, a notice was published in the Federal Register. However, the White House intervened, stipulating that only active military test pilots would be used as the source of selection of astronauts. It was also decided that, since the individuals who were volunteering for duty as an astronaut would be subjected to stringent medical examinations, it would be inappropriate to jeopardize their military careers if some anomalies were discovered. Therefore, such information was not to be entered on candidates' permanent medical records. Additionally, the candidates were required to be graduates of test pilot school and to have at least 1,500 hours of flying time in high-performance jet aircraft.

This decision simplified the selection process considerably since it provided a known pool of applicants, each with a complete file of information on medical status at the time of application and a medical history spanning his service career. The records of 500 individuals were subsequently reviewed, including candidates from the Air Force, the Navy, the Marine Corps, and the Army. The National Aeronautics and Space Administration announced in January of 1959 that 110 had met the basic requirements, and on Monday, 2 February 1959, 69 candidates out of this pre-screened group reported to Washington under special military orders.

In Washington, these individuals were further screened through interviews with NASA project officials, psychiatric evaluations by two Air Force medical officers, reviews of their medical records and background, and a battery of written tests. On the basis of this initial evaluation, 32 pilots were selected to undergo the second phase of evaluation to become astronauts.

The 32 finalists were screened with an extensive battery of medical and psychological tests, some of which were standard and some of which were devised specifically for this purpose on the basis of the best medical judgment and state-of-the-art medical practice. Some of the most up-to-date procedures were included in the test battery; for example, exercise capacity was determined using a bicycle ergometer.

The first element of the medical evaluation program was conducted for NASA by the Lovelace Clinic in Albuquerque, New Mexico. Each candidate spent seven and a half days at the Lovelace facility undergoing detailed medical examinations. The medical examinations were followed by a series of provocative physiological stress tests administered at the USAF Aerospace Medical Laboratory at Wright-Patterson AFB, Ohio. The stress tests were designed to simulate, as far as could be anticipated, all of the stresses an astronaut might encounter during a Mercury mission. The testing provided tolerance information obtained through use of a centrifuge, low pressure chambers, an anechoic chamber, thermal exposure units, and aircraft modified to fly "Keplerian" trajectories producing a brief period of weightlessness. As a final step, peer ratings were obtained.

On 2 April 1959, NASA announced that seven astronauts had been chosen for Project Mercury. According to the Lovelace report, the seven ultimately selected were chosen because of their exceptional resistance to mental, physical, and psychological stresses and because of their particular scientific discipline or specialty. The average age was 35.2 years.

Subsequently, eight additional groups were selected to join the astronaut corps, for a total (as of 1982) of 127 individuals, including eight women. The medical evaluations for astronaut selections from 1962 through 1969 were conducted at Brooks AFB, San Antonio, Texas, using the protocol shown in Table 1. In 1977 and 1980 medical evaluations using new NASA standards were conducted at the Johnson Space Center, Houston, Texas. Table 2 shows the nine selection groups and statistics relating to each group.

It is interesting to note that all the astronaut selections through 1969 were conducted without a specific set of "pass-fail" medical standards. Instead, a ranking system was applied so that individuals in each selection group were competing in terms of health and physical fitness against other individuals in that same group. Since these examinations were conducted by experts in the field of aerospace medicine, it is clear that individuals with disqualifying defects for duties as high-performance test pilots had a lower probability of receiving a high ranking in the medical examination process. However, since all candidates had been repeatedly screened during their careers as pilots of high performance aircraft, there was little likelihood that such medical defects would be found. As a result, there were uncertainties about what was and what was not disqualifying.

Introduction of Formal Medical Standards

In anticipation of providing crewmen for the forthcoming Space Shuttle missions, in 1976 NASA made two major changes in the medical evaluation process for astronaut selection. First, it was decided that the medical evaluations would be performed at NASA's Lyndon B. Johnson Space Center in Houston and that they would be conducted by NASA personnel, with outside consultation as desired. The second change was in the development of a formal set of medical evaluation standards specifically tailored to individual categories of crewmen flying in space. These standards would reflect the growing body of knowledge in aerospace medicine based on the successful completion of five major programs of manned space flight, and would reflect the requirements of the Space Shuttle era.

Table 1
Medical and Psychological Tests Administered
to Astronaut Candidates

1. Medical history and review of systems.

2. Physical examination.

3. Electrocardiographic examinations, including routine electrocardiographic studies at rest, during hyperventilation, carotid massage, and breath holding, a double Master exercise tolerance test, a cold pressure test, and a precordial map.

4. Treadmill exercise tolerance test.

5. Vectorcardiographic study.

6. Phonocardiographic study.

7. Tilt table studies.

8. Pulmonary function studies.

9. Radiographic studies, including cholecystograms, upper GI series, lumbrosacral spine, chest, cervical spine, and skull films.

10. Body composition study, using tritium dilution.

11. Laboratory examinations, including complete hematology workup, urinalysis, serologic test, glucose tolerance test, acid aklaline phosphatase, BUN, sodium, potassium, bicarbonate, chloride, calcium, phosphorus, magnesium, uric acid, bilirubin (direct and indirect), thymol turbidity, cephalin flocculation, SGOT, SGPT, total protein with albumin and globulin, separate determination of Alpha 1 and Alpha 2, Beta and Gamma globulins, protein bound iodine, creatinine, cholesterol, total lipids and phospholipids, hydroxyproline, and RBC intracellular sodium and potassium. Stool specimens were examined for occult blood and microscopically for ova and parasites. A urine culture for bacterial growth was done, and a 24-hour specimen analyzed for 17-ketosteroids and 17-hydroxycorticosteroids.

12. Detailed examination of the sinuses, larynx, and Eustachian tubes.

13. Vestibular studies.

14. Diagnostic hearing tests.

15. Visual fields and special eye examinations.

16. General surgical evaluation.

17. Procto-sigmoidoscopy.

18. Dental examination.

19. Neurological examination.

20. Psychologic summary, including Weschler Adult Intelligence Test, Bender Visual-Motor Gestalt Test, Rorschach Test, Thematic Apperception Test, Draw-A-Person Test, Gordon Personal Profile, Edwards' Personal Preference Schedule, Miller Analogies Test, and Performance Testing.

21. Electroencephalographic studies.

22. Centrifuge testing.

From Hawkins & Zieglschmid, 1975

Table 2
Chronology of Selection of
Space Personnel in U.S. Program
(to August 1982)

Year	Category	Number Selected	Average Age (Yrs)	Number Who Have Flown	Resigned (Never Flown)
1959	Mercury Pilots	7	35.2	7	0
1962	Pilots/Astronauts	9	32.3	8	0
1963	Pilots/Astronauts	14	31.0	10	0
1965	Scientist/Astronauts	6	30.3	4	2
1966	Pilots/Astronauts	19	32.3	15	1
1967	Scientist/Astronauts	11	31.4	0	4
1969	Pilot/Astronauts	7	29.1	4	0
1978	Pilots	15	32.3	0	0
	Mission Specialists	20		0	0
1980	Pilots	8	33.5	0	0
	Mission Specialists	11		0	0
		127		48	7

In January 1977, NASA published the *Medical Evaluation and Standards for Astronaut Selection: NASA Class I-Pilot Astronaut*, and *NASA Class II-Mission Specialist* (NASA, 1977a, 1977b). These were followed in April of that year by corresponding standards for the category NASA Class III-Payload Specialist (NASA, 1977c). These standards served as a basis for selection cycles in 1978 and 1980. During these selections, 54 candidates were approved for the ratings of Pilot or Mission Specialist. Among those selected were eight women.

Tables 3 and 4 compare values for some physiological health indices for those selected prior to 1970 and those chosen in the two most recent selections. Table 5 shows the principal differences in the medical standards for Class I, II, and III crew members. Medical risk factors for acute and chronic diseases are weighted equally for all three classes.

Table 3
Medical Parameters Describing
Astronauts Selected Before 1970

Selection Group	Selection Year	Weight Mean	SD	Sys BP Mean	SD	Dia BP Mean	SD	Chol. Mean	SD	Trig. Mean	SD	Glucose Mean	SD	O$_2$ (Max) Mean	SD
1	1959	166	13	121	8.3	77	1.1	194	26.8	—		103	4.6	35.0	3.2
2	1962	162	13	122	11.8	75	8.4	177	33.8	—		93	7.1	NA*	
3	1963	161	13	127	12.9	79	11.8	174	16.7	116	36.3	102	9.3	42.2	5.9
4	1965	158	13	123	15.3	77	11.6	151	42.5	117	24.7	84	14.0	46.3	5.5
5	1966	166	15	124	13.4	76	7.6	197	45.7	99	18.5	109	12.3	43.8	4.6
6	1967	164	23	128	10.5	76	9.0	194	72.0	73	28.7	99	8.9	44.4	5.4
7	1969	162	15	127	10.0	68	7.7	194	44.4	93	36.2	112	7.9	43.7	2.7

*Not Available.

Table 4
Medical Parameters Describing
Crews Selected for Shuttle Programs

Selection Group	Selection Year	Weight Mean	SD	Sys BP Mean	SD	Dia BP Mean	SD	Chol. Mean	SD	Trig. Mean	SD	Glucose Mean	SD	O$_2$ (Max) Mean	SD
						Males									
8	1977	164	20	122	10.3	78	7.3	192	43.8	82	37.9	95	8.9	46.4	7.7
9	1980	160	14	118	8.8	79	5.1	167	30.6	74	21.0	94	4.5	49.4	5.8
						Females									
8	1977	131	22	107	11.3	76	3.8	191	47.2	67	24.3	86	6.9	35.0	5.9
9	1980	113	13	118	0.0	75	9.9	150	36.8	54	.7	86	1.4	40.8	2.6

Table 5
Difference in Standards Between NASA
Medical Class I, II, and III Spaceflight Evaluations

Item	Astronauts Pilots (class I)			Mission Specialist (class II)			Payload Specialist (III)
Distant Vision	20/50 or better uncorrected; correctable to 20/20 each eye			20/100 or better uncorrected; correctable to 20/20 each eye			corrected less than 20/40, better eye
Near Vision	uncorrected <20/20 each eye			uncorrected <20/20 each eye			corrected less than 20/40, better eye
Hearing Loss (db) Per 150, 1964 Standard	500 Hz 30	1000Hz 25 each ear	2000Hz 25	better ear: 500Hz 30 / worse ear 35	1000Hz 25 / 30	2000Hz 25 / 30	500Hz 35 / 1000Hz 30 / 2000Hz 30 better ear
Height (inches)	64-76 in.			60-76 in.			not specified
Refraction/Astigmatism	specified			specified			not specified
Contraction Visual Field (better eye)	15°			15°			30°
Phorias	specified			specified			not specified
Depth Perception	specified			specified			not specified
Color Vision	specified			specified			not specified
Joints Motion Range	specified			specified			not specified
Radiation Exposure	<5 rem/year			<5 rem/year			not specified

The medical evaluation procedures and standards for astronaut selection established in 1977 parallel requirements developed by the Department of Defense for high-performance aircraft pilots. They have been augmented to the extent necessary to include tests to evaluate ability to withstand the stresses of space flight. Certain stress and endurance tests, initially used as part of the selection process for astronauts, are now included in the training curriculum. These stress tests include Keplerian parabolic flights, altitude chamber indoctrination, water tank immersion training, survival training, and workload tolerance evaluations. These tests were moved from the selection process to the training phase since pass/fail selection criteria cannot readily be applied. They are retained in training under the philosophy that physiological stress testing enhances endurance for the environmental factors of space flight. It is believed that these tests, given during the one-year indoctrination period prior to formal acceptance into the astronaut corps, help to prepare a candidate to function as an astronaut.

Medical Issues in Space Crew Selections

With the formulation of medical standards, NASA has established individual medical evaluation programs for Pilot and Mission Specialist astronauts, as well as for the non-astronaut Payload Specialist. Of the 329 applicants examined in 1977 and 1980, 25 percent were disqualified for medical reasons. Of those, 75 percent of the rejections were due to visual, psychiatric, and cardiovascular problems. Vision problems alone accounted for 54 percent of those rejected, with visual acuity accounting for 60 percent of all ophthalmological problems. Other diagnoses showed a wide degree of variability. Table 6 shows the disqualifying medical conditions encountered during the different selection processes dating back to 1959. Results from the different selections prior to 1969 have been pooled since no standardized evaluation process was used during that period.

Table 6
Disqualifying Medical Conditions Found
for Astronaut Candidates

| | Selection Category and Time Period | | | | |
| | Pilot | | | Mission Specialist | |
Standard	1959-67	1977	1980	1977	1980
Visual	21	0	2	28	7
Surgical	9	0	3	4	4
Neurological	10	0	1	3	6
Cardiovascular	3	2	1	11	4
Hematological	2	0	0	1	3
Otolaryngological	16	0	2	1	1
Metabolic/Endocrine	5	0	0	4	3
Gastrointestinal	10	0	0	1	0
Genito-Urinary	10	0	0	2	1
Pulmonary	1	0	0	1	0
Dental	1	0	0	0	0
Psychological (Human Factors)	5	0	1	1	12

The successful development and application of selection standards is demonstrated by the fact that, to date, all of the newer astronaut candidates have completed their prescribed one-year training course without developing untoward health problems which could have resulted in rejection. Another important consequence of the publication of NASA medical standards is that, in combination with the unique operational aspects of the STS, they provide the field of space medicine with the long-awaited opportunity to establish longitudinal health trend studies which were difficult to conduct in the past.

Longitudinal Medical Surveillance

Biomedical results from Mercury, Gemini, and Apollo missions were based largely on comparisons of pre- and postflight physiological status, and on observation of postflight physiological readaptation. Inflight medical monitoring provided data only on certain functions which were essential to the assessment of crew health and safety and the evaluation of basic physiological response to activities such as EVA. It was not until the advent of Skylab that provisions were made for extensive experimentation in space to determine the rate and time-course of adaptation to spaceflight conditions, based on scheduled analyses of metabolic, cardiovascular, and biochemical parameters under a variety of stressed and non-stressed conditions (Johnston, 1977).

Early biomedical studies were mostly observational in nature, and were often dictated by unexpected physiological manifestations exhibited by the small, highly select group of individuals who participated in space flight. However, these studies did contribute to our understanding of the gross physiological changes associated with single exposures to space flights of varying duration. As the number of missions increased and more biomedical data were accrued, and as space endurance records were established, a clustering of events was seen which permitted specialists to approximate physiological trends and formulate newer hypotheses (Nicogossian, Pool, & Rambaut, 1980).

Medical Programs in the STS Era

The development of the Space Transportation System represents a significant change of character and emphasis for the space program. The Space Transportation System is being developed with a view to maximum utility and practical benefit. Therefore, the philosophy is that most individuals with an approved experiment or other legitimate basis for flight will be given permission to fly on the Space Shuttle provided there is no reason to suspect that such flight will jeopardize the health and well-being of the individual or threaten the success of the mission. There are to be no arbitrary age limits or similar barriers. In this context it has been estimated that, through 1990, more than 400 individuals will participate in as many as 200 Shuttle orbital flights.

The Space Transportation System will offer new opportunities to obtain flight-related biomedical data. To support the increased number of flight crew personnel, the high launch and landing rate, and the short turnaround times required for STS operations, an efficient biomedical system must be established for the orderly and timely collection of standardized biomedical data from selection, retention, and flight medical evaluations. An essential requirement for this biomedical system is for the initiation of longitudinal studies of large groups of space travelers. The publication of NASA medical standards in 1977, along with the novel operational features of the Space Shuttle, afford the opportunity to establish longitudinal follow-up studies, and thus to begin to bring space biomedical research into accord with standard public health practices. Although these studies will be of limited pathophysiological and etiological value for the foreseeable future, they will provide an invaluable feedback mechanism for (1) designing high-quality experiments addressing specific hypotheses to be tested, and/or (2) developing more accurate modeling capabilities.

Current Status of NASA's Longitudinal Studies

In 1976, a retrospective survey dealing with aging and the development of coronary artery disease among U.S. astronauts selected prior to 1970 was presented (Degioanni et al., 1976). The conclusion was that the incidence of coronary artery disease among U.S. astronauts was not significantly different from that of the general population. In the context of the length of observation (15 years) and the rate of occurrence of the disease under investigation (N = 1 out of a sample size of 73), this conclusion may have been premature. The pre-1977 astronaut population is a relatively homogenous, prescreened group of men with a low index for cardiovascular risk factors. This is supported by a review of trend data for weight, recumbent blood pressure, fasting blood sugar, and triglycerides for this group (Figure 1). All of these health indices remained remarkably stable in the time frame of the observation (1969 through 1979). The small variations in weight recorded in the years 1975 and 1976 probably represent fluctuations in the number of astronauts examined, and also changes in lifestyle incurred in the absence of foreseeable space missions during this period of time. Changes in blood cholesterol levels are presented in Figure 2. A steady and significant increase in blood cholesterol values with progression of time is noted, which is comparable to trends observed in the general population. However, values for high density lipoprotein (HDL) and the persistently high maximum oxygen consumption levels (VO_2 = 42 ± 8 ml/kg/body) as well as a lower-than-expected percent fat for age (15.7), all point to the fact that the cardiovascular risk profiles of astronauts remain below the national average.

Constructing Longitudinal Studies in the STS Era

In order to assess systematically the effects of space flight on humans, several types of studies, differing primarily in the length of time devoted to data collection, are planned by NASA:

1. Short-term studies based on analyses of data from single flights. These will add to the information collected describing physiological changes during space flight. The ongoing STS space motion sickness study is an example. This study involves routine collection of pre-, in-, and postflight space motion sickness-related data from Space Shuttle crew members and passengers through at least the mid-1980s. The goals are to (1) validate ground-based predictive tests of susceptibility to space motion sickness and (2) define operationally acceptable countermeasures.

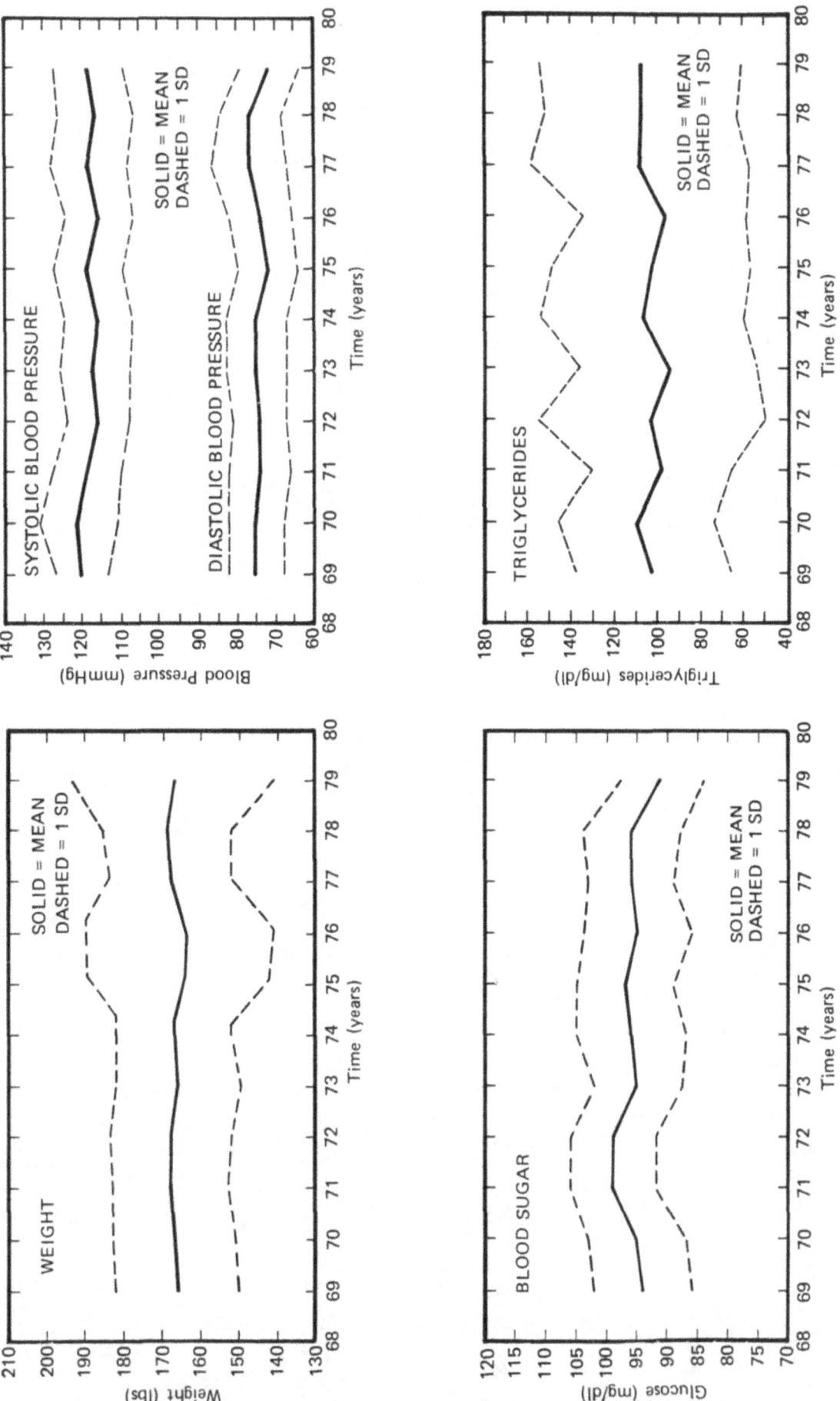

Figure 1. Ten-year review of four health indices for astronauts selected before 1970 (N = 57).

243

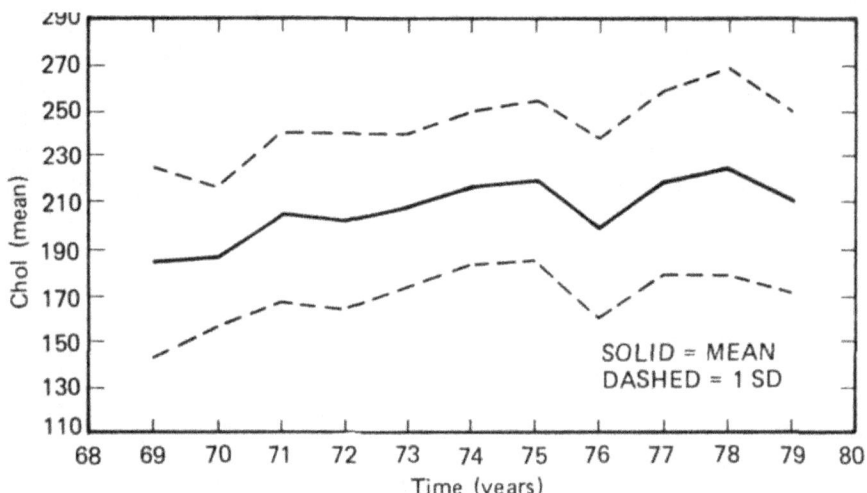

Figure 2. Changes in blood cholesterol levels
of astronauts selected before 1970 (N = 57).

2. Mid-term longitudinal studies spanning several missions. These studies will measure the effect, if any, of repeated exposures to the spaceflight environment. Attention will be given to issues such as the cumulative effect of radiation exposure and the rate of bone mineral loss through repeated missions.

3. Long-term longitudinal studies aimed at collecting career data through a medical surveillance program to develop occupational information based on diseases or injuries incurred during or following space flight.

Much of the data in NASA longitudinal studies will come from standardized annual medical evaluations. The resulting biomedical data base will be divided into Astronaut-Pilot, Astronaut-Mission Specialist, and Payload Specialist categories. This will allow studies to be made concerning effects of different work requirements in space as well as differential length and frequency of exposure. These longitudinal studies will help to validate medical standards, to better define medical risk factors associated with space flight, to characterize the "health profile in zero gravity," and, as appropriate, to define new problems associated with space flight which might result from exposure of population groups other than astronauts.

References

Colton, T. *Statistics in medicine*. Boston: Little, Brown & Co., 1974.

Degioanni, J., Nicogossian, A., Karstews, A.I., Burchard, E.C., & Zieglschmid, J.F. A survey of fifteen years of medical experience with astronauts. In *Preprints of the 48th ASMA Annual Scientific Meeting*, Bal Harbour, FL, 1976.

Hawkins, W.R., & Zieglschmid, J.F. Clinical aspects of crew health. In R.S. Johnston, L.F. Dietlein, and C.A. Berry (Managing Eds.), *Biomedical results of Apollo* (NASA SP-368). Washington, D.C.: National Aeronautics and Space Administration, 1975.

Johnston, R.S. Skylab Medical Program overview. In R.S. Johnston & L.F. Dietlein (Eds.), *Biomedical results from Skylab* (NASA SP-377). Washington, D.C.: National Aeronautics and Space Administration, 1977.

Link, M.M. *Space medicine in Project Mercury* (NASA SP-4003). Washington, D.C.: U.S. Government Printing Office, 1965.

NASA. *Medical evaluation and standards for astronaut selection: NASA class I—Pilot Astronaut* (JSC-11569). Prepared by Space and Life Sciences Directorate, NASA Lyndon B. Johnson Space Center, Houston, TX, January 1977a.

NASA. *Medical evaluation and standards for astronaut selection: NASA class II—Mission Specialist* (JSC-11570). Prepared by Space and Life Sciences Directorate, NASA Lyndon B. Johnson Space Center, Houston, TX, January 1977b.

NASA. *Medical evaluation and standards: NASA class III—Payload Specialist* (JSC-11571). Prepared by Space and Life Sciences Directorate, NASA Lyndon B. Johnson Space Center, Houston, TX, April 1977c.

Nicogossian, A., Pool, S., & Rambaut, P. The Shuttle and its importance to space medicine. Presented at the 31st IAF Congress, Tokyo, Japan, 22-25 September 1980.

CHAPTER 16

BIOMEDICAL TRAINING OF SPACE CREWS

Biomedical training, which has always been part of the total training program for space crews, is designed for two purposes. The first is to familiarize crew members with the unique features of the space environment, especially weightlessness, as these features produce physiological change or affect performance. The second is to prepare crew members to manage medical events, such as an unforeseen illness or injury, which might occur during a mission.

Physiology training in space medicine is a direct outgrowth of that used in military aviation medicine. Aviation programs are concerned with the traditional problems of the high-speed, high-altitude flight environment. Physiological stresses include low pressure and temperature, lack of oxygen, vibration, high linear and angular acceleration forces, and many others. Altitude chambers and human centrifuges are used extensively in aviation physiology training programs. In the early stages of manned space flight, much of this training technology was applied, with some modification, to the issues of man in space. Astronauts were given training in high altitude physiology, human response to thermal extremes, survival in different recovery environments, tolerance to different acceleration forces, and the effects of psychological stressors such as confinement and isolation. To the extent that it could be accomplished, an introduction to the physiology of weightlessness was added.

The value of zero-gravity simulation for training purposes was demonstrated dramatically following the first attempts at extravehicular activity during the Gemini program. The extravehicular tasks attempted in both the Gemini 9 and 11 missions were not particularly successful. In Gemini 9, the task was to check out and don the Astronaut Maneuvering Unit to be used as an assist during EVA. Foot restraints proved to be inadequate, causing the pilot to exert a continuous high workload to maintain body position. Heat and perspiration were produced at a rate exceeding the removal capability of the life-support system, and fog accumulated on the spacesuit visor until the astronaut's vision was almost totally blocked (Schultz et al., 1967).

Shortly after the Gemini 9 mission, the pilot made use of an underwater zero-gravity simulation to test the use of various restraint systems. It was concluded at that time that this type of simulation very nearly duplicated the actual weightless condition and could be used profitably for study of problems in extravehicular operations. As a result of these tests, it was decided that the Gemini 12 flight crew would use an underwater zero-gravity simulator as part of their training for extravehicular activity. Two periods of intensive underwater simulation and training were added to the program. During these simulations the pilot followed scheduled flight procedures and duplicated the planned EVA systematically. Among other benefits, the pilot was able to condition himself to relax completely within the suit. All movements were slow and deliberate, using only those muscles required for task performance. An effort was made to relax all other muscles not in use. This technique then was transferred to the actual EVA of Gemini 12.

246

Results of the Gemini 12 EVA showed that all tasks attempted were feasible, using the techniques perfected in the zero gravity simulator as well as improved body restraints. EVA workload could be controlled within the desired limits by application of proper procedures and indoctrination. It was concluded that any task which could be accomplished readily in a valid underwater simulation would have a high probability of success during actual extravehicular activity (Machell et al., 1967). Use of this type of training for performance in weightlessness was included in all subsequent training programs.

A second training issue results from problems encountered in returning a spacecraft to Earth prior to mission completion. When a spacecraft cannot be recovered quickly, emergency medical services, except in the unusual event of a physician crew member, must be provided through an onboard combination of trained crewmen and stowed medical supplies. While medical advice can be provided by Mission Control, the direct treatment of a serious illness or injury must be provided by another crew member.

During orbital flight, a spacecraft can make an emergency return to Earth within a matter of hours, given prior preparations, but this is not desirable in terms either of mission objectives or economy. Under other circumstances, an emergency return simply may not be possible, however dire the circumstances. This was amply demonstrated during the Apollo 13 mission. After insertion into a lunar trajectory, the spacecraft experienced an explosion and rupture of an oxygen tank, with serious damage to other components. Oxygen required for breathing and for the electricity-producing fuel cells was rapidly depleted. It was initially calculated that only about 38 hours of power, water, and oxygen were available, and that this was about one-half as much as would be needed to complete circumnavigation of the moon and return to Earth. By transferring to the Lunar Module, greatly reducing power consumption, and improvising on-board carbon dioxide removal systems, the crew was able to accomplish a successful return. While the lowered cabin temperature made the crew uncomfortably cold, no serious medical problems were experienced with the exception of a urinary tract infection developed by one of the crewmen. Under conditions of extreme hazard, a space flight was completed successfully. Apollo 13 served a very useful purpose in demonstrating the need for meticulous planning and training for inflight emergencies. Medical training for crew members, supported by simulator practice, is an important part of this emergency planning.

Training for the Stresses of Space

An astronaut can encounter any number of stresses, including weightlessness, noise and vibration, unusual acceleration forces, possible temperature extremes, and isolation. Every effort is made to minimize these stresses through appropriate design of the spacecraft and attention to mission parameters. However, the possibility always remains that through an unusual combination of circumstances, one or more of these stresses might suddenly increase. The Apollo 13 inflight emergency provides an excellent example. In this mission, the crew operated under very low thermal conditions, with suit inlet temperatures dropping to as low as 5° C (41° F).

The philosophy of environmental stress training is that repeated exposure to a stress factor results in a better understanding of its effects and an increased ability to cope with it. This type of training, which in one sense might better be termed "conditioning," was given considerable emphasis in the early preparations for manned space flight both by the United States and the Soviet Union. Since then, emphasis has abated somewhat in the American program while continuing at much the same level in the Soviet Union.

In planning for Project Mercury, there was considerable concern over the possible effect of space flight environmental factors on crew performance. Consequently, training emphasized crew exposure to such conditions as high acceleration forces, zero-gravity conditions, heat, noise, and spacecraft tumbling motion (Link & Gurovskiy, 1975). Particular attention was given to the capability of the crew to control the spacecraft manually during the high acceleration loads imposed during launch and entry. In order to examine this capability, and to provide training in dealing with acceleration, a number of runs were made in the centrifuge at the Naval Air Development Center, Johnsville, Pennsylvania. These training programs were considered quite valuable in providing practice in the operation of spacecraft systems and in allowing astronauts time to develop their adaptive capabilities to accelerative forces. Training in the centrifuge has continued in Soviet programs but was discontinued in the U.S. prior to the Apollo missions.

As experience in manned space flight increased, more attention was given to enhancing performance in weightlessness and less to environmental stress factors per se. While indoctrination concerning the consequences of hypoxia and exposure to reduced pressure is still a practice through use of altitude chambers, more attention is directed to the NASA KC-135 aircraft which, when flying a Keplerian trajectory, can present short periods of weightlessness. Figures 1 and 2 show use of this aircraft to train astronauts in tasks such as handling of space food and maneuvering in space suits during weightlessness. Even though an astronaut has only a few seconds of zero gravity in which to experiment with such tasks, this practice is judged useful for later mission performance.

Use of water immersion techniques to simulate zero gravity has a distinct advantage over the use of an aircraft flying a zero-gravity profile, since there is no time constraint. Under conditions of neutral buoyancy in a water tank, astronauts can practice the tasks required during extravehicular activity over a long period of time, or until the desired proficiency is achieved. Since the Gemini 12 flight, all astronauts have been trained in this manner. Figures 3 and 4 show training operations at the two NASA facilities used for this purpose.

Medical Training

All spacecraft carry a medical kit tailored to mission duration, availability of physicians, and training level of crew members. The purpose of the crew medical training is to provide crews with the knowledge and skills necessary to respond to inflight illnesses and injuries in an appropriate and timely manner. This objective is met through both the general medical training which is part of each astronaut's initial training and mission-specific training in the month prior to a Space Shuttle flight (Vanderploeg & Hadley, 1982).

Figure 1. Astronaut Richard Truly practicing use of the Space Shuttle food system while in weightlessness in the KC-135 aircraft.

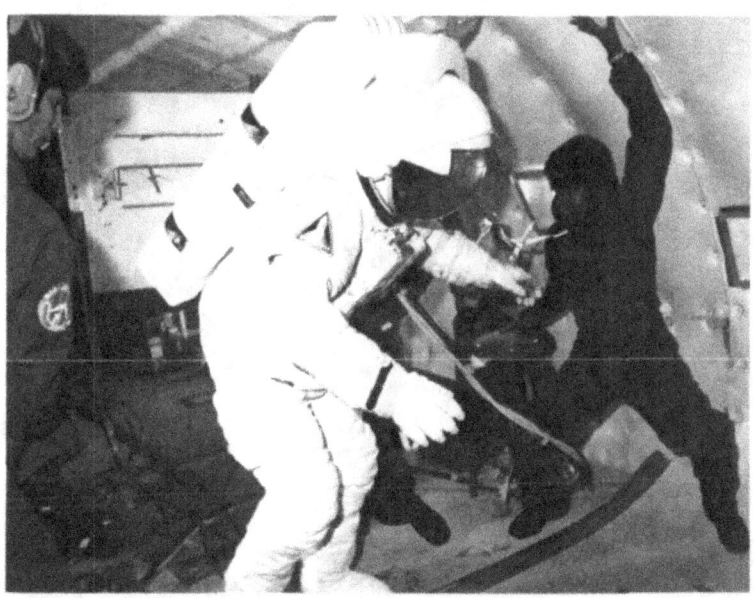

Figure 2. Astronaut Gordon Fullerton practicing with the Space Shuttle Extravehicular Mobility Unit in the KC-135 aircraft.

Figure 3. Astronaut Richard Truly entering the water immersion training facility at the Johnson Space Center.

Figure 4. Astronaut Shannon Lucid using the Neutral Buoyancy Simulator at the Marshall Space Flight Center.

The initial medical training given an astronaut involves 16 hours of instruction during the first year after selection. The training effort covers altitude physiology and includes training in composition of the atmosphere; the gas laws; signs, symptoms, and treatment of hypoxia; operation of life support equipment; effects of increased gravity; use of anti-G maneuvers (Valsalva manuever, etc.); use of the anti-G suit, and a demonstration of hypoxia.

Medical training for management of inflight episodes typically begins after an astronaut has been given a crew assignment. Topics covered include location and use of (1) the Shuttle Orbiter Medical System (SOMS), (2) the Operational Bioinstrumentation System, (3) the anti-gravity suit, and (4) the radiation monitoring equipment. The medical training given in support of SOMS is shown in Table 1.

Table 1

STS-1 Crew Medical Training

VITAL SIGNS:		Pulse, Blood Pressure, Temperature, Respiratory Rate, Pupil Size and Reaction
PHYSICAL EXAMINATION AND TREATMENT:		
	EYE —	Ophthalmoscopy, Lid Eversion, Foreign Body Reaction and Treatment, Fluorescein Staining
	EAR —	Otoscopy
	NOSE —	Control of Nose Bleeds
	THROAT —	Examination, Oral Airway Insertion
	AUSCULTATION —	Heart, Lung, and Bowel Sounds
EMERGENCY PROCEDURES:		One-man CPR, Heimlich Maneuver, Cricothyrotomy
HEMORRHAGE CONTROL:		Direct Pressure, Pressure Points, Tourniquets, Pressure Bandaging
BANDAGING:		Extremities, Chest, Abdomen
SPLINTING:		Neck, Fingers, Upper and Lower Extremities
LACERATION TREATMENT:		Bleeding Control, Steristrip Application
DENTAL PROCEDURES:		Temporary Fillings, Gingival Injections
EKG:		Use of OBS
MOTION SICKNESS:		Prophylactic Medications, Treatment, Head Positioning and Movement
SOMS-A:		Organization, Drug Usage, Medical Checklist Organization and Use

From Vanderploeg, 1981

251

The training in emergency medical procedures is supported through use of checklists dealing with the kinds of episodes anticipated in an onboard emergency. An example of such a checklist is shown in Table 2.

Table 2

Checklist Used by Space Shuttle Crewmen
for Medical Emergency

CUTS:

MBK[1]
1. Stop bleeding with 2x2 sponges (E2-1)[2], pressure over wound

2. If bleeding uncontrolled, apply tourniquet (EMK C1-2) above cut and apply pressure, bandage over cut

3. After bleeding controlled, cleanse wound, adjacent skin with Betadine (E1-9) or alcohol swabs (E1-7)

4. Consult surgeon regarding use of steristrips

5. Dress wound
   Betadine (Povidone) ointment (F1-5) sufficient to prevent dressing from adhering to wound
   Cover with adaptic (E2-3) or 2x2 sponges (E2-1), Kling (E1-2), Kerlix (F1-4), Ace (E2-5)

---

[1] MBK = Medications and Bandage Kit

[2] Refers to location of supplies in the kit.

From Degioanni et al., 1979

Training also is given in the handling of biomedical experiments that might be carried in the payload bay of the Space Shuttle Orbiter. This training focuses on Mission Specialists and Payload Specialists.

Emergency Medical Support

Launch and landing operations of the Space Shuttle are supported by an Emergency Medical Services System (EMSS) designed to provide an ill or injured crewman with rapid access to the appropriate level of medical care. The success of this system depends on availability of medical personnel and facilities, rapid transportation procedures, trained rescue personnel, and knowledgeable space crewmen. Training for space crewmen is directed principally toward use of the system. Eight major egress modes have been identified for launch and landing emergencies of the Space Shuttle (Pool, 1981). Figure 5 shows rescue operations being practiced. Figure 6 shows one of the NASA rescue vehicles.

Figure 5. Rescue swimmers assisting an "astronaut" during recovery from an Orbiter mock-up which supposedly has landed in water.

Figure 6. Rescue helicopter used by NASA for transport of injured crewmembers to appropriate medical facilities.

Training System Development

Training of space crews is a team effort under continuous development. The physician, as one member of the team, provides inputs concerning environmental stress effects and medical requirements. Lessons learned from completed missions, combined with prospective activities for future missions, are analyzed in terms of the feasibility of dealing with inflight medical episodes through a combination of trained crewmen and spacecraft equipment. The proposed improvements in procedures and equipment are verified under laboratory conditions, in spacecraft simulators, and during parabolic aircraft flights. Experienced astronauts review the procedures and provide an invaluable feedback process as to their feasibility for use inflight. At the completion of this process, new training procedures and new onboard medical equipment are incorporated into the space program. At some future time, as space stations become operational and as physicians-astronauts are included in crew complements, it may be that more advanced biomedical training procedures can be developed which will use the actual space platform as a training base, rather than relying heavily on ground-based simulation, as is now the case.

References

Degioanni, J., Smart, R.C., & Snyder, R.D. Shuttle operational medical system (SOMS): A medical checklist—preliminary, rev. A (JSC-14798; LA-B-11110-5Q). Houston, TX: Lyndon B. Johnson Space Center, July 13, 1979.

Link, M.M., & Gurovskiy, N.N. Training of cosmonauts and astronauts. In M. Calvin and O.G. Gazenko (General Eds.), *Foundations of space biology and medicine* (Vol. III) (NASA SP-374). Washington, D.C.: U.S. Government Printing Office, 1975.

Machell, R.M., Bell, L.E., Khyken, N.P., & Prim, J.W., III. Summary of Gemini extravehicular activity. In *Gemini summary conference* (NASA SP-138). Washington, D.C.: U.S. Government Printing Office, 1967.

Pool, S.L. Emergency medical services system (EMSS). In S.L. Pool, P.C. Johnson, Jr., and J.A. Mason (Eds.), *ST-1 medical report* (NASA TM-58240). Washington, D.C.: National Aeronautics and Space Administration, Scientific and Technical Information Branch, December 1981.

Schultz, D.C., Ray, H.A., Jr., Cernan, E.A., & Smith, A.F. Body positioning and restraints during extravehicular activity. In *Gemini summary conference* (NASA SP-138). Washington, D.C.: U.S. Government Printing Office, 1967.

Vanderploeg, J.M. Crew medical training. In S.L. Pool, P.C. Johnson, Jr., and J.A. Mason (Eds.), *STS-1 medical report* (NASA TM-58240). Washington, D.C.: National Aeronautics and Space Administration, Scientific and Technical Information Branch, December 1981.

Vanderploeg, J.M., & Hadley, A.T., III. Crew medical training and Shuttle Orbiter medical system. In S.L. Pool, P.C. Johnson, Jr., and J.S. Mason (Eds.), *STS-2 medical report* (NASA TM-58245). Houston, TX: Lyndon B. Johnson Space Center, May 1982.

CHAPTER 17

GROUND-BASED MEDICAL PROGRAMS

The principal goal of the medical program of the National Aeronautics and Space Administration is to assure the health and well-being of space crews. This program, dealing as it does with healthy individuals, is one of preventive medicine in the fullest sense. Medical personnel must anticipate health hazards and potential medical problems, adverse physiological reactions, and possible injury conditions which might occur during flight. The task then is to establish an environment and procedures to preclude such occurrences. Should a medical problem or an injury in fact occur, prompt and proper attention must be available to minimize the consequences for the mission and the crewman.

The success enjoyed by the U.S. and USSR space programs has brought with it a confidence in the ability of space crewmen to remain healthy and to work productively for long periods and under stressful conditions. There have been illness episodes and health problems, to be sure, but the overall record is good. In achieving this success, space medicine personnel have developed principles and guidelines of demonstrated effectiveness for the medical surveillance and care of space crews. Today, in comparing procedures used for Apollo missions with those for the Space Shuttle, the medical oversight of space crews is seen to be less demanding of crew time, to require a smaller medical support establishment, and to operate from a broader base of medical data.

Routine Medical Care

Health maintenance for space operations includes routine monitoring of the general health of crew members, their families, and, during periods of preparing for and conducting missions, mission-control and support personnel. The Johnson Space Center provides health care services and maintains health records for all flight crew personnel and their dependents. This care is administered through the JSC Flight Medicine Clinic and includes routine physical examinations, dental care, and provision of medical consulting services as required. Through this routine care and oversight, medical officers insure a general state of good health and, in addition, build a medical file concerning space crews to aid in dealing, at some later time, with any medical issues which might affect the conduct of a mission.

Medical oversight requirements increase significantly as a mission draws near. All sites supporting Space Transportation System operations maintain clinical facilities for routine treatment and emergency medical care on an on-going basis. These care services are provided to NASA employees, on-site contractor employees, and personnel temporarily assigned for mission duties. Part of these services include assuring that all flight control team members meet medical certification requirements. This is in recognition that an illness of a key flight controller would have negative mission impact just as with a member of the space crew.

The Crew Physician and his staff are responsible for monitoring flight crew training in the days before a mission to assess potential hazards. This coverage includes tests in the Weightlessness

Environment Training Facility (Figure 1), altitude chamber runs (which routinely have a physician in attendance), and any other tests thought to have an element of medical hazard. An abbreviated physical examination is performed on all participants, including the crew, other subjects, and support personnel, prior to a test or training operation.

Figure 1. Weightless Environment Training Facility used for underwater neutral buoyancy training, simulating zero gravity, at the Lyndon B. Johnson Space Center.

The NASA Occupational Health Program operates principally to assure the health and safety of all NASA employees. It also provides important services in direct support of manned spaceflight activities. These support efforts include training employees who work with toxic chemical or physical agents, assisting with direct medical support of mission activities, and playing a key role in the emergency medical support system used during periods of launch and recovery. Personnel skilled in occupational medicine are of considerable help in insuring the safety of such tasks as handling of hypergolic fuels, testing of rocket engines, and general monitoring of spacecraft safety.

Health Stabilization Program

Early flights in the Mercury and Gemini programs were so brief that there was little concern over the possibility of an infectious disease occurrence. In Gemini, as flight durations increased, medical personnel acknowledged the possibility of disease and were successful in obtaining a measure of reduction in the number of personal contacts by crew members prior to launch. While there were no serious inflight episodes, Gemini crews did experience some minor illnesses, such as colds and influenza, during the period of pre-launch preparation.

As the Apollo program developed, concern increased over the likelihood of an infectious disease and its consequences. Certainly, there would be considerable difficulty in successfully accomplishing a lunar landing mission should one of the crew members become seriously ill during the early stages. During the Apollo 8 preflight period, all crew members suffered viral gastroenteritis. Treatment appeared to be successful, and the spacecraft was launched on schedule. However, viral gastroenteritis reoccurred in one crew member inflight (Wooley & McCollum, 1975). This episode greatly increased awareness of the need for stricter preflight measures to insure the health of astronauts during a mission.

The Apollo 14 mission was the first to be conducted under a formalized Flight Crew Health Stabilization Program. The purpose of the program was to minimize or eliminate the possibility of adverse alterations in the health of flight crews during the immediate preflight, inflight, and postflight periods. The four key elements in this program were (1) clinical medicine, (2) immunology, (3) exposure prevention, and (4) epidemiological surveillance. The essence of the program was to institute, for a three-week period prior to launch, strict control over locations to which flight crew members had access, number of personal contacts for the astronauts, and careful monitoring of the health of individuals required to be in contact with flight crew members (Wooley & McCollum, 1975). The program was quite successful, producing an immediate improvement in the number of crew members experiencing illness during the mission period.

The need for health stabilization became even more critical during the Skylab missions for two reasons. First, the greatly extended periods of orbital flight time increased the probability of an inflight illness. Second, an inflight illness would have contaminated the results of a number of biomedical experiments designed to examine in detail human physiological response to the space environment. For these reasons, the Health Stabilization Program developed in Apollo was continued as a formal part of the Skylab mission sequence, again using a 21-day isolation period prior to launch. A seven-day postflight isolation period also was used. This allowed extensive postflight medical examinations and assessment of recovery from the physiological changes noted inflight.

A summary of the illness occurrences in the Apollo and Skylab missions is presented in Table 1 (Ferguson et al., 1977). Prior to the initiation of the Health Stabilization Program in Apollo, 57 percent of the prime crew members experienced some illness during the 21 days prior to launch, as well as illness events both inflight and postflight. The illnesses included a number of upper respiratory infections, viral gastroenteritis, and one rubella exposure. These infections are notably absent from the beginning of the Health Stabilization Program on Apollo 14 through the three Skylab missions. On Apollo 17 and Skylab missions 3 and 4, minor skin infections, or rashes, were seen. Medical monitoring personnel concluded that these occurred for reasons other than preflight exposure. The Health Stabilization Program introduced in Apollo and continued through Skylab was remarkably successful in reducing the number of illness events for flight crewmen.

Table 1

Effect of the Flight Crew Health Stabilization Program
on the Occurrence of Illness in Prime Crewmen

Health Stabilization Program Absent				Health Stabilization Program Operational			
Mission	Illness Type[1]	Number of Crewmen Involved	Time Period[2]	Mission	Illness Type[1]	Number of Crewmen Involved	Time Period[2]
Apollo 7	URI	3	M	Apollo 14			
8	VG	3	P,M	15			
9	URI	3	P	16			
10	URI	2	P	17	SI	1	P
11				Skylab 2			
12	SI	2	M	3	SI	2	M
13	R	1	P	4	SI	2	M

[1] Illness type:
    URI, Upper respiratory infection.
    VG, Viral gastroenteritis.
    SI, Skin infection.
    R, Rubella exposure.

[2] Time period:
    M, During mission.
    P, Premission.

From Ferguson et al., 1977

Advent of the Space Transportation System offers new problems for health stabilization. An inflight illness, again, would be of real consequence. The problem, however, is that many more people will be involved in launch preparation and, as missions become increasingly frequent, more flight crew personnel will be under medical surveillance than was true for any previous program. There also will be an increased number of personal contacts prior to a mission as preparations are completed for a variety of payloads to be carried into space.

The Health Stabilization Program for the Space Transportation System identifies three levels of health coverage (NASA, 1981). Level I is the least stringent program. This is a voluntary program based on health education, increased health awareness by flight crew members and contact personnel, and involves no special medical surveillance or examination beyond the initial health screening program usually completed two to three weeks before the flight. Level II provides limited personal contact with flight crew members and requires medical examinations and surveillance of contact personnel. This is essentially the Health Stabilization Program used during previous space programs, but with reduced isolation time. Level III is a true isolation (quarantine) program which would provide the greatest amount of health protection.

The particular level of health stabilization to be employed is decided for each STS mission by a Health Stabilization Board established by the Johnson Space Center. For the STS-1 mission, a Level II program was used. Personnel required to be in work areas were identified and given medical examinations. Those found medically qualified were identified as primary contacts. Security was placed over the training building as well as the principal work area for crewmen, and only primary contacts were allowed to enter. Primary contacts were instructed to wear surgical masks when within six feet of crewmen. Each contact voluntarily reported his or her illnesses to the NASA Clinic. Table 2 shows the number, type, and location of personnel given medical examinations and approved as primary contacts for STS-1. The illness rate in this population during the program was 28 illnesses per 1,000 persons per week (Ferguson, 1981). A summary of these illnesses is presented in Table 3. The STS-1 program effectively kept 38 known ill persons out of crew work areas and thereby prevented exposure and possible illness.

The Health Stabilization Program for the STS-2 mission was reduced from a Level II to a Level I effort (Ferguson, 1982). This level, as noted, focuses on an educational program creating health awareness among personnel entering crew work areas. Posters, signs, and information sheets were placed in these areas. Information sheets also were distributed to contacts. Special crew travel routes were established to prevent accidental exposures. Only three illness reports were received from the 164 primary contacts during the 14 days of program operation. Crew members experienced no illness from infectious disease.

Mission Procedures

Pre- and Postflight Medical Examinations

All space missions have included a schedule of detailed medical examinations of flight crew members prior to launch and following recovery. The preflight physical examinations, typically scheduled over the month preceding launch, are intended to detect any medical problems which might require remedial or preventive intervention and to provide a baseline for postflight comparison. The purpose of the final examination, just prior to launch, is to certify the crewmember medically for flight and to document physical status at the onset of the mission.

In the Orbital Flight Test (OFT) program for the Space Shuttle, comprising the first four orbital missions, there were four preflight medical examinations and three conducted postflight. In addition, all flight crew members were given a vestibular testing program beginning 180 days prior to the mission. This was to establish a need for personal treatment regimens. This testing, which continues today, is not given to crewmen with previous space flight experience who were not subject to space motion sickness, unless such testing is specifically ordered by the Crew Physician.

259

Table 2

Number, Type, and Location of Personnel Given Medical Examinations
and Approved as Primary Contacts for STS-1

Type	Location					Subtotal
	JSC	KSC	DFRC	ARC	Headquarters	
NASA	216	35	7	1	5	264
Contractor	643	42	12	0	0	697
Others	10	1	0	0	0	11
Subtotal	869	78	19	1	5	972 Total

From Ferguson, 1981

Table 3

Types of Illnesses Reported by STS-1 Primary Contacts

Illness	Location			Percent of Total*
	JSC	KSC	DFRC	
Upper Respiratory Infection	24	3	3	81
Bronchitis	1	0	0	3
Pneumonia	0	0	0	0
Upper Enteric Illness	3	0	0	8
Lower Enteric Illness	2	0	0	5
Fever Present	4	0	0	11
Headache Present	1	0	0	3
Skin Infection Present	0	0	0	0
Other Infectious Illness	1	1	0	5

*Percentages total more than 100% because one illness may contain more than one symptom complex.
From Ferguson, 1981

In the operational STS program, only one or two preflight and postflight medical evaluations are planned in order to have the least impact on crew training between missions and ground turn-around time for the Space Shuttle.

The schedule and contents of pre- and postflight evaluations used for OFT missions are shown in Table 4. The duration of each examination listed in this table is that used for the STS-1 mission. Accompanying the physical examination given immediately following recovery was a formal medical debriefing. Crew members were debriefed using a standardized protocol to ensure completeness of coverage and to conserve time during the postflight period.

Table 4
STS-1 Pre- and Postflight Medical Evaluations

Examination	F-30 Days	F-10 Days	F-2 Days	F-0	L+0	L+3 Days	L+3 to L+7 Days
Location	JSC	JSC	JSC	KSC	DFRC	JSC	JSC
Approximate Duration	1:05	1:55	0:15	0:10	0:30	1:45	
Examination Components	LM PX D V T	LM ST PX A	LM	PX	LM STW PX	ST D V T A PX	T-38 Check Out

LEGEND:

A — Audiometry
D — Dental Examination
LM — Laboratory—Microbiology
PX — Physical Examination
ST — (Cardiovascular) Stress Tests, including 80% treadmill
STW — Stand Test—Weight
T — Tonometry
V — Visual Examination

F-30 = 30 days before flight
L+3 = 3 days after landing

From Fischer and Degioanni, 1981

The STS-1 recovery medical examination showed little effect of space flight, as was anticipated for a mission lasting only two days and six hours. The crew did exhibit the expected hyperreflexia and dependent venous stasis (Fischer & Degioanni, 1981). The STS-2 mission, abbreviated to two days because of fuel cell problems, also showed nominal change. In this case, the most notable finding was that both crewmen appeared fatigued and under-hydrated (LaPinta & Fischer, 1982). No residual mission effects were observed during the next medical examination three days later.

Inflight Monitoring

An important feature of medical surveillance during space missions is the inflight monitoring of telemetered biomedical data. During the course of the U.S. and Soviet space programs, sophisticated and highly reliable biotelemetry devices have been developed. These allow real-time comprehensive monitoring of biomedical status and provide critical data for medical personnel in mission control centers. Table 5 shows the medical information collected, some of which was telemetered directly and some held for later transmission, and the techniques used to obtain this information.

Table 5

Inflight Biomedical and Performance Measures

Parameter Measured	Techniques Employed	
	American Missions	Soviet Missions
Cardiac activity and circulation	2-lead ECG with synchronous phonocardiography, vector-cardiography, cardiotachography, blood pressure-pressure cuffs and automatic measurement of tones, leg plethysmography during LBNP tests, venous compliance	Continuous 1-lead ECG, periodic 12-lead ECG, seismocardiography (myocardial contractility, kineto-cardiography, blood pressure-pressure cuffs, tachooscillography and other measurements, sphygm-ography, rheoencephalography, and cardiac output (Bremser-Ranke)
Hematology	Hemoglobinometry, venus blood collection with separation and preservation	
Respiration	Impedance pneumography, spirometry, gas exchange	Perimetric pneumography, pulmonary volumes, gas exchange
CNS, sensory function, and performance	EEG, sleep analysis with EOG, voice communication, vestibular tests in rotating chair, overall task performance, otolith test goggles, rod and sphere (for spatial orientation)	EEG, EOG, voice communication, vestibular tests, psychophysiological tests, overall task performance
Metabolism	Body mass measurement, biosampling, bicycle ergometry, metabolic analyzer (O_2 con-sumption, CO_2 production), body temperature (ear probe)	Biosampling

From Berry, 1975

The value of telemetered biomedical data as an aid in evaluating crew member status was demonstrated vividly during the Apollo 15 mission. At that time, arrhythmias were noted while the astronauts were on the lunar surface and, again, during the return flight to Earth. Bigeminies and premature auricular and ventricular contractions were seen. These arrhythmias have been linked to potassium deficits and excessive workloads (Johnston & Hull, 1975). The finding of these unexpected cardiac events led to a comprehensive review of nutritional requirements and work scheduling for subsequent lunar exploration missions.

The inflight medical monitoring of crews of the Space Transportation System continues but at a lesser level. Concern over man's ability to withstand the stresses of space flight has diminished following the success of earlier programs. For this reason, minimum biomedical instrumentation is used during STS missions. Crew members are supplied with an Operational Bioinstrumentation System, which provides two channels for electrocardiographic data. Monitoring during initial flights of the Space Shuttle is done by recording the ECG during pre-launch, launch, early orbit time, entry, and landing. Other information is obtained by monitoring crew voice transmissions and through use of a daily private crew medical communication channel.

As STS missions mature, ECG data will be monitored routinely for any crew member performing extravehicular activities. For at least the initial flights using expanded crews, the ECG channels will be used to monitor the Mission Specialist and Payload Specialist positions during ascent and recovery.

Program Development

The Space Transportation System, with multiple Space Shuttle Orbiters in a continuing rotation pattern with frequent flights, is ambitious and will continue to grow over the next decade. As this program increases in scope, so will the need for a carefully tailored medical support program. To this end, procedures used for medical oversight and for the provision of health care services are in a program of continuing review and update.

A computerized medical management and data storage system has been installed by NASA. With each STS flight, additional medical information is added to the medical data base as a means of initiating as well as verifying selected changes in STS health care services.

References

Berry, C.A. Medical care of spacecrews (Medical care, equipment, and prophylaxis). In M. Calvin and O.G. Gazenko (General Eds.), *Foundations of space biology and medicine* (Vol. III) (NASA SP-374). Washington, D.C.: U.S. Government Printing Office, 1975.

Ferguson, J.K. Health stabilization program. In S.L. Pool, P.C. Johnson, and J.A. Mason (Eds.), *STS-1 medical report* (NASA TM-58240). Washington, D.C.: National Aeronautics and Space Administration, Scientific and Technical Information Branch, December 1981.

Ferguson, J.K. Health stabilization program. In S.L. Pool, P.C. Johnson, and J.A. Mason (Eds.), *STS-2 medical report* (NASA TM-58245). Houston, TX: Lyndon B. Johnson Space Center, May 1982.

Ferguson, J.K., McCollum, G.W., & Portnoy, B.L. Analysis of the Skylab flight crew health stabilization program. In R.S. Johnston and L.F. Dietlein (Eds.), *Biomedical results from Skylab* (NASA SP-377). Washington, D.C.: U.S. Government Printing Office, 1977.

Fischer, C.L., & Degioanni, J. Evaluation of crew health. In S.L. Pool, P.C. Johnson, and J.A. Mason (Eds.), *STS-1 medical report* (NASA TM-58240). Washington, D.C.: National Aeronautics and Space Administration, Scientific and Technical Information Branch, December 1981.

Hawkins, W.R., & Zieglschmid, J.F. Clinical aspects of crew health. In R.S. Johnston, L.F. Dietlein, and C.A. Berry (Managing Eds.), *Biomedical results of Apollo* (NASA SP-368). Washington, D.C.: U.S. Government Printing Office, 1975.

Johnston, R.S., & Hull, W.E. Apollo missions. In R.S. Johnston, L.F. Dietlein, and C.A. Berry (Managing Eds.), *Biomedical results of Apollo* (NASA SP-368). Washington, D.C.: U.S. Government Printing Office, 1975.

LaPinta, C.K., & Fischer, C.L. Evaluation of crew health. In S.L. Pool, P.C. Johnson, and J.A. Mason (Eds.), *STS-2 medical report* (NASA TM-58245). Houston, TX: Lyndon B. Johnson Space Center, May 1982.

National Aeronautics and Space Administration, *Health stabilization program for the Space Transportation System* (JSC 14899). Houston, TX: Lyndon B. Johnson Space Center, May 1981.

Wooley, B.C., & McCollum, G.W. Flight crew health stabilization program. In R.S. Johnston, L.F. Dietlein, and C.A. Berry (Managing Eds.). *Biomedical results of Apollo* (NASA SP-368). Washington, D.C.: U.S. Government Printing Office, 1975.

Ferguson, J.K. Health stabilization program. In S.L. Pool, P.C. Johnson, and J.A. Mason (Eds.), *STS-1 medical report* (NASA TM-58240). Washington, D.C.: National Aeronautics and Space Administration, Scientific and Technical Information Branch, December 1981.

Ferguson, J.K. Health stabilization program. In S.L. Pool, P.C. Johnson, and J.A. Mason (Eds.), *STS-2 medical report* (NASA TM-58245) Houston, TX: Lyndon B. Johnson Space Center, May 1982.

CHAPTER 18

COUNTERMEASURES TO SPACE DECONDITIONING

Previous chapters have described physiological adaptive changes which occur in both short- and long-duration space missions. Some of these adaptations, such as motion sickness, are self-limiting; others produce progressive changes in different body systems. These changes occasionally affect crew performance inflight. However, with continuing technological advances in space systems design, more concerns have been raised regarding the adaptation of individuals to weightlessness and their ability to cope with complex and demanding tasks while physiologically stressed. Therefore, a search for countermeasures is considered to be of high priority in order to enhance performance inflight or upon return to one gravity. This chapter is an attempt to summarize progress made in the development of countermeasures to counteract space deconditioning by both the U.S. and USSR manned programs.

Maintenance of Physical Condition

Two of the most immediate and significant effects of weightlessness are the headward shift of fluids, and the removal of weight forces from bone and muscle. These changes lead to a progressive degradation of cardiovascular and musculoskeletal system conditioning by Earth's standards. Cardiovascular deconditioning is manifested by a postflight orthostatic intolerance, decreased cardiac output, and reduced exercise capacity. Musculoskeletal system changes, brought about by hypodynamia and the absence of gravitational forces, are mostly reversible, but they contribute substantially to weakness and poor gravitational tolerance in the postflight period. Both forms of deconditioning may also severely impair the ability of a weightlessness-adapted individual to function and perform adequately during the critical phases of reentry and landing.

Because the underlying factor producing the changes leading to both cardiovascular and musculoskeletal deconditioning is the absence of gravity, the effort to reduce these deconditioning effects has been primarily focused on restoring weight forces on the body and simulating Earth-normal physical movements, stresses, and system interactions to the greatest extent possible. The most direct approach would be the generation of artificial gravity inside the spacecraft. Thus far this has not been deemed practical, for both technical and economic reasons. On a smaller scale, the use of onboard centrifugation has been considered (Vil'-Vil'yams & Shulzhenko, 1980), but this requires significant technological and engineering modification to spacecraft systems, and probably offers far less physiological advantage. The single approach which so far has received wide operational acceptance in the U.S. and USSR space programs is exercise, particularly if combined with the periodic application of low-level lower body negative pressures (LBNP) (up to -30 torr). The provision of adequate caloric and nutrient content in the space diet constitutes another significant countermeasure, particularly for counteracting the losses of muscle mass and electrolytes that are important for the general metabolic regulation. Another potential countermeasue is the use of drugs to help maintain homeostasis. However, this approach has not yet received wide acceptance, since only minimal supporting research in space has been conducted in this area.

Exercise

Exercise Techniques

Over the course of both the Soviet and American space programs, a variety of exercise techniques and devices have been employed. The development of exercise as a countermeasure has been primarily a search for the best possible compromise among efficacy, equipment size, ease of use, and operational time requirements. Since complete maintenance of one-gravity physical condition is not possible, the identification of functional capabilities that are compatible with different missions is also an important factor. It has been suggested that different types of exercise offer different degrees of protection and efficacy (Thornton, 1981). Thus, isotonic and isokinetic exercises work well in preventing mass and strength losses in certain muscle groups. Quiet standing under gravitational loading (by means of straps) might afford some protection against loss of bone mineral. Walking under the same applied force appears to reverse muscle atrophy and improve coordination. Endurance exercises such as pedaling a bicycle ergometer prevent reductions in heart size and mass as well as respiratory capacity; and may also increase circulating blood volume. However, the latter approach (which has been used whenever possible on long-term flights since the early 1970's) did not appreciably prevent mineral or muscle mass losses.

In a recent review, Thornton (1981) has suggested that utilizing a combination of exercise techniques and devices when applying one-gravity equivalent forces (e.g., on a passive treadmill) is the most effective way to reduce muscle, joint, and bone atrophy; minimize reductions in heart size/mass; and maintain coordination and exercise capacity. Table 1 summarizes the adaptation of different bodily systems/functions to weightlessness and the recommended preventive exercise techniques.

Although the type of exercise best suited to the maintenance of different physiological capabilities can be fairly well specified, what is more uncertain is the optimum amount of daily exercise. Table 2 gives an estimate of time required to maintain various one-gravity capacities, based on maintaining average non-athletic performance on return.

Experience Gained From Inflight Exercise Programs

Exercise regimens prescribed for space missions have required gradually longer and more frequent periods of exercise—particularly as the length of missions has increased. On the first prolonged (18-day) Soviet manned flight, Soyuz 9, physical exercises were performed by the cosmonauts for two one-hour periods each day. In a subsequent 24-day flight, 2.5 hours of exercise per day was employed, including walking/running on a treadmill. By 1975, the standard program involved three exercise periods per day, with a variety of equipment for a total of 2.5 hours, with the selection of exercises on the fourth day being optional (Dodge, 1976). Over the three missions of the Skylab program a similar increase in exercise quantity was imposed (0.5, 1.0, and 1.5 hours/day), although the total amounts were less than those used by the Soviets. On the third mission, Skylab 4, a treadmill was provided which allowed more vigorous exercise (Hordinksy, 1977).

266

Adaptation	Prevention

MUSCLE

Loss of Muscle Strength and Mass

Arms

Arms of normal strength are relatively protected by usual activities in flight. There may be emergency situations on return to Earth that require significant strength.

Two or three sets of 6 to 12 maximum repetitions every other day should maintain or increase strength depending upon prior training of crewman. The forces should be equivalent to weight and, of course, must be properly oriented as to direction. Isometric exercises have not proven realistic for training for isotonic work. Isokinetic loading may be acceptable.

Legs

Relatively rapid loss of strength with expected decrease to levels incompatible with ambulation in 1-g. There is an associated loss of muscle mass and probably of resting tension (tone).

Legs and Trunk

The primary normal functions of legs are standing and walking, hence, walking/running under properly distributed 1-g loads can be considered exercise par excellence for legs and trunk. A series of other exercises might be devised to maintain strength in various groups, but this would be inefficient. It has been amply demonstrated that strength cannot be maintained by repetitive low level exercises, especially in a non-walking mode, such as the bicycle ergometer.

Endurance

Arms

The arms will typically not be required to exert prolonged isometric or isotonic forces except in EVA suit work, in which high counter forces to many motions are generated by the suit, and in EVA work situations which may include repetitive manual actuation of mechanisms, e.g., secondary erection mechanism of the space telescope.

In situations requiring endurance training, the training motions chosen should be those required by the task at hand. In EVA suit work, the forces are not those of weight in 1-g ($F = Km + mX$), but are primarily of the form, $F = K_r X^n$ or $F = K_r$ when $X > 0$, i.e., resistive forces with continuous reaction against motion qualitatively similar to movement through a viscous medium, such as movement of a paddle through molasses. The body responds specifically to the type of load trained with; hence, load magnitude, type, duration, and direction must be specified. Further, this training must be done preflight. On short missions, no inflight training will be required.

Legs

Although not measured specifically, leg and muscle endurance may be expected to decrease more rapidly than the strength function. Walking in 1-g is an endurance exercise for the average individual.

Duration of walking required to maintain 1-g capacity is unknown but obviously will be at the level of 1-g strength capacity required. An estimate of duration required to maintain the typical endurance level in 1-g is 30 minutes per day maximum at an equivalent of 6-7 mph, with 1 to 2 days per week without such exercise. Walking/running are endurance exercises,

From Thornton, 1981.

267

and while it might be theoretically possible to devise a series of synthetic exercises for various muscle groups, this is not nearly as practical as walking/running itself. Again, muscle endurance at a given load cannot be maintained by much smaller loads no matter how many repetitions are made.

SKELETON

Significant calcium and bone losses occur in weightlessness. There are reports and indications that this is self-limited, but possibly at a much reduced strength level. Evidence from 1-g studies shows that 1-g weight-bearing activity with *intact muscle mass* will prevent such losses.

Although unproven, the preponderance of clinical evidence is that weight-bearing 1-g equivalent exercise will prevent mineral loss in weightlessness and that anything less will allow some degree of adaptation. Again, the archetype of such exercise is walking/running. The critical question is how much time per day is required, and this cannot be answered. Extrapolating from the bedrest standing studies, it could be as low as 1 hour per day.

JOINTS AND TENDONS

There is no direct evidence available on weightless effects; however, if these areas follow the usual trend, both could be compromised. Again, the areas which would be expected to suffer most are the lower back and legs, along with the principal functions served by these areas.

Walking/running under 1-g equivalent loads should prevent any loss of integrity in these areas. The levels and duration required to maintain leg strength and endurance should also be adequate to maintain these closely related functions.

NEUROMUSCULAR COORDINATION

The chief area of concern is walking, and while virtually all areas of neuromuscular coordination may be expected to suffer, walking is the most obvious and probably the most likely to cause injury.

Walking under 1-g equivalent loads for periods of minutes per day should be adequate to maintain coordination for normal purposes.

CARDIOVASCULAR

Blood and Fluid Shifts

Liters of fluid are shifted from the lower body and limbs within 48 to 72 hours. A small portion of this fluid is excreted and the remainder sequestered in the upper body. Plasma volume is reduced by several hundred ml within the same time frame and red cell mass is then adjusted over a period of weeks such that blood volume is reduced. Some of these effects are reversed within 24 to 72 hours on exposure to 1-g but are present on reentry.

Whether or not blood/fluid shift and loss can be *partially* compensated by endurance exercises of major muscle masses is an open question. If one is willing to expand the concept of exercise to include any physical modality to train or stress the body, then an exercise/technique capable of shifting blood and fluid from the upper to the lower body for appreciable time is required.

Adaptation	Prevention

Cardiorespiratory Endurance

The reduced blood/fluid volume, increased compliance, and changes in muscle circulatory and metabolic patterns all combine to reduce maximum work capacity and endurance. These are further degraded by reductions in heart capacity and size and by changes in neuro-endocrine responses.

Any system which can cause prolonged large elevations of oxygen intake will generally provide the stress for cardiorespiratory endurance. Treadmills and bicycle ergometry are the two most commonly used means. The time required to maintain 1-g equivalent endurance will obviously be a function of the condition of the individual. For example, a 50 miles per week runner will need at least that in weightlessness, clearly an impractical situation.

Musculovascular

Although unproven, evidence indicates that the tone of a muscle is related to its strength and endurance or possibly to changes in strength/endurance; i.e., a muscle losing strength and mass may have decreased tone, and vice versa. In addition, if venous and somatic muscle reflexes play a role, then some method of distending veins might be useful.

Strength and presumably tone can be maintained by walking exercises. If one extends the definition of exercise, then the same schemes to shift blood into the lower body and leg vessels could be used here, except that techniques should be varied to stimulate any rate-of-change receptors that might be present.

Table 2

Exercise Time Required to Maintain One-Gravity Capacities

Physical Parameter	Exercise Modality	
	Walk	Jog
Leg Strength	15-20 min/day	5 min/day
Leg Endurance	25-30 min/day	15-20 min/day
Cardiorespiratory Endurance	30 min/day	15-20 min/day
Bone Strength	1-2 hrs/day	30-90 min/day
Coordination	10 min/day	5 min/day

1-g Equivalent Hydrostatic Forces on Leg Vessels

Blood & Fluid Redistribution & Loss	1-3 hrs/day in addition to above exercises. This should be a static exercise which would not interfere with other duties.

From Thornton, 1981.

Throughout the Skylab missions, successive improvements were seen in postflight leg strength and volume changes, orthostatic tolerance and recovery time, and cardiac output and stroke volume, even though each mission lasted four weeks longer than the last (Henry et al., 1977; Thorton & Rummel, 1977). Results of exercise on Soviet missions have shown a similar pattern of reduced physiological deconditioning in response to more strenuous exercise programs (Gazenko, Genin, & Yegorov, 1981; Yegorov, 1980). These have become so successful that, in flights of six months' duration aboard the Salyut 6 station some parameters, such as exercise capacity, did not change from their preflight values. Body mass has tended to increase linearly during such missions— a result which can be attributed partly to exercise, particularly in view of the fact that leg volumes have typically increased toward the end of the missions. In some instances, the response of heart rate and arterial pressure to lower body negative pressure tests inflight has remained essentially unchanged from baseline values. Although other cardiovascular parameters, such as cardiac ejection and filling times, output and pressure both at rest and during stress, have continued to be affected by weightlessness, overall physical fitness has nevertheless been adequately maintained.

Anti-Deconditioning Devices and Technology

A wide variety of onboard equipment has been developed for use in weightlessness. Most of these are related to exercise, and all are directed at the maintenance of physical conditioning.

Ergometer

This device is a bicycle-like apparatus which can be pedaled with either the hands or the feet (Figure 1). During work a variety of parameters of cardiovascular and metabolic activity can be monitored. Electrodes and a blood pressure cuff are attached to the person using this device, and ventilation and aerobic capacity can be estimated (Sawin et al., 1975). Similar devices were present onboard Skylab and all Soviet space stations from Salyut 4 on.

Treadmill

Both Soviet and American space stations have also featured a treadmill (Figure 2) on which astronauts and cosmonauts could exercise the lower limbs more thoroughly than on the ergometer (Dodge, 1976; Hordinsky, 1977). In the Soviet program this device was present on all Salyut space stations; on Skylab it was present only on the final mission, Skylab 4, but it is included onboard the Space Shuttle (Figure 3). In both the Soviet and American versions, elastic "bungee cord" straps are used to press the individual down onto the treadmill and provide simulated gravitational force (1.1 g on Skylab; 0.62 g on Salyut).

"Penguin" Suit

Soviet cosmonauts are provided with a special elasticized suit which provides passive stress on antigravity muscle groups. This constant-loading, or "Penguin" suit (Figure 4), is worn by cosmonauts not only during exercise (the tie-down straps used with the treadmill can be attached to it), but also during the entire working day. It provides partial compensation for the absence of gravity by opposing movement, and functions as a constant gravitational load on muscles of the legs and trunk (Yegorov, 1981).

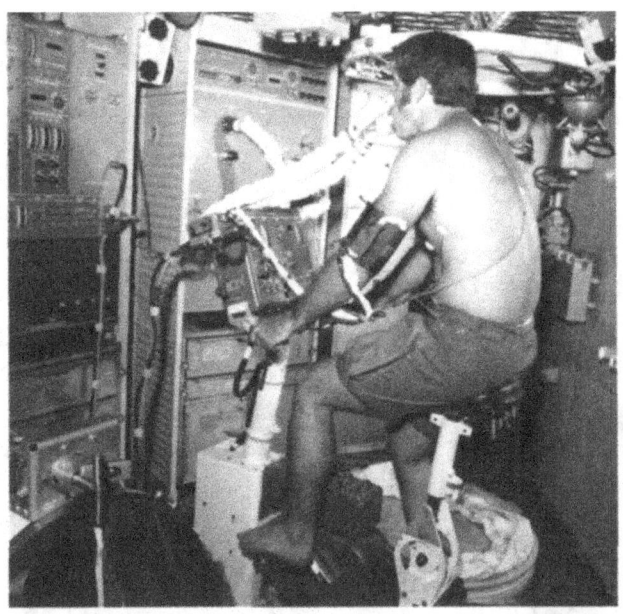

Figure 1. The ergometer device used aboard Skylab. This apparatus permits both exercise and metabolic/cardiorespiratory analysis.

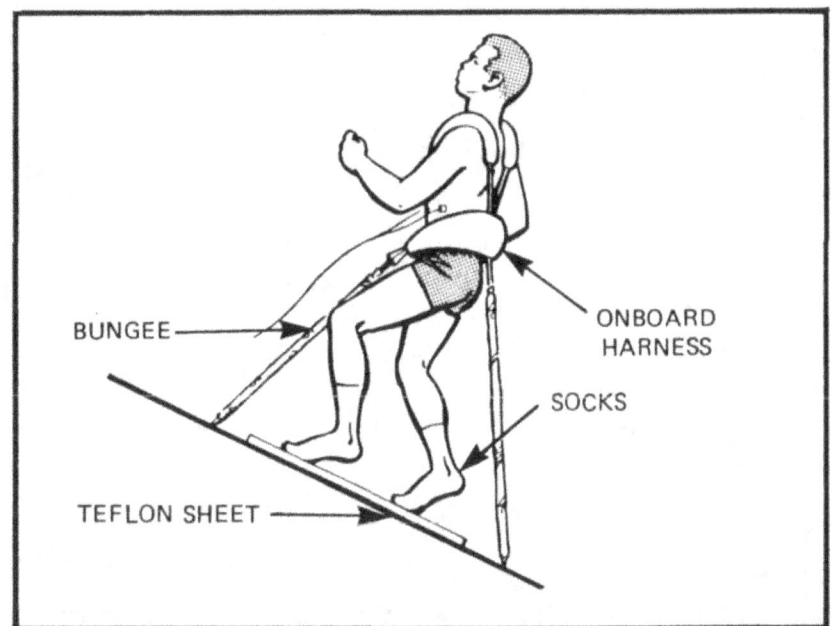

Figure 2. Representation of a passive treadmill used aboard Skylab 4, and similar to that used aboard Salyut. Elastic bungee cords provide simulated gravitational loading on the user. (From Thornton, 1981)

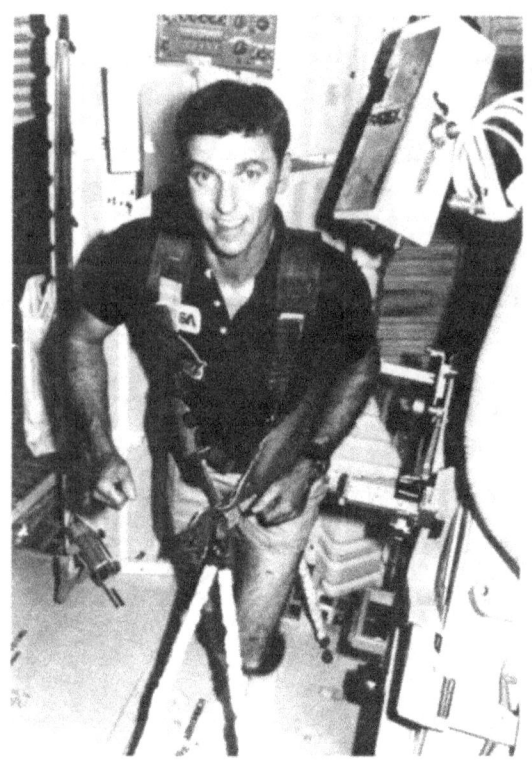

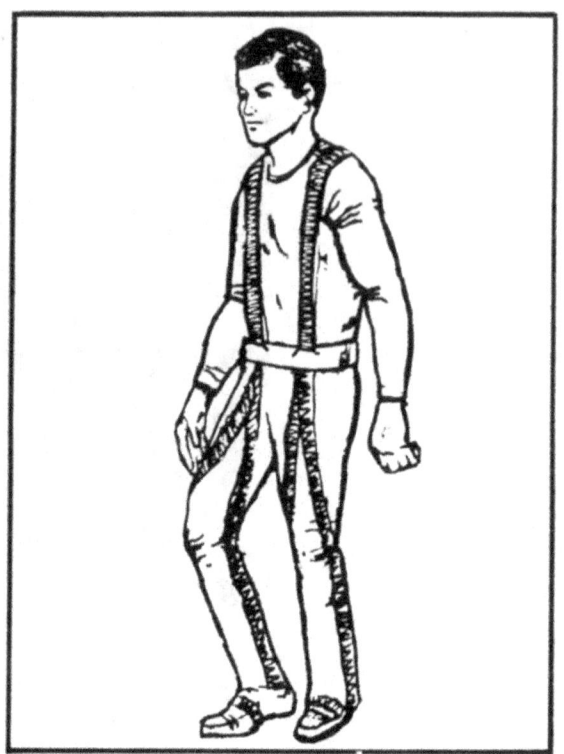

Figure 3. The STS-2 Commander using the
Space Shuttle treadmill.

Figure 4. The "Penguin" constant-loading
suit worn by cosmonauts.

Lower Body Negative Pressure

Lower body negative pressure, or LBNP, partially reverses headward fluid shifts, thus minimizing cardiovascular deconditioning. LBNP can also be used to assess the level of deconditioning during a mission, therefore predicting the degree of postflight orthostatic intolerance. The LBNP chamber on Skylab (Figure 5) was used for this purpose (Johnson et al., 1977). The Soviets use LBNP for this purpose and as an inflight countermeasure to cardiovascular deconditioning. The "Chibis" vacuum suit (Figure 6) can be used while sitting or standing, thus freeing a cosmonaut for other tasks. LBNP sessions in the Chibis are scheduled toward the end of long-term missions in order to subject the cardiovascular system to sudden fluid shifts, and thus to help restore orthostatic tolerance upon return to Earth. Typical usage is once every four days for 20 minutes (five minutes at each of four pressure levels), and for 50 minutes per day in the final two days of a mission (Gazenko et al., 1981). At the end of a mission, about 300 ml of water is consumed during the application of LBNP to restore fluid volume lost as a result of space flight.

Figure 7 shows an ambulatory LBNP suit developed by NASA for possible future use. It has not been evaluated in flight as yet.

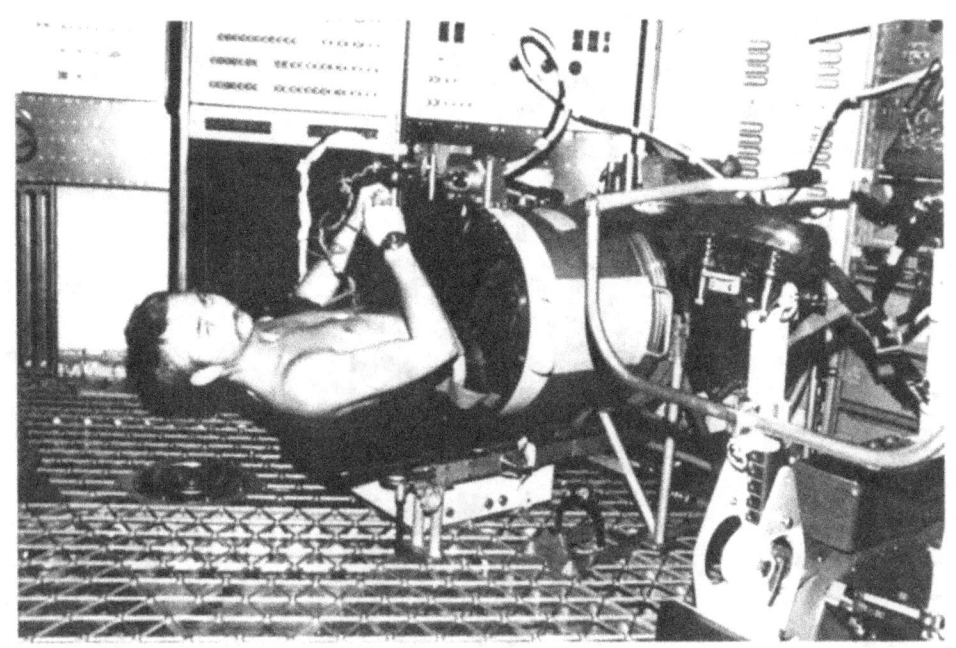

Figure 5. The LBNP apparatus in use aboard Skylab.

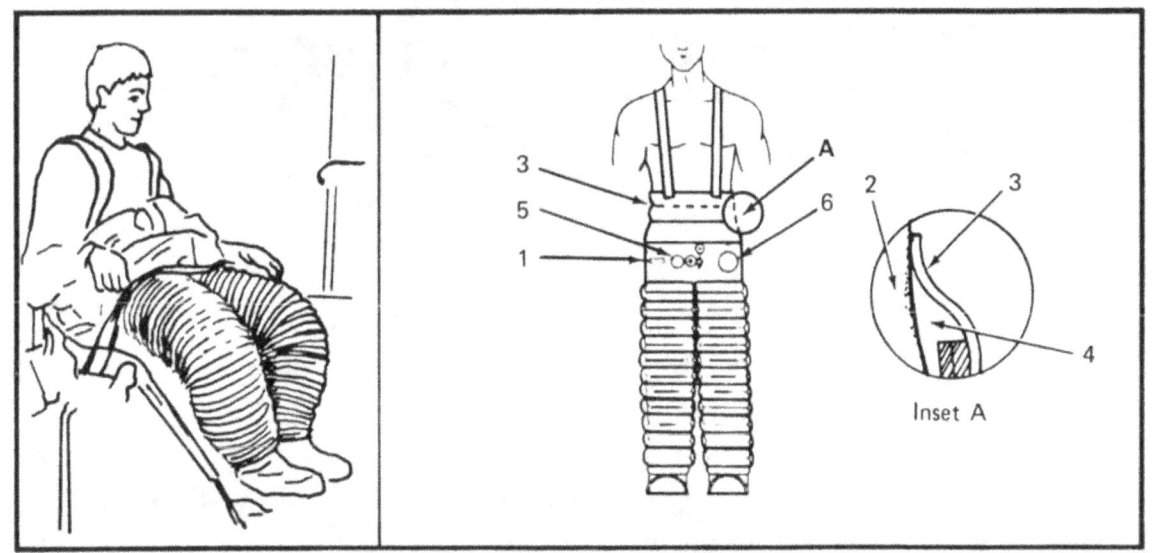

Figure 6. Subject wearing Chibis garment. At right is a general diagram of the Chibis garment. (1) Micropump, (2) body, (3) rubber shutter, (4) air space, (5) air intake valve, (6) manometer.

Figure 7. The NASA experimental ambulatory LBNP suit.

Antigravity Suit

Another countermeasure for minimizing postflight orthostatic intolerance used both by the U.S. and USSR is the antigravity suit (G-suit). Similar in appearance to the Chibis suit, it has somewhat the opposite function, as it increases pressure on the lower body, preventing blood from pooling in the extremities during reentry and postflight and disrupting circulation. The G-suit is found to improve venous return and orthostatic stability postflight (Gazenko et al., 1981).

Electrostimulation

The Soviets employ a variety of procedures in their medical program that have not received wide acceptance by Western space medicine specialists. One of these is electrostimulation of selected muscles, a technique which relies partly on biofeedback mechanisms. Various muscle groups are stimulated by an apparatus called "Tonus," in an effort to reduce disuse atrophy of muscle tissue.

Other Devices

A great many other exercise devices of lesser importance or more limited application have been employed in space. Most of these can be categorized into two standard types: the capstan-type "stretcher" and the elastic "chest expander." Representations of these are shown in Figure 8. Although such devices are extremely lightweight, inexpensive, and easy to use, because of the nature of the forces they generate and the limited groups of muscles they exercise they are considered to be not especially effective (Thornton, 1981). They are useful as accessory exercisers, however.

Figure 8. Two simple forms of loading devices for arm exercise in weightlessness. On the left (A) is a rope/capstan device that allows development of high forces; on the right (B) is a simple elastic device.

Nutrition and Pharmacological Support

Nutritional and pharmacological procedures have been investigated in a number of studies as possible countermeasures to space-induced physiological changes such as fluid shifts, cardiovascular deconditioning, muscle strength and mass losses, and bone mineral losses. Studies dealing with nutritional and pharmacological benefit are difficult to conduct and interpret, however, because of the potential broad spread of effect across many body functions, the many drug side effects, and the fact that the physiological changes themselves are interrelated. Nonetheless, these studies are important in the search for effective countermeasures.

Nutritional Considerations

The need for food—more precisely, the need for nutrient substances—is driven by the body's total energy expenditure as expressed in the activity of muscles, organs, systems, and mental/ nervous processes. This need is regulated by thirst, appetite, digestion, and metabolism, and also by physical work output. Weightlessness results in a substantial loss of the fluids and electrolytes that govern many functions (Leach, 1981). The much-reduced physical work requirement results not only in an altered energy output, but also in a loss of protein nitrogen through muscle atrophy (Ushakov, 1980). The reduction in electromechanical stresses and other factors brings about a loss of calcium from bone. It has been suggested by some investigators that metabolic and digestive processes undergo substantial changes, partly as a result of the altered stress environments and physical confinement (Popov, 1975).

Much effort has been devoted in both the U.S. and Soviet space programs to determining the optimum diet for consumption during space missions. Considerations of storage time, size/weight restrictions, and practicality for consumption in weightlessness at first led to the use of freeze-dried food bars (Figure 9) and purees and juices packages in squeeze tubes. The palatability of these early food items left much to be desired, which meant that the intended quantities were sometimes not consumed. In addition, early estimates of energy requirements on space missions were unrealistically low; nor were metabolic changes adequately taken into account. For these reasons, space diets have undergone a considerable evolution.

Food Variety. The most significant change has been a large increase in the diversity of foods and an improvement in packaging and presentation methods. For example, foods available to Skylab crews included 70 different food items presented as freeze-dried rehydratables, thermostabilized foods, dry and moist bite-sized foods, and a variety of beverages (Johnston, 1977). Hot and cold water were available for rehydration, as well as an oven for heating. Utensils could be used for eating solid foods covered by a plastic membrane flap (Figure 10). Spice packets were provided for the preparation of food to individual tastes. In the Space Shuttle, the Skylab innovations have been augmented by an airliner-like food galley and a pantry, stocked according to crew preference, to supplement the menu (Sauer & Rapp, 1981). The Soviets have taken even larger steps toward the satisfaction of individual preferences, customizing menus on an individual basis from the time of the early Soyuz flights (Dodge, 1976). The advent of the Progress cargo ships in the later Salyut

Figure 9. Typical space food developed for early Apollo missions.

Figure 10. Astronaut reconstituting food during Skylab 3. Note use of normal utensils.

missions meant that fresh fruits, vegetables, and condiments could be periodically supplied to supplement the diet. Crews were encouraged to "order out" for items they wished to eat. In all, the tendency has been to attempt to establish an Earth-normal pattern and quality of meals while meeting energy and metabolic requirements.

Energy Content. On the basis of experience, and particularly with the advent of longer flights and extensive inflight exercise programs, total energy content of the diet in U.S./USSR space programs has been progressively increased (Dodge, 1976; Popov, 1975). In the first Soviet flights, for example, daily caloric intake was about 2,600 kcal. In the first phase of the Soyuz program, it was about 2,800 kcal. By the time of Salyut 1, this had been raised to 2,950 kcal. The Salyut 4 diet provided 3,000 kcal. On Salyut 6, the caloric allowance was 3,150 kcal. Energy content of American diets in space has been somewhat lower, on average—about 2,500 kcal (except in the Apollo lunar landing missions, where it was 2,800-3,000 kcal). Table 3 shows the caloric and nutrient content of a typical Apollo meal. By the time of Skylab, the energy content of the space

Table 3
Typical Composition and Caloric Content
of Apollo Daily Meal

Food Composition of Daily Menu

Meal A	Meal B	Meal C
Fruit cocktail	Chicken salad	Beef stew
Bacon squares	Beef with vegetables	Potato salad
Strawberry cubes	Butterscotch pudding	Sweet pastry cubes
Cocoa	Fruitcake	Grapefruit drink
Orange drink	Pineapple-grapefruit drink	

Food Values

Constituents	Meal A	Meal B	Meal C	Total
Energy (kcal)	759	1123	911	2793
Protein (g)	28.5	45.2	28.7	102.4
Fat (g)	25.4	42.0	32.4	99.8
Carbohydrate (g)	106.4	140.0	125.7	372.1
Ash (g)	7.0	6.8	7.3	21.1
Ca (mg)	176.0	505.0	486.0	1168.0
P (mg)	342.0	712.0	592.0	1646.0
Fe (mg)	3.3	4.8	4.9	13.0
Na (mg)	1659.0	1526.0	1916.0	5101.0
K (mg)	818.0	863.0	1047.0	2728.0
Mg (mg)	64.3	89.5	95.3	249.1
Cl as NaCl (g)	4.30	3.05	3.94	11.29

From Popov, 1975.

diet was equivalent to that of the normal pre-mission diet. The daily menu aboard the Space Shuttle provides 3,000 kcal per person, although individual intake is not necessarily this large. Table 4 shows estimated daily intake aboard STS-1.

Table 4
STS-1 Estimated Mean Daily Inflight Nutrient Consumption
Per Crewman

Meal*	Calories	Protein (g)	Cho (g)	Fat (g)	Ca (mg)	Phos (mg)	Na (mg)	K (mg)	Mg (mg)	Fe (mg)	Zn (mg)
Day 1-L	1006	33.7	114.6	45.8	256	521	2368	706	137	11.4	6.6
D	662	38.8	83.6	18.6	318	458	1219	754	84	5.6	6.5
2-B	1021	33.2	170.4	22.8	766	804	1180	1418	209	17.8	6.7
L	810	45.9	107.2	22.3	275	476	1262	726	123	5.8	7.0
D	884	32.8	112.1	33.8	310	484	1656	1431	116	5.2	6.0
3-B	930	29.1	129.2	22.9	494	670	1328	1440	105	8.4	2.5
Mean/Man/Day	2656	106.8	358.6	83.1	1210	1706	4506	3238	387	27.1	17.6
Recommended Levels:											
JSC	3000	56			800	800	3450	2737	350	18	
FDA		56			800	800			350	10	15

*B = Breakfast, L = Lunch, D = Dinner.

From Sauer and Rapp, 1981.

Dietary Supplements. Currently, attention is being paid to the diet and its potential impact on higher cerebral function (via the neurotransmitters) and performance. This research is in the early stage of experimentation, and further development and validation of this approach utilizing specially devised diet composition with respect to amino acids needs to be carried out (Fernstrom, 1981). Soviet scientists feel that administration of vitamins, amino acids, and minerals promotes the retention of fluids (and, thus, electrolytes). These preparations are administered in larger doses just prior to reentry in order to facilitate the readaptation process (Yegorov, 1981). A lower level of potassium in the diet has been implicated in the cardiac arrhythmias and long postflight recovery period observed on Apollo 15; potassium supplements appeared to prevent these problems on subsequent missions (Berry, 1981).

Drugs

Pharmaceutical compounds are used in a variety of ways as countermeasures to the physiological effects of space flight. The most significant of these is their use in preventing the symptoms of space motion sickness. (For a detailed treatment of this subject, see Chapter 8.) Medical kits carried aboard manned spacecraft also include supplies of routine medications such as aspirin, antihistamines, stimulants, and sedatives, as well as a variety of emergency medications; but these are not intended for use as countermeasures to the problems discussed in this chapter. Although their research programs in this area are similar, American and Soviet biomedical specialists have adopted somewhat different approaches to use of drugs as countermeasures in space flight.

Cardiovascular Deconditioning. More so than their American counterparts, Soviet biomedical researchers favor the use of drugs as a countermeasure to cardiovascular deconditioning. In conjunction with the Penguin, Chibis, and G-suits, and with exercise, cosmonauts have used such drugs as ephedrine, papazol, isoptin, ritmodan, novocainamid, cytochrome C, Aldactone, Nitrol, and mexitil to control cardiovascular disturbances due to weightlessness. At different times drugs such as tropaphen, phenoxybenzamine, anapriline, octadine, carbocromene, and papaverine, separately or in combinations, have been used during missions to improve adrenergic regulation of blood circulation. Attempts have been made to normalize the water-sodium metabolism by administration of vasopressin, pitressin, desoxycorticosterone acetate, and nerobol (Shaskov & Yegorov, 1979). Antidiuretic hormone is sometimes administered to control fluid loss. In addition, as was mentioned earlier, isotonic solution loading prior to reentry is used to reduce postflight orthostatic intolerance.

Bone Mineral Loss. In view of the seemingly progressive loss of calcium, considerable research has been directed at finding pharmaceutical means of halting this process. Soviet efforts in this area have so far been unsuccessful. However, American researchers have recently reported promising results with clodronate disodium, a diphosphonate compound which appears to prevent hypercalciuria during bedrest (Dietlein & Johnston, 1981).

References

Berry, C.A. Physiological and psychological parameters of life in space stations. *Spaceflight*, 1981, *23*(1): 23-26.

Dietlein, L.F., & Johnston, R.S. U.S. manned space flight: The first twenty years. *Acta Astronautica*, 1981, *8*(9-10):893.

Dodge, C.H. The Soviet space life sciences (Ch. 4). In *Soviet space programs, 1971-1975: Overview, facilities, and hardware, manned and unmanned flight programs, bioastronautics, civil and military applications, projections of future plans* (Vol. 1). Washington, D.C.: Library of Congress, Science Policy Research Division, Congressional Research Service, 1976.

Dodge, C.H. The space life sciences (Ch. 8). In *United States civilian space programs, 1958-1978*. Washington, D.C.: Congressional Research Service, Science Policy Research Division, January 1981.

Fernstrom, J.D. Effects of the diet on brain function. *Acta Astronautica*, 8(9-10): 1035-1041.

Gazenko, O.G., Genin, A.M., & Yegorov, A.D. Major medical results of the Salyut 6-Soyuz 185-day space flight. Preprints of the XXXII Congress of the International Astronautical Federation, Rome, Italy, 6-12 September 1981.

Henry, W.L., Epstein, S.E., Griffith, J.M., Goldstein, R.E., & Redwood, D.R. Effect of prolonged space flight on cardiac function and dimensions. In R.S. Johnston and L.F. Dietlein (Eds.), *Biomedical results from Skylab* (NASA SP-377). Washington, D.C.: National Aeronautics and Space Administration, 1977.

Hordinsky, J.R. Skylab crew health—crew surgeon's report. In R.S. Johnston and L.F. Dietlein (Eds.), *Biomedical results from Skylab* (NASA SP-377). Washington, D.C.: National Aeronautics and Space Administration, 1977.

Johnson, R.L., Hoffler, G.W., Nicogossian, A.E., Bergman, S.A., Jr., & Jackson, M.M. Lower body negative pressure: Third manned Skylab mission. In R.S. Johnston & L.F. Dietlein (Eds.), *Biomedical results from Skylab* (NASA SP-377). Washington, D.C.: National Aeronautics and Space Administration, 1977.

Johnston, R.S. Skylab medical program overview. In R.S. Johnston & L.F. Dietlein (Eds.), *Biomedical results from Skylab* (NASA SP-377). Washington, D.C.: National Aeronautics and Space Administration, 1977.

Leach, C.S. An overview of the endocrine and metabolic changes in manned space flight. *Acta Astronautica*, 1981, 8(9-10): 977-968.

Sauer, R.L., & Rapp, R.M. Food and nutrition. In S.L. Pool, P.C. Johnson, Jr., and J.M. Mason (Eds.), *STS-1 medical report* (NASA TM-58240). Washington, D.C.: National Aeronautics and Space Administration, 1981.

Sawin, C.F., Rummel, J.A., & Michael, E.L. Instrumented personal exercise during long duration space flights. *Aviation, Space, and Environmental Medicine*, 1975, 46: 394-400.

Shashkov, V.S., & Yegorov, B.B. Problems of pharmacology in space medicine. *Farmakologiya i Toksikologiya*, 1979, 42(4), 325-339.

Thornton, W. Rationale for exercise in spaceflight. In J.F. Parker, Jr., C.S. Lewis, & D.G. Christensen (Eds.), Conference proceedings: *Spaceflight deconditioning and physical fitness*. Prepared for National Aeronautics and Space Administration under Contract NASW-3469 by BioTechnology, Inc., 1981.

Thornton, W.E., & Rummel, J.A. Muscular deconditioning and its prevention in space flight. In R.S. Johnston and L.F. Dietlein (Eds.), *Biomedical results from Skylab* (NASA SP-377). Washington, D.C.: National Aeronautics and Space Administration, 1977.

Toscano, W.B., & Cowings, P.S. Transference of learned autonomic control for symptom suppression across opposite directions of Coriolis acceleration. Preprints of the 1978 Annual Scientific Meeting of the Aerospace Medical Association, New Orleans, Louisiana, 8-11 May 1978.

Ushakov, A.S. Nutrition during long flight (NASA TM-76436). Washington, D.C.: National Aeronautics and Space Administration, October 1980.

Vil'-Vil'yams, I.F., & Shul'zhenko, Ye.V. Functional condition of the cardiovascular system after 3 day immersion and prophylactic rotation on a short radius centrifuge (NASA TM-76299). *Fiziologiya Cheloveka*, 1980, 6(2): 323-327.

Yegorov, A.D. Results of medical research during the 175-day flight of the third prime crew on the Salyut 6-Soyuz orbital complex (NASA TM-76450). Washington, D.C.: National Aeronautics and Space Administration, January 1981.

SECTION VI

Medical Problems of Space Flight

CHAPTER 19
TOXIC HAZARDS IN SPACE OPERATIONS

Introduction

It has long been known that human exposure to trace levels of many gases can present a threat to health. Submarine operations and certain industrial workplace environments were the traditional spheres of concern with respect to toxic gas contamination. The closed-loop environment of spacecraft presented a new focus for such concerns. As early as the Apollo Program, with the advent of longer space missions, steps were taken to provide adequate protection for crews by eliminating, or at least minimizing, crew exposures to possibly harmful levels of trace contaminant gases and other hazardous material contained in the spacecraft cabin (Rippstein, 1975). Accordingly, from the Apollo Program onward each NASA manned spaceflight program has incorporated a toxicology program as an element of biomedical support. Such programs have had two primary objectives: (1) identification and control of sources of contaminant gases, and (2) control or removal of the gases themselves. This chapter summarizes the toxic hazard issues that are of concern for the Shuttle Toxicology Program, and presents some initial results of Shuttle inflight toxicological sampling and analysis.

Whenever possible or appropriate, the chemicals described in this section include a Threshold Limit Value (TLV), expressed in parts per million (ppm). The TLV specifies the concentration of a particular substance in air which is believed to present no danger of undesirable side effects; it is derived from observations of industrial workers under conditions of repeated daily exposure. TLVs are not poison thresholds, and can be exceeded for short periods under certain conditions. Nevertheless, TLVs do not take into account the conditions of space flight, during which crew members are exposed to potentially toxic substances continuously, rather than repeatedly and for only eight hours each day.

Sources of Toxic Substances

Toxic substances can come from a number of different sources, including: (1) leaks or spills from storage tanks, (2) volatile metabolic waste products of the crew, (3) particulate pollutants which are not easily removed from the air under conditions of weightlessness, (4) volatile components of spilled food, (5) leaks from environmental or flight control systems, (6) thermal reaction products produced by small electrical fires or contaminated removal systems, and (7) outgassing of cabin construction materials. All of the nonmetallic materials used in the interior of the Orbiter crew compartment are known to outgas contaminant compounds. Some specific sources are electrical insulation, paints, lubricants, adhesives, and the degradation of nonmetallic console and equipment structures. The heat produced by equipment operation increases outgassing. However, outgassing presents less of a problem in the Shuttle than in previous spacecraft because the atmospheric pressure is higher (14.7 psi). Earlier spacecraft had reduced-pressure atmospheres, a condition which accelerates outgassing. The Environmental Control and Life Support Systems

(ECLSS) of both the Space Shuttle and Spacelab have been designed so as to reduce trace contaminant gases to acceptable levels (Rippstein, 1981).

Fluids and Gases Aboard the Space Shuttle

A novel feature of the Space Shuttle in terms of U.S. manned spaceflight experience is its ground landing capability. This is significant because the transition from reaction jet to aerodynamic control and then to landing depends on spacecraft systems using fluids and gases with toxic potential. These materials are present in storage tanks and could present problems because of leaks during normal operations or from container rupture during a crash and rescue episode. Figure 1 shows the location of storage tanks on the Orbiter. The amount of material in each tank at landing depends upon whether or not the Orbiter completed the full flight schedule. Exposure to each of these substances presents individual toxicological problems. A description of those which present the most significant potential health hazards follows:

Ammonia. Storage tanks are located in the aft-fuselage, and are a part of the Shuttle ECLSS. A colorless vapor with a pungent odor, ammonia is a powerful irritant to the eyes and to mucous membranes of the upper respiratory tract. Symptoms of exposure include irritation of the eyes, conjunctivitis, swelling of the eyelids, irritation of the nose and throat, coughing, dyspnea, and vomiting. Liquid anhydrous ammonia produces severe burns on contact (TLV = 25 ppm, 18 mg/m^3).

Liquid and Gaseous Oxygen. The main health hazard associated with oxygen is fire. An oxygen-enriched atmosphere can be ignited by a spark. In the liquid state, oxygen can produce extensive "burns" on skin contact. Oxygen tanks are located in the mid-fuselage, and are part of the ECLSS. (No TLV—however, upper limit is 100 percent for six hours at one atmosphere, lower limit = 19%).

Freon 21. Storage tanks for this colorless, odorless, nonflammable gas are located in the mid- and aft-fuselage. Moderate concentrations in the atmosphere can produce lightheadedness, giddiness, shortness of breath, and narcosis. The TLV is <1000 ppm. Concentrations above this value may cause cardiac arrhythmia. If the exposed individual is having difficulty breathing, adrenaline or similar drugs should not be administered because of the possibility of initiating irregular heartbeat.

Flourinert FC-40 (Fluorocarbon). This fluorinated liquid is used as a dielectric coolant in the fuel cells of the Electrical Power System. At normal ground temperatures and pressures it presents little or no health hazard, but exposures to temperatures of 600° F may produce toxic products.

Helium. Helium tanks are located in several places on the Orbiter. The inert, nonflammable, colorless and odorless gas acts as a simple asphyxiant in concentrations where the oxygen level is reduced to less than 15 percent.

Hydrazine (N_2H_4) and Monomethyl Hydrazine (MMH). Part of the auxiliary power unit, hydrazine tanks are located in the aft-fuselage. At room temperature, hydrazine is a clear, oily liquid with an odor resembling ammonia. In the liquid or vapor form, it is extremely toxic. Liquid

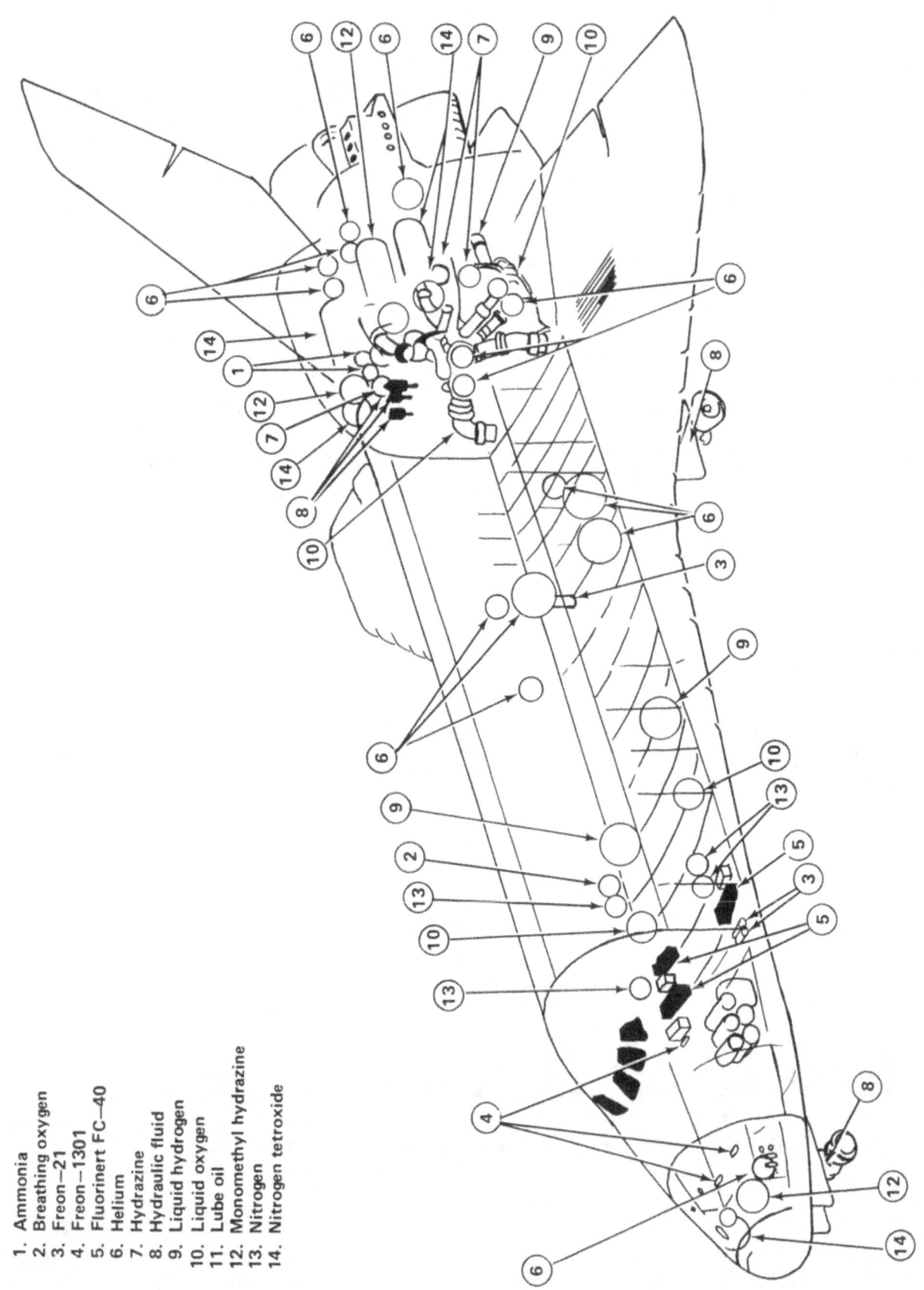

1. Ammonia
2. Breathing oxygen
3. Freon—21
4. Freon—1301
5. Fluorinert FC—40
6. Helium
7. Hydrazine
8. Hydraulic fluid
9. Liquid hydrogen
10. Liquid oxygen
11. Lube oil
12. Monomethyl hydrazine
13. Nitrogen
14. Nitrogen tetroxide

Figure 1. Location of potentially hazardous fluids and gases aboard the Orbiter.

C—4

hydrazine produces severe burns on contact with the skin or eyes; it can penetrate the skin to cause systemic effects similar to those produced when swallowed or inhaled. The vapor causes local (irritation of the eye and respiratory tract) and systemic effects. Hydrazines produce effects on the central nervous system which result in convulsions and death. Rapid therapeutic treatment using barbiturates or barbiturates in combination suppresses the convulsive seizures and provides protection through the acute intoxication phase until the chemical has metabolized (Azar et al., 1970). Repeated to prolonged exposure to hydrazine may cause toxic damage to the liver and kidneys. Pyridoxine, a form of vitamin B, prevents fatty liver changes from hydrazine. The TLV for hydrazine is .1 ppm; the TLV for MMH is .2 ppm. Table 1 presents the emergency exposure limits (EEL) for military and space operations for hydrazine and MMH.

Table 1
Emergency Exposure Limits (Military and Space Operations)

	10 min.	30 min.	60 min.	24 hr.
Hydrazine	30 ppm	20 ppm	10 ppm	—
MMH	90 ppm	30 ppm	15 ppm	1 ppm

From Back et al., 1978.

Liquid Hydrogen. Part of the main propulsion system, liquid hydrogen tanks are located in the aft-fuselage. In the gaseous form, hydrogen acts as a simple asphyxiant. Liquid hydrogen produces serious "burns" on skin contact because of its low temperature.

Nitrogen. Tanks are located in the mid-fuselage, and are part of the ECLSS. At room temperature, nitrogen is an inert, nontoxic, colorless, odorless, nonflammable gas. It acts as a simple asphyxiant when the oxygen level has been reduced to less than 15 percent.

Nitrogen Tetroxide. This highly toxic chemical is used in the Reaction Control Subsystem and in the Orbital Maneuvering System. Skin contact with liquid nitrogen tetroxide results in a yellow stain if momentary, and severe chemical burns if prolonged. In the eyes, the liquid can produce blindness. Inhalation results in irritation of the respiratory tract, and may cause pulmonary edema (TLV = 5 ppm). An unfortunate operational experience with exposure to this chemical occurred when the three U.S. crewmembers of the Apollo-Soyuz Test Project inhaled gaseous nitrogen tetroxide upon reentry. Some pulmonary edema occurred, forcing cancellation of many of the postflight medical tests (Nicogossian et al., 1977).

288

Spacecraft Contaminants

A wide variety of contaminants has been found in the atmosphere of confined spaces during actual or simulated space flights. Analyses of samples taken from Apollo cabin atmospheres indicated the presence of some 300 different compounds (Rippstein, 1981); approximately 400 compounds are outgassed in the Shuttle Orbiter cabin (Rippstein, 1982). It has been very difficult to judge the toxic hazards of these contaminants because their concentrations are usually not known. Many of these substances are commonly found in air, but in the confined atmosphere of the spacecraft they may build up to toxic levels. Table 2 presents some of the contaminants which may pose health hazards, and which have been found in the confined atmospheres of actual or simulated space flights (MacEwen, 1973).

Initial Results of Shuttle Missions

The Shuttle Toxicology Program provides for preflight and inflight atmospheric samplings and analyses during each operational mission. In the case of STS-1, the first Operational Flight Test, atmospheric samples were obtained throughout the 54-hour flight. Four whole gas samples were taken during the mission: at the beginning; two evenly spaced periods in the middle; and at the end. Absorbed gas samples were taken on a continuous basis. In postflight chemical analysis of the samples, a total of 84 compounds were detected. Extrapolation of the data obtained in this mission to a seven-day mission assessment showed that the Orbiter's cabin environment was safe for a flight of that duration (Rippstein, 1981). After the 54-hour STS-2 mission, analysis revealed a total of 99 compounds in the whole gas atmospheric samples (Rippstein, 1982).

Toxicological Aspects of Skylab

Toxicological support for the Space Shuttle benefited from the experience gained during the Skylab missions. Following the loss of Skylab 1's micrometeoroid shield, the interior wall insulation material of the Orbital Workshop overheated. This created a potential toxic hazard because the decomposition of the rigid polyurethane foam produced trace quantities of toluene diisocyanate, and because the overheating might have produced accelerated offgassing of all nonmetallic materials in the cabin. To protect the Skylab 2 crew from potentially toxic air contaminants, several measures were taken. Before the crew entered the space station cluster, a series of pressurization-depressurization cycles designed to discharge and dilute any contaminating gases were conducted. The crew also sampled the air for carbon monoxide and toluene diisocyanate by means of two types of gas analysis detector tubes. The analysis revealed no detectable toluene diisocyanate, and an extrapolated 5 mg/m^3 level of carbon monoxide. The crew energized the ECLSS, which was designed to remove trace levels of contaminants by means of activated carbon filtration. Thirty minutes later the crew entered the space station, and the mission was completed with no more atmospheric trace gas problems.

The Skylab 4 crew took three samples of the atmosphere, on mission days 11, 46, and 77. Analysis of the three samples revealed the presence of more than 300 compounds in the Skylab

Table 2

Potentially Toxic Air Contaminants

Contaminant	Description	TLV (ppm)*	Symptoms of Exposure	Treatment
Allyl alcohol	Limpid liquid, pungent odor	2	Eye and nose irritation, lacrimation, retrobulbar pain, photophobia, blurring of vision; dermatitis.	Flush eyes with water, wash skin with soap and water. If dermatitis occurs, see Chap. 5, Proctor & Hughes (1978).
Ammonia	Colorless gas, extremely pungent odor	25	Eye, nose & throat irritation; dyspnea, bronchospasm, chest pain, pulmonary edema, pink frothy sputum; skin and eye burns and vesiculation.	If liquid is splashed on skin, or eyes, flush with water. If inhaled, consult Chap. 6 (Treatment of Respiratory Irritants), Proctor & Hughes (1978). Obtain chest x-rays and examine for infiltrates. Perform analysis of arterial blood gases. Maintain P_{O_2} above 60 mm Hg by instituting the following measures: 1) administration of 60-100% O_2 by mask or cannula, 2) intubation and mechanical ventilation, 3) positive end expiratory pressure breathing. Fluid balance must be maintained; use diuretic if required. Steroids may be administerd (2-4 days) to decrease the inflammatory response of the lung.
Benzene	Clear, colorless liquid	10	Vertigo, headache, nausea, and coma; increase or decrease in erythrocytes, leukocytes, or platelets; aplastic anemia; coagulation disturbances; excessive bleeding.	Remove from exposure, wash skin areas and irrigate eyes with water. Treat CNS effects symptomatically. If dermatitis occurs, see Chap. 5 (Proctor & Hughes, 1978).
Carbon disulfide	Clear, colorless liquid, nearly odorless	20	Dizziness, headaches, sleep disturbances, fatigue, nervousness, anorexia, weight loss, psychosis; peripheral neuropathy; Parkinson-like syndrome, ocular changes, including enlargement of blind spot and contraction of peripheral field of vision, diminished pupillary reflexes; coronary heart disease, diastolic hypertension; cardiac arrhythmias; gastritis; signs of kidney and liver damage; eye and skin burns from splashes; dermatitis.	Remove from exposure, wash skin areas and irrigate eyes. In severe cases of peripheral neuropathy, institute bedrest and discourage use of affected limb. Prevent contractures by splints and passive stretching. Treat coma symptomatically; give artificial resuscitation and administer O_2 if indicated.

*TLV's are published by the American Conference of Governmental Industrial Hygienists. Values are based on a time weighted average for an 8 hour day. Compiled from Proctor and Hughes (1978).

Table 2 (Continued)

Potentially Toxic Air Contaminants

Contaminant	Description	TLV (ppm)	Symptoms of Exposure	Treatment
Carbon monoxide	Colorless, odorless gas	50	Headache, tachypnea, nausea, weakness, dizziness, mental confusion, hallucinations, cyanosis, depression of the S-T segment of EKG; syncope.	Aggressive administration of O_2 within the limits of toxicity. With 100% O_2, and a tightly fitting mask, the elimination half time is 80 min. Use hyperbaric chamber if possible. Expose to 3 atmospheres absolute, for no longer than 90 min to avoid convulsions. At this level, elimination half time is 23 min. Treat acidosis with intravenous sodium bicarbonate. Cerebral edema may be treated with corticosteroids and diuretics.
1, 2-Dichloroethane (ethylene dichloride)	Colorless liquid pleasant odor, sweet taste	50	Irritation of the eyes and mucous membranes; anorexia, vomiting and epigastric pain; signs of liver and kidney dysfunction such as increased serum bilirubin, albuminuria, and positive Takata-Ara test; dizziness, weakness, and coma; dyspnea, rapid pulse, and cyanosis.	Remove from exposure, wash skin areas, irrigate eyes with water. Perform artificial resuscitation if needed.
Hydrofluoric acid (hydrogen fluoride)	Clear, colorless corrosive fluid or gas	3	Severe eye, nose and throat irritation, delayed fever, cyanosis, and pulmonary edema; severe and painful skin and eye burns from splashes.	Remove clothing and drench individual with copious water. Dress affected areas with soft dressing soaked with iced zephiran or hyamine chloride. (See further procedures under "ammonia".) For severe acid burns, inject small quantities of 10% calcium gluconate into affected tissue.
Hydrogen sulfide	Colorless, flammable gas, offensive odor	10	Apnea, coma, convulsions; irritation of the eyes, including conjunctivitis, pain, lacrimation, photophobia, keratoconjunctivitis, corneal vesiculation; irritation of the respiratory tract, including rhinitis, pharyngitis, bronchitis, and pneumonitis.	Remove from exposure; flush eyes with water. Observe for 72 hrs for signs of pulmonary edema. The use of nitrite-induced methemoglobinemia has been advocated as a treatment for H_2S poisoning. Refer to "Treatment of Respiratory Irritants", Chap. 6, Proctor & Hughes (1978).
Nitrogen dioxide	Colorless solid to yellow liquid	5	Cough, mucoid or frothy sputum production, increasing dyspnea; chest pain; signs and symptoms of pulmonary edema with cyanosis, tachypnea, tachycardia; eye irritation.	Refer to "Treatment of Respiratory Irritants", Chap. 6, Proctor & Hughes (1978). Observe for 72 hrs for signs of pulmonary edema. See further procedures under "ammonia".
Sulfur dioxide	Colorless gas or liquid, pungent odor.	5	Irritation of the eyes, nose, throat, rhinorrhea; choking, cough; reflex bronchoconstriction; eye and skin burns.	Flush eyes and skin with water; if dermatitis occurs, see Chap. 5, Proctor & Hughes (1978). See procedure under "ammonia" for inhalation exposure.

atmosphere. The contaminants in each of the three samples were very similar, indicating that the amounts of individual trace contaminants were not rising throughout the mission. The gas generation rates of the contaminant sources and the removal rate by the ECLSS had attained a state of equilibrium.

The experience in Skylab demonstrated that certain specific toxic hazards can be ascertained and dealt with as they arise, and that a spacecraft's onboard ECLSS is capable of establishing an equilibrium between the generation of air contaminants and the removal rate, at least under normal operating conditions (Rippstein & Schneider, 1977).

References

Azar, A., Thomas, A.A., & Shillito, F.H. Pyridoxine and phenobarbital as treatment for aerozine-50 toxicity. *Aerospace Medicine*, 1940, *41*, 1.

Back, K.C., Carter, V.L., & Thomas, A.A. Occupational hazards of missile operations with special regard to the hydrazine propellants. *Aviation, Space, and Environmental Medicine*, 1978, *49*, 591-598.

MacEwin, J.D. Toxicology. In J.F. Parker, Jr., & V.R. West (Eds.), *Bioastronautics data book, (2nd Ed.)* (NASA SP-3006). Washington, D.C.: U.S. Government Printing Office, 1973.

Nicogossian, A.E., LaPinta, C.K., Burchard, E.C., Hoffler, G.W., & Bartelloni, P.J. Crew health. In A.E. Nicogossian (Ed.), *The Apollo-Soyuz Test Project medical report* (NASA SP-411). Springfield, VA: National Technical Information Service, 1977.

Proctor, N.H., & Hughes, J.P. *Chemical hazards of the workplace*. Philadelphia: J.B. Lippincott, 1978.

Rippstein, W.J. The role of toxicology in the Apollo space program (Ch. 7). In R.F. Johnston, L.F. Dietlein, & C.A. Berry (Eds.), *Biomedical results of Apollo* (NASA SP-368). Washington, D.C.: National Aeronautics and Space Administration, 1975.

Rippstein, W.J. Shuttle toxicology. In S.L. Pool, P.C. Johnson, Jr., & J.A. Mason (Eds.), *STS-1 medical report* (NASA TM-58240). Washington, D.C.: National Aeronautics and Space Administration, December 1981.

Rippstein, W.J. Shuttle toxicology. In S.L. Pool, P.C. Johnson, Jr., & J.A. Mason (Eds.), *STS-2 medical report* (NASA TM-58245). Houston, TX: Lyndon B. Johnson Space Center, May 1982.

Rippstein, W.J., & Schneider, H.J. Toxicological aspects of the Skylab program. In R.S. Johnston & L.F. Dietlein (Eds.), *Biomedical results from Skylab* (NASA SP-377). Washington, D.C.: U.S. Government Printing Office, 1977.

CHAPTER 20
RADIATION EXPOSURE ISSUES

A major concern for space medicine is the radiation exposure experienced by space travelers. Over the past 25 years, we have developed a wealth of data describing the radiation environment in orbital space (Chapter 2) as well as through much of the rest of the solar system. Results from numerous space probes present a picture of heightened radiation levels, changing character of radiation, and even "radiation storms" as the level of solar activity waxes and wanes.

Radiation in space has been described as "the primary source of hazard for orbital and inter-planetary space flight" (Petrov et al., 1981). The successful flights made by the United States and the Soviet Union prove clearly that comparatively brief flights, when there are no solar flares and when flight trajectories are carefully planned, do not present a radiation hazard for space crews. However, as the space programs of both nations expand and as greater numbers of specialists spend more time in space, the problem of providing radiation protection grows in importance. Space medicine personnel must be knowledgeable concerning the extent of the danger; means of providing an appropriate measure of protection; and, most important, proper medical procedures for use in the event of a radiation emergency.

Measurement of Exposure

Consideration of radiation hazard and exposure risks for space crews requires an understanding of the units used for expressing radiation levels and the effect on biological systems. There are several interrelated units, as follows:

Roentgen

Radiation cannot be measured directly; therefore its magnitude is determined by the ionization produced by the passage of radiation through a medium. The roentgen, the basic unit of quantity, refers to the ionization produced by the passage of X- or gamma radiation in air resulting in 0.001293 grams of air ions carrying one electrostatic unit of electricity of either sign. The roentgen as a measure is little used in describing exposure.

Rad

One rad (radiation absorbed dose) of any type of radiation corresponds to the absorption of 100 ergs per gram of any medium. This is the most common unit describing exposure. For biological systems, one rad is equivalent to 10^{-5} joules of energy absorbed per gram of tissue.

Rem

The rem (roentgen equivalent, man) is the most accurate unit for expressing exposure of man. One rem equals the absorbed dose of any ionizing radiation which produces the same biological effect in man as that resulting from the absorption of one roentgen of X-rays. The rad measure is qualified by an RBE (relative biological effectiveness) measure in order to arrive at an expression in rems. The RBE differs for various types of radiation, from a value of one for X-rays and gamma rays, to values as high as 15 to 20 for alpha particles at different energy levels (U.S. Atomic Energy Commission, 1962). For the type of radiation encountered in space, particularly in low-Earth orbit, one rem may be considered as roughly equal to one rad.

LET

The biological effects of heavy particles, such as found in galactic radiation, are not described adequately by a single dose parameter such as rad. For these purposes, the linear energy transfer (LET) is used, which is the cosmic ray energy absorbed per micrometer of cells or living tissues penetrated. The LET is larger for higher-energy, highly charged particles.

Investigations of radiation levels in space are accomplished with specially instrumented unmanned vehicles which telemeter information concerning radiation levels to ground stations on Earth. Radiation measuring devices also are carried on manned space vehicles. The purpose in this case is two-fold. First, there is a need to gather information concerning personnel exposures and to monitor the radiation hazard. Second, these missions add to the radiation mapping of near-Earth space.

In the U.S. program, radiation is monitored in flight through use of two types of dosimeter, passive and active (Barnes, 1981). Precise radiation information is obtained from the passive dosimeters, which are composed of thermoluminescent dosimeter (TLD) chips, plastic sheets, and metal foils, each affected by different kinds and energies of radiation. These dosimeters are sealed units which must be processed in a laboratory following the flight. Active dosimeters are used as a means of determining in real-time the radiation danger being encountered. They may be read by a crewmember at any time. These are integrating dosimeters consisting of pen-sized ion chambers which measure three ranges of radiation exposure. The pocket dosimeter, low range (PDL), measures accurately in the range of zero to 200 millirads. The pocket dosimeter, high range (PDH), measures accurately in the range of zero to 100 rad. There also is included a contingency high rate dosimeter (HRD), provided for measurement of doses from zero to 600 rad. The combination of passive and active dosimeters allows accurate monitoring and recording of the unique radiation in space, including the electron, proton, and heavy cosmic rays encountered.

Human Response to Radiation

Human response to radiation can be quite varied, depending on the operation of a number of variables. Key variables affecting the response include:

1. **Radiation Variables.** The type of radiation, the possibility of combined types, and the time-intensity relationship.
2. **Subject Variables.** The location of exposure (whole body versus specific organs or tissues), the condition of the subject, his age, and complicating separate medical factors.
3. **Situational Variables.** Stress factors such as weightlessness, acceleration, and body temperature.

The bulk of the information in the medical literature on human response to radiation is drawn from clinical exposures and can be extrapolated to space flight only with caution. Also, little information is available concerning whole-body exposures of healthy subjects and the effects of exposure to heavy ions. For these reasons, it is difficult to specify with precision the medical consequences of a particular spaceflight exposure situation.

For a discussion of clinical effects of space-type radiations, the reader is referred to Tobias and Grigor'yev (1975). A review of radiation effects following whole-body exposure and exposure of specific organ and tissue systems is presented in Warren and Grahn (1973).

It is characteristic of radiation exposures that all body organs and systems are not affected equally. Knowledge of the differential sensitivity of body structures is important both for protection purposes and for evaluation of clinical syndromes. Table 1 presents three groupings of body elements in terms of their relative radiosensitivity. The lymphoid system is considered to be the most sensitive of all body components to radiation. This is followed closely by bone marrow elements and the gastrointestinal epithelium. The key characteristic of this most sensitive group is that all of the tissues are growing. Tissue elements which are rapidly dividing are the most sensitive to any radiation exposure.

Table 1
Differential Radiosensitivity of Human Tissue

High Sensitivity
- Lymphoid Tissue
- Bone Marrow Elements
- Gastrointestinal Epithelium
- Gonads (Testis and Ovary)
- Embryonic Tissues

Moderate Sensitivity
- Skin
- Vascular Endothelium
- Lung
- Kidney
- Liver
- Lens (Eye)

Low Sensitivity
- Central Nervous System
- Endocrine Glands
  (Except Gonads)
- Muscle
- Bone and Cartilage
- Connective Tissue

From Tyler, 1982.

Tissues that show moderate sensitivity to radiation include the skin, the vascular endothelium, and internal structures such as the lung. These systems all have slow growth characteristics. Of these, the lens of the eye is somewhat different in that it shows a particular sensitivity to neutron radiation, which can produce cataracts and lenticular opacities at lower levels of exposure than would be predicted. In general, by the time significant damage is received by any of these tissue systems, severe damage has occurred within the hematopoietic system.

Those tissues which show low radiosensitivity, such as the central nervous system, the endocrine glands, muscle, and bone and cartilage, reproduce slowly and are considered as very stable.

Exposures During Space Flight

The Radiobiological Advisory Panel, Committee on Space Medicine of the National Academy of Sciences issued in 1970 a set of recommended radiation exposure limits for use by NASA in planning and developing protection for future manned space missions. These limits are presented in Table 2. The mission limit (30 days maximum for bone) was established at 25 rem in view of research which indicated this level of exposure would have an extremely low probability of episodes such as diarrhea, anorexia, and vomiting, any of which might impact the conduct of a mission. The career limit of 400 rem was established on the basis of doubling the dose considered a risk for leukemia. Of course, this value of 400 rem refers to a career exposure and not an acute episode. Other intermediate values were calculated using the career dose as a reference.

Table 2

Radiation Exposure Limits Recommended
for Spaceflight Crewmembers

Constraint	Primary Reference Risk (Rem at 5 cm depth in tissue)	Ancillary Reference Risks			
		Bone Marrow* (Rem at 5 cm)	Skin (Rem at 0.1 mm)	Ocular Lens (Rem at 3 mm)	Testes (Rem at 3 cm)
1-year average daily dose		0.2	0.6	0.3	0.1
30-day maximum		25	75	37	13
Quarterly maximum**		34	105	52	18
Yearly maximum		75	225	112	38
Career limit	400	400	1200	600	200

*Whole body exposure.

**May be allowed for two consecutive quarters followed by 6 months of restriction from further exposure to maintain yearly limit and includes all occupational exposures.

Table adapted from *Radiation Protection Guides and Constraints for Space Mission and Vehicle-Design Studies involving Nuclear Systems*, Radiobiological Advisory Panel, Committee on Space Medicine, National Academy of Sciences, Washington, D.C., 1970.

The radiation levels encountered by both the Soviet Union and the United States during manned missions have been low. Table 3 presents the radiation exposures for crewmembers in all of the U.S. manned space missions to date. All radiation levels are seen to be acceptable. The highest exposure (7.8 rad) was recorded during the 84-day Skylab 4 mission. To put this risk into perspective, it has been stated that the Skylab 4 crewmen could fly a mission comparable to one 84-day flight per year for 50 years before exceeding the NASA career limit (Bailey et al., 1977).

Table 3
Measured Radiation Exposures
in U.S. Manned Spaceflight Programs

Mission	Mean Dose (Rad)	Mission	Mean Dose (Rad)
Mercury 9	0.027	Apollo 10	0.480
Gemini 3	.020	Apollo 11	.180
Gemini 4	.045	Apollo 12	.580
Gemini 5	.177	Apollo 13	.240
Gemini 6	.025	Apollo 14	1.140
Gemini 7	.150	Apollo 15	.300
Gemini 8	.010	Apollo 16	.510
Gemini 9	.018	Apollo 17	.550
Gemini 10	.840	Skylab 2	1.980
Gemini 11	.025	Skylab 3	4.710
Gemini 12	.015	Skylab 4	7.810
Apollo 7	.160	STS-1	.020
Apollo 8	.160	STS-2	.015
Apollo 9	.200	STS-3	.461

From Nachtwey, 1982.

It is of interest that four nuclear devices were detonated by France at the Murora Test Site during the flight of Skylab 3. No radiation attributed to these tests was recorded in Skylab. Of the radiation which was received, however, it was calculated that galactic cosmic radiation contributed roughly 20 to 30 percent.

An indication of the risk associated with the exposures recorded during manned space missions can be obtained by comparing radiation levels encountered in space with those found in a number of other situations on Earth. Table 4 presents radiation exposures from such diverse sources as chest X-rays and transcontinental jet flights. This table indicates that routine medical procedures (upper G.I. series, barium enema. etc.) produce radiation exposures many times greater than that recorded during the two-day Space Shuttle flight of STS-1. Living in Kerala, India, for one year would be considerably more dangerous, due to the increased level of thorium in the local sands. Even here, however, the radiation presents little if any risk over the lifetime of a human.

Table 4
Comparisons

	Average Rad to Total Red Marrow
Transcontinental Round Trip, Jet	0.004
Chest X-Ray (Photofluoroscopic)	0.045
Chest X-Ray (Radiographic, Low Dose)	0.010
Living in Houston 1 Year	0.100
Living in Denver 1 Year	0.200
Upper G.I. series	0.535
Barium Enema	0.875
Living in Kerala, India, 1 Year	5.-----

From Nachtwey, 1982.

Table 5 compares the radiation exposure of six astronauts from their space missions with the radiation these individuals were calculated to have received from all other sources, such as diagnostic X-ray procedures. With the exception of the two astronauts who had flown missions on the order of two and three months, radiation received from the more normal sources on Earth consistently exceeded that received in space. This low level of radiation during space flights has resulted from the combination of careful trajectory planning plus the good fortune to have avoided any unpredicted periods of heightened solar activity.

Table 5
Representative Career Exposures for Astronauts

Astronaut	Career Time	Space Exposure Time	Space Rad Dose	Diagnostic X-Ray and Other Rad Dose
X	15 years	9 days, 15 min.	.64	2.29
Y	10 years	12 days, 11 hours, 4 min.	.18	2.01
Z	18 years	22 days, 2 hours, 45 min. (4 Flights)	1.83	2.76
A	15 years	28 days	1.66	2.82
B	15 years	59 days	3.71	2.36
C	11 years	84 days	8.01	2.65

From Nachtwey, 1982.

Were astronauts to encounter a sudden burst of radiation from the sun, such as occurred in August 1972, or a nuclear detonation exploded at high altitude, the radiation picture would change dramatically. A nuclear explosion would be particularly dangerous. There would be prompt intense radiation in the form of X-rays and neutron and gamma radiation. The spacecraft also might be unfortunate enough to fly through the debris cloud itself. Should such an event occur, crew exposures would be recorded and the data immediately transmitted to ground stations. The mission management team, with medical consultation, then would decide whether circumstances would allow the crew to continue or if an immediate return to Earth was indicated.

A review of space crew exposures must include an interesting visual phenomenon, obviously related to space radiation, which was first noted in the Apollo program. Crewmembers of Apollo 11, during the time of trans-Earth coast, reported seeing faint spots or flashes of light when the cabin was dark and they had become dark-adapted (Osborne et al., 1975). From these reports and more systematic studies on later Apollo flights, it was concluded that the light flashes resulted from high-energy, heavy cosmic rays penetrating the spacecraft structure and the crewmembers' eyes. The fact that dark adaptation is necessary for this experience indicates that the phenomenon occurs at the retina rather than through direct stimulation of the optic nerve.

Measures made during the Skylab program corroborated the light flash findings of Apollo. Light flash observations were made in two particular orbits to provide data on the effects of both latitude and the South Atlantic Anomaly. During each session, the astronaut donned a blindfold, allowed ten minutes for adaptation to darkness, and then recorded his observation of each flash. Results are shown in Figure 1. Two conclusions were made (Lundquist, 1979). The occurrence

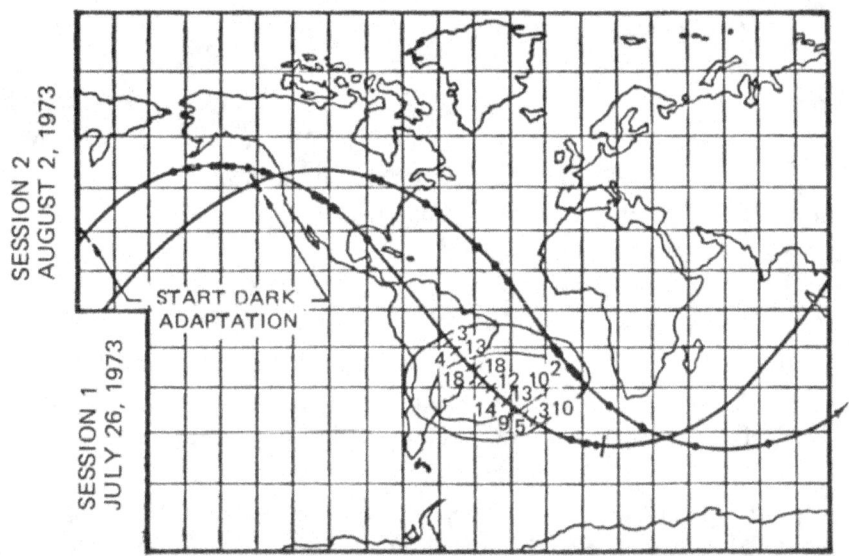

Figure 1. Light flashes observed during two particular orbits selected
to provide data on the effects of latitude and the South Atlantic anomaly.
(Lundquist, 1979)

of light flashes correlates with the flux of cosmic particles. The greatly increased number of flashes in the South Atlantic Anomaly, however, probably results either from trapped protons in this region or from trapped heavier nuclei.

Protection and Treatment Principles

The protection of spacecraft crews from radiation injury can follow any or all of four directions, as outlined by Grigor'yev (1976):

1. Passive physical protection—attenuation of radiation by increasing thickness of the skin of the spacecraft, arranging the equipment, creating shelter or shadow protection;

2. Active protection—creating protection against charged particles with the aid of magnetic and electric fields;

3. Local protection of critical organs and systems;

4. Pharmaco-chemical anti-radiation protection and methods of combined protection.

The feasibility of each of these avenues of protection will be discussed briefly. Passive physical protection simply means increasing the thickness of a protective coating, generally the exterior of the spacecraft, in order to reduce the level of radiation from solar flares and from the Van Allen belts. However, the relationship is not straightforward. When the thickness of protection is increased to a certain limit, the dose from galactic cosmic radiation can lead to increased secondary radiation and thereby defeat the purpose of the protection system. The biggest problem with this approach, however, is that the additional weight required for proper protection is extremely high. For this reason it is generally considered better to use passive protection from spacecraft components such as onboard equipment, fuel reserves, etc.

It has been known, at least in theory, for a number of years that spacecraft can be protected from charged particles through the use of magnetic and electric fields developed around the vehicle (Johnson & Holbrow, 1977). Such protection is promising and has a number of advantages, one being its comparatively low weight. In one approach, magnetic protection is based on the use of super-conducting materials which, being cooled to liquid-helium temperatures (-269° C), have zero electrical resistance and are able to provide circulation of a current, once established, for a practically unlimited time (Grigor'yev, 1976). The energy consumption for magnetic protection of this type is not great and is determined basically by the power required for the cryogenic unit. Another advantage is that the magnetic field can be increased, by raising the working current level, when the radiation situation worsens in comparison with that expected.

Local protection of critical organs and systems implies the availability of specific garments to be worn in the event of an anticipated increase in radiation. With dogs, it has been found that protective effect is highest with screening of the abdomen. Following radiation, when such abdominal screening is used, changes in peripheral blood, bone marrow, and immunobiological reactivity of the blood are not as severe as in other animals not similarly protected (Grigor'yev,

300

1976). Shielding a portion of the body, particularly if it is the abdominal area, appears therefore not only to increase survival rate, but also to reduce the severity and frequency of appearance of the primary syndromes of radiation injury.

Pharmaco-chemical protection is a topic which has been given considerable attention in recent years, principally for basic medical purposes. This is a broad field of endeavor. Saksonov (1975) notes that more than 15,000 different chemical substances, having widely dissimilar physico-chemical properties and pharmacological effects, have been tested. These include vitamins, antibiotics, nitrites, cyanides, amino acids, alkaloids, flavonoids, polysaccharides, sulphur-containing substances, analeptics, narcotics, central nervous system stimulants, colline and acridine derivatives, local anesthetics, and others. Many substances have been found which, when given to test animals at a specific time before irradiation, result in some reduction of the damaging effects of the radiation. There are many problems, however. First, dose levels of the protective agent frequently must be quite high. Second, any number of side effects can be produced through the administration of the protective substance. For these reasons, the practical use of pharmacological or chemical substances for protection of space crews from radiation will not occur until some point in the distant future.

References

Bailey, J.V., Hoffman, R.A., & English, R.A. Radiological protection and medical dosimetry for the Skylab crewmen. In R.S. Johnston & L.F. Dietlein (Eds.), *Biomedical results from Skylab* (NASA SP-377). Washington, D.C.: U.S. Government Printing Office, 1977.

Barnes, C.M. Radiological health. In S.L. Pool, P.C. Johnson, Jr., and J.A. Mason (Eds.), *STS-1 medical report* (NASA TM-58240). Washington, D.C.: National Aeronautics and Space Administration, Scientific and Technical Information Branch, December 1981.

Grigor'yev, Yu.G. *Radiation safety of space flights* (NASA TT F-16, 853). Translated by SCITRAN under contract NASW-2791. Washington, D.C.: National Aeronautics and Space Administration, 1976.

Johnson, R.D., & Holbrow, C. (Eds.). *Space settlements—a design study* (NASA SP-413). Washington, D.C.: U.S. Government Printing Office, 1977.

Lundquist, C.A. (Ed.). *Skylab's astronomy and space sciences* (NASA SP-404). Washington, D.C.: U.S. Government Printing Office, 1979.

Nachtwey, S. Radiation exposure, detection, and protection. Paper presented at the 53rd Annual Scientific Meeting of the Aerospace Medical Association, Bal Harbour, FL, 10-13 May 1982.

National Academy of Sciences (Radiobiological Advisory Panel, Committee on Space Medicine). *Radiation protection guides and constraints for space mission and vehicle-design studies involving nuclear systems.* Washington, D.C.: Author, 1970.

Osborne, W.Z., Pinsky, L.S., & Bailey, J.V. Apollo light flash investigations. In R.S. Johnston, L.F. Dietlein, and C.A. Berry (Managing Eds.), *Biomedical results of Apollo* (NASA SP-368). Washington, D.C.: U.S. Government Printing Office, 1975.

Petrov, V.M., Kovalev, E.E., & Sakovich, V.A. Radiation: Risk and protection in manned space flight. *Acta Astronautica*, 1981, *8*(8-10), 1091-1098.

Saksonov, P.P. Protection against radiation (biological, pharmacological, chemical, physical). In M. Calvin and O.G. Gazenko (General Eds.), *Foundations of space biology and medicine* (Vol. III). Washington, D.C.: U.S. Government Printing Office, 1975.

Tobias, C.A., & Grigor'yev, Yu. G. Ionizing radiation. In M. Calvin and O.G. Gazenko (General Eds.), *Foundations of space biology and medicine* (Vol. II, Book 2). Washington, D.C.: U.S. Government Printing Office, 1975.

Tyler, P. Health effects of ionizing radiation. Paper presented at the 53rd Annual Scientific Meeting of the Aerospace Medical Association, Bal Harbour, FL, 10-13 May 1982.

Warren, S., & Grahn, D. Ionizing radiation. In J.F. Parker, Jr. and V. West (Managing Eds.), *Bioastronautics data book* (NASA SP-3006). Washington, D.C.: U.S. Government Printing Office, 1973.

CHAPTER 21

MEDICAL CARE AND HEALTH MAINTENANCE IN FLIGHT

The demands of space flight require that crew members be healthy and at peak proficiency at all times. The preventive medicine program used in preparation for a mission is designed to ensure that an astronaut is in excellent health at the moment of launch. The extensive preflight precautions, however, do not guarantee that there will be no medical episodes during a mission. Indeed, the history of space flight to date indicates that there will inevitably be episodes, for the most part minor but with occasional events of real medical consequence. While the principal effort of space medicine is preventive, there must be sufficient planning and preparation for inflight problems so that these relatively rare events can be dealt with effectively at the time.

Medical problems that arise during a mission are handled through a combination of procedures. These procedures include:

1. **Ground-Based Monitoring.** A space mission is monitored continuously by a Flight Control Team, which includes medical personnel as team members. The medical team obtains health-related information via spacecraft telemetry. This is supplemented through use of a private medical conference, as necessary, with the crew. The information obtained during monitoring is intended to deal with direct medical problems and also to evaluate circumstances which appear to be leading toward such problems. This information includes data concerning status of environmental control systems, radiation exposure, food supply, water condition, and personal hygiene. Any malfunction of a spacecraft life-support system having a potential effect on crew health can be examined.

Additional health monitoring can be accomplished using data from onboard biomedical experiments and from instrumented physical exercise. Such data has been used in both the U.S. and USSR space programs. In the Skylab era, inflight motion sickness studies, cardiovascular studies, and metabolic load experiments provided invaluable information to medical teams via biotelemetry. Instrumented physical exercise sessions both in Skylab and more recently on the Shuttle flights were useful adjuncts in the implementation of an inflight health monitoring and maintenance program. Environmental data, such as partial pressures of oxygen and carbon dioxide, as well as cabin temperature, can be obtained as required. In addition, noise and radiation exposure surveys are conducted at select points of a mission to evaluate long-term effects of potential environmental hazards. Finally, during critical operations inflight, such as extravehicular activities, biomedical monitoring can be achieved through voice communications supported by one-lead electrocardiography and information on metabolic expenditure during work performed.

2. **Medical Training.** Each astronaut is given training concerning both the prevention and the management of inflight medical problems. This training, described at greater length in the chapter on biomedical training, includes general medical training, augmented by special courses dealing with the handling of onboard medical systems and biomedical experiments. As part of mission preparation, additional briefings are designed to review pre- and postflight medical procedures, discuss crew preventive medicine measures, instruct the crew in the contents and use of the onboard

medical kit, demonstrate the configuration and operation of the biomedical harness, and familiarize the crew with toxicological considerations (Vanderploeg, 1981). Finally, on certain missions, if a Physician-Astronaut is included in the crew complement, he can direct the management of an inflight emergency.

3. **Onboard Medical Kit.** The provision of a medical kit for use by trained crewmen is designed to provide treatment for life-threatening emergencies through stabilization and to permit diagnosis and treatment of all minor injuries and illness events.

Inflight Disease and Injury

Space crews are given detailed medical examinations during the month prior to launch. Included is a physical examination administered on the day of launch. These examinations, combined with the use of a Health Stabilization Program to minimize disease contact prior to flight, insure, to the extent possible, that the health of a space crew about to be launched is excellent.

The extensive preflight preventive programs have held inflight medical problems to a minimum. Table 1 lists the illnesses and injuries experienced by U.S. space crews to date. The top entries in this table list those episodes which occurred inflight and which would require management by the flight crew. The instances of trauma and toxic pneumonia encountered during landing were dealt with by members of the medical recovery team. Events of space motion sickness are not included in this table since they were discussed in a previous chapter.

Table 1

Illness/Injury Occurrence in U.S. Space Crews

Inflight Illnesses/Injuries	Number
Dysbarism	2
Eye-skin irritation (fiberglass)	3
Skin infection	2
Contact dermatitis	2
Urinary tract infection	2
Arrhythmias	2
Serious otitis	1
Eye and finger injury	1
Sty	1
Boil	1
Rash	1
Illness/Injury During Recovery and Landing	
Trauma (scalp laceration from detached camera)	1
Toxic pneumonia (inadvertent atmosphere contamination by N_2O_4)	3
Injury Discovered Postflight	
Back strain (due to lifting of heavy object)	1

From data of Furukawa et al., 1982

Some caution should be observed when using the results in Table 1 to define medical care requirements for future space missions. The data in this table are based on findings from the Gemini program through the Apollo-Soyuz Test Project. Each new space program brings a new spacecraft environment and different crewmember work demands. In Space Shuttle missions, for example, astronauts work for the first time in a sea-level atmosphere. They also perform a variety of new work activities related to the different payloads that can be carried. These features may well alter the potential for injury from that found, for example, in the Skylab program.

Onboard Medical Equipment

All U.S. spacecraft have included an onboard medical kit as part of the supplies. In the Mercury program, this kit was rudimentary, containing only an anodyne, an anti-motion sickness drug, a stimulant, and a vasoconstrictor for treatment of shock (Link, 1965). As Project Mercury evolved into Project Gemini, more medications were included in the medical kit. The contents for the 14-day Gemini 7 kit are listed in Table 2. In the subsequent Apollo, Skylab, and Space Shuttle programs, drugs included in the medical kit have been quite similar. The onboard medical supplies for Skylab were more elaborate, including surgical instruments and diagnostic equipment, in view of the length of these flights.

Table 2
Gemini 7 Inflight Medical and Accessory Kits

Medication	Dose and Form	Label	Quantity
Cyclizine HCl	50-mg tablets	Motion sickness	8
d-Amphetamine sulfate	5-mg tablets	Stimulant	8
APC (aspirin, phenacetin, and caffeine)	Tablets	APC	16
Meperidine HCl	100-mg tablets	Pain	4
Triprolidine HCl	2.5-mg tablets	Decongestant	16
Pseudoephedrine HCl	60-mg tablets	Decongestant	16
Diphenoxylane HCl	2.5-mg tablets	Diarrhea	16
Tetracycline HCl	250-mg film-coated tablet	Antibiotic	16
Methyl cellulose solution	15 cc in squeeze-dropper bottle	Eyedrops	1
Parenteral cyclizine	45 mg (0.9 cc in injector)	Motion sickness	2
Parenteral meperidine HCl	90 mg (0.9 cc in injector)	Pain	2

From Berry, 1975

The Space Shuttle Orbiter carries with it a specially designed medical system to allow crewmen to deal with medical emergencies occurring inflight. This package, known as the Shuttle Orbiter Medical System (SOMS), has three configurations, called SOMS A, SOMS B, and SOMS C. The smallest unit, SOMS A, is designed for use on flights of 21 man-days or less. The next unit, SOMS B, is designed for flights of up to 44 man-days. The largest unit, SOMS C, is designed for flights extending beyond this period. Since the nominal Space Shuttle mission in the immediate future will be in the order of seven days, only the smallest system, SOMS A, will be described here.

The Shuttle Orbiter Medical System includes two medical kits, a Medications and Bandage Kit (MBK) plus an Emergency Medical Kit (EMK); a medical checklist for crew member use; and the occasional use of other Orbiter systems such as the Portable Oxygen System (Vanderploeg, 1981). Each kit contains three pallets with items stowed on both sides of each pallet. The general contents and purpose of each pallet are shown in Table 3.

Table 3
SOMS Medical Kit

Pallets	Pallet Contents	General Purpose of Equipment in the Pallet
A	Injectables	Medications to be administered by injection
B	Emergency Items	Minor surgery equipment
C	Diagnostic/therapeutic items	Instruments for measuring and inspecting the body
D	Oral medications	Pills, capsules, and suppositories
E	Bandage items	Material for covering or immobilizing body parts
F	Non-injectables	Medications to be administered by topical application

The SOMS medical kits contain an array of drugs for use as dictated by illness and injury conditions. As part of the preparation for a mission, an evaluation is made of the sensitivity of each astronaut to each of the drugs present in the medical kit. The drug sensitivity evaluation is carried out in two steps (Vanderploeg, 1981). First the health record of each crew member is reviewed and a listing made of all medications received. Any reported reactions or side effects are noted. In the second step, each crew member is given sensitivity testing for those medications felt to have a high likelihood for use inflight. Any reactions are noted in a listing maintained at the Mission Operations Control Room as reference information for use as required during a mission by medical personnel.

Figure 1 shows the two medical kits included in SOMS A. The three pallets can be seen stored within each of these kits. Figures 2 and 3 show the contents of one pallet from the Emergency Medical Kit and one from the Medications and Bandage Kit, respectively.

306

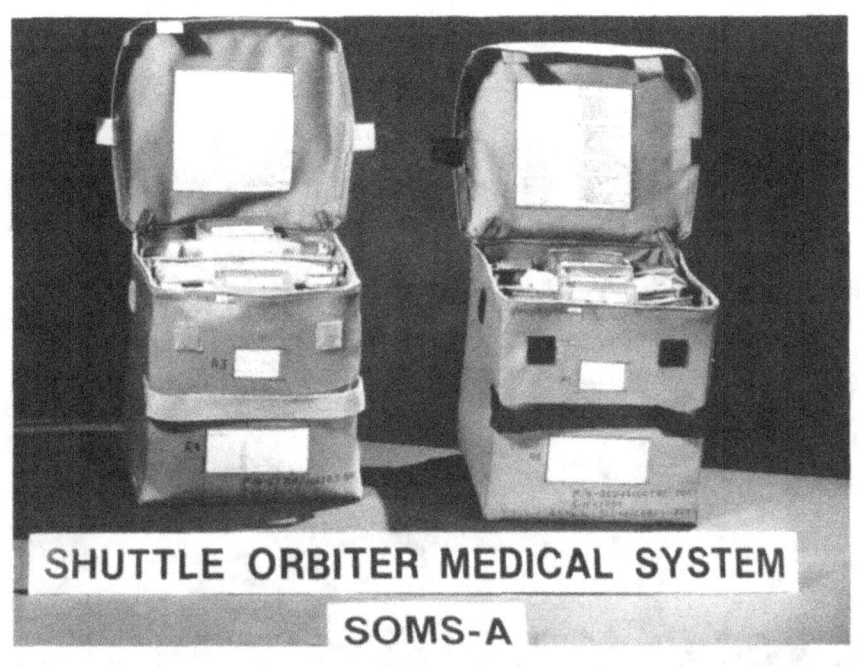

Figure 1. The two kits comprising the Shuttle Orbiter Medical System (SOMS-A).

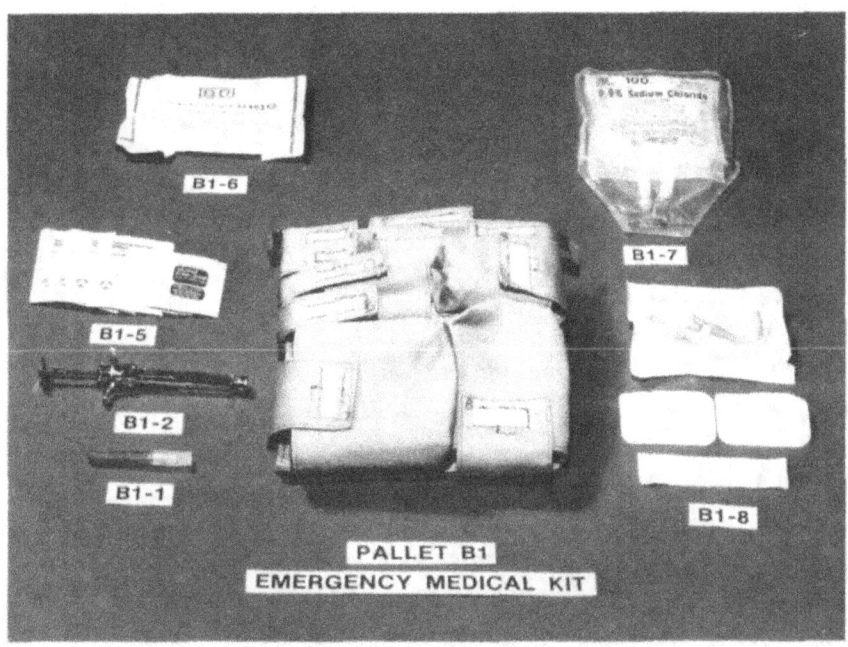

Figure 2. Contents of Pallet B1 of the Emergency Medical Kit.

307

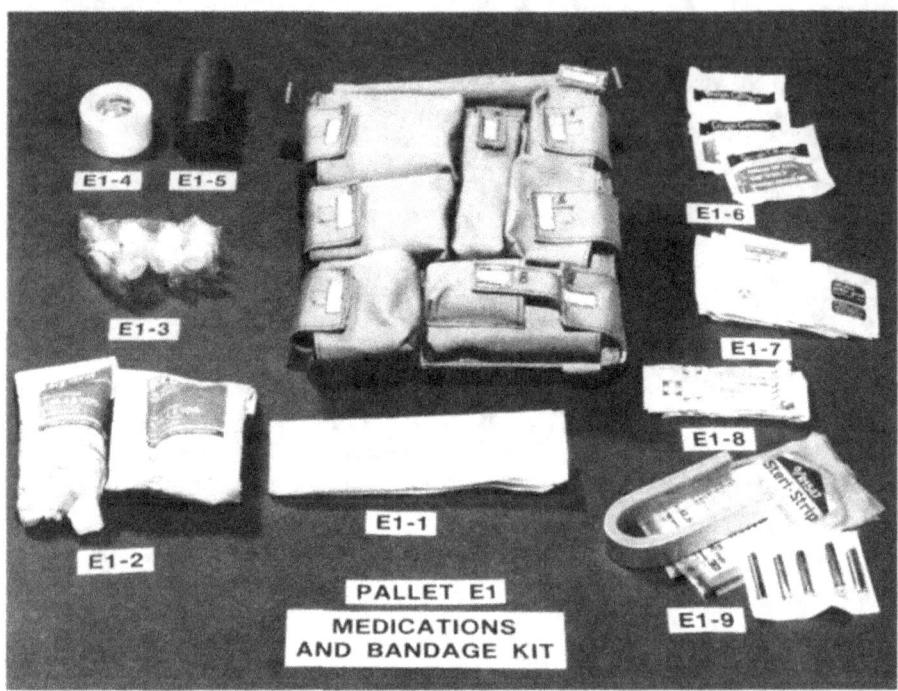

Figure 3. Contents of Pallet E1 of the Medications and Bandage Kit.

A listing of the medicine and supplies contained in each of the six pallets of the two medical kits is presented in Tables 4 through 9. Periodically, the items in these pallets are reviewed in the light of growing flight experience and the contents are revised as appropriate. In addition, periodic refurbishment of the medications is performed based both on a consideration of the manufacturer's specification for the drug's shelf life and possible effects of the space environment on pharmacological activity of the drugs.

In the event of a serious mishap, such as a cervical spine injury or an illness requiring immobilization of a crew member in the weightless environment for the purpose of rendering care, special procedures utilizing available Orbiter equipment have been devised. Figures 4 and 5 show these procedures as developed by the medical support team.

308

Table 4
Contents of Pallet A (SOMS)

Aramine 10 mg/cc	Lidocaine HCl 20 mg/cc
Epinephrine 1:1000	Lidocaine 20 mg/cc
Atropine 0.4 mg/cc	Valium 5 mg/cc
Phenergan 25 mg/cc (IV)	Xylocaine 2% with Epinephrine
Compazine 5 mg/cc	Xylocaine 2% without Epinephrine
Pronestyl 500 mg/cc	Demerol 25 mg/cc
Morphine Sulfate 10 mg/cc	Benadryl 50 mg/cc
Decadron 4 mg/cc	Vistaril 50 mg/cc

Table 5
Contents of Pallet B (SOMS)

Needles, 22 g, 1.5 inch	Suture 4-0 Dexon with needle
Tubex injector	Bili-lab Stix and chart
Syringes, 10 cc	Forceps, small point
Needle, 21 g, butterfly (IV)	Needle holder and small hemostat
Normal Saline, 100 cc	Tweezers, fine point
Tubing, IV, without chamber	Scalpel No. 13, Scalpel No. 11, and
Suture 4-0 Dexon with needle	curved scissors

Table 6
Contents of Pallet C (SOMS)

BP cuff and stethoscope	Foley catheter, No. 11 Fr., with
Tourniquet and cotton balls	30 cc balloon
Oral airway and cricothyrotomy set	Ophthalmoscope head
Disposable thermometers	Penlight
Tongue depressors	Fluorescein strips
Otoscope specula	Binocular loupe
Otoscope	Sterile drape
Cobalt light	Sterile gloves

Table 7
Contents of Pallet D (SOMS)

Actifed tablets

Dexedrine tablets, 5 mg

Donnatal tablets

Lomotil tablets

Dalmane capsules, 15 mg

Ampicillin capsules, 250 mg

Erythromycin tablets, 250 mg

Tetracycline capsules, 250 mg

Throat Lozenges, Cepacol

Scop/Dex capsules, 0.4/5 mg

Pyridium capsules, 200 mg

Valium tablets, 5 mg

Tylenol tablets, No. 3

Benadryl capsules, 25 mg

Nitroglycerin tablets, 0.4 mg

Periactin tablets, 4 mg

Sudafed tablets, 30 mg

Pen VK tablets, 250 mg

Phenergan/Dexedrine, 25/5 mg

Compazine suppositories, 25 mg

Keflex capsules, 250 mg

Aminiphyllin suppositories, 500 mg

Digoxin, 0.25 mg

Phenergan suppositories

Dulcolax tablets

Parafon Forte

Aspirin

Table 8
Contents of Pallet E (SOMS)

Q-tips

3 inch Kling

Mylanta tablets

Dermicel tape, 1 inch

Gauze

Robitussin Cough Calmers

Wipes, alcohol

Wipes, Betadine

Bandaids

Steri-strips

Finger splint

Benzoin wipes

2x2 sponges

Eye patch

Adaptic dressing

Toothache kit

Ace bandage

Table 9
Contents of Pallet F (SOMS)

Afrin Nose Spray

Methylcellulose eyedrops

Kerlix dressing

Povidone-Iodine Ointment, tube

Anusol-HC cream

Kenalog cream, 0.1%

Mycolog cream

Sulfacetamide

Neocortef

Halotex

Neosporin

Pontocaine, 15 ml bottle

Blistex

Cortisporin otic solution

Triangular bandage

Surgical masks

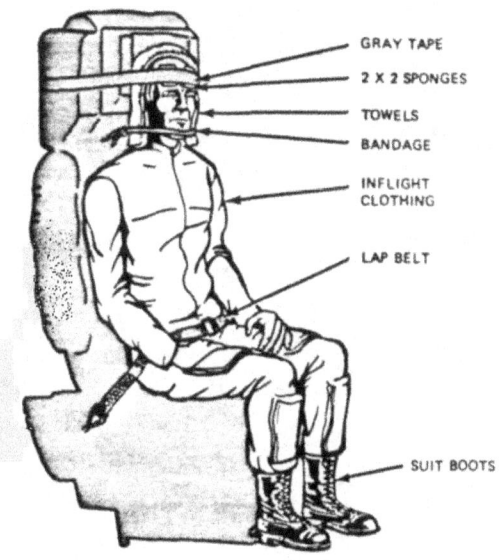

Figure 4. Procedure for immobilizing a Space Shuttle crew member in the event of serious neck injury.

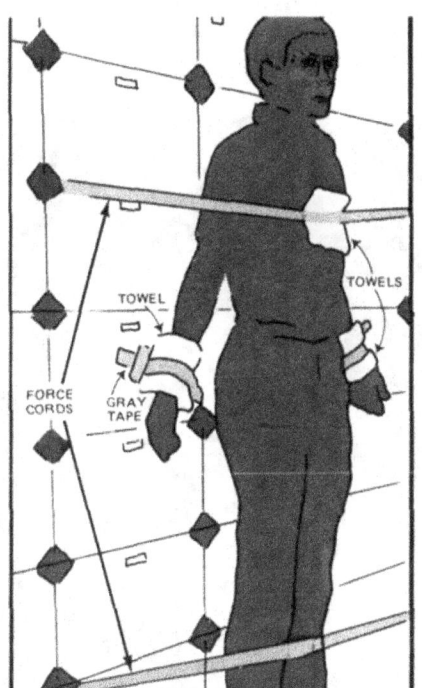

Figure 5. Procedure for restraining Space Shuttle crew member at the first aid station in the event of spinal injury or illness requiring immobilization.

The medical system carried on the Space Shuttle Orbiter includes a limited capability to perform laboratory tests inflight, such as microbial analyses. There is always the possibility of an infectious disease going undetected during preflight examinations, so that an inflight diagnosis must be made. Equipment is carried to allow inflight microbiological sampling and analyses of samples from the throat, urine, and wounds or obvious areas of infection. For later flights, a hematology kit is being designed. This kit will allow drawing of blood and limited clinical analyses of hematocrit, hemoglobin, and, with a microscope, white-cell counts.

In the event of a minor illness occurring in space, a crewman, in consultation with the Crew Physician, will be able to use the medical kit to administer aid. If the medical condition is more serious and requires definitive treatment in order that it not become life-threatening, the mission can be aborted and an emergency rescue made at a landing facility. An orbital rescue also is possible, although more complicated. When the Shuttle fleet is augmented, it is conceivable that a second vehicle could be used for rescue. In this event, astronauts who are trained in EVA procedures would transfer using the space suit. Other crew members or passengers, not trained for EVA, would use the personal rescue system shown in Figure 6. This is an inflatable sphere, currently under evaluation, with a diameter of 86 centimeters which allows personnel to be transferred to another spacecraft. The system includes the personal rescue enclosure, constructed of a gas-tight restraint and thermal protection garment; a rescue support umbilical to provide life support functions and communications capability during the pre- and post-rescue modes; and a portable oxygen system. The oxygen system provides pressurized gas for the rescue system sufficient for a one-hour rescue period.

Figure 6. The personal rescue system used for transferring personnel to a rescue vehicle.

312

Health Practices

Medical personnel are primarily responsible for the direct medical issues arising during manned space flight. However, they share with spacecraft designers and engineers a responsibility to see that spacecraft procedures and living conditions are conducive to good health and performance. Health maintenance is a broad topic and includes more than simply the prevention of disease. For example, in order for crewmen to be at peak mental and physical condition during a launch period, circadian rhythms are adjusted as needed. In the event of an early morning or late night launch, circadian rhythms may be adjusted by altering the work schedule by several hours during the days immediately prior to launch. During missions, the general rule is that crew schedules will be tied to either Houston or Cape Canaveral time. Adjustments in schedules also may be made before landing but, if biomedical experiments are being conducted, the adjustments are made so that these experiments are not affected.

An important factor in health maintenance is food palatability, quality, nutritional value, and the manner of preparation. In operations of the Space Shuttle, every attempt has been made to provide food service in a relatively normal manner. Prior to flight, a list of available foodstuffs is discussed with individual crew members and, based on their preferences, daily menus are established. Table 10 shows a typical menu plan for a seven-day mission. (The nutritional quality of these menus is discussed in Chapter 18, dealing with countermeasures to the effects of the space environment.) Since unforeseen events might occur which would extend a mission beyond the planned time lines, contingency rations also are included in the pantry of the Orbiter (Table 11). The foods listed in these tables require special containers and utensils for use in the space environment. Figure 7 shows a Space Shuttle astronaut preparing beverages in weightlessness. While every attempt has been made to minimize any problems associated with food service, and thereby to make dining in space a positive experience, it is obvious that this activity remains different from the accustomed practices on Earth.

Housekeeping in orbit and provisions for hygiene are considered important from the standpoint of both preventive medicine and the morale of space crews. Considerable effort has been expended to devise efficient and acceptable waste management systems. These systems are much improved over earlier bag collection devices and, in the Space Shuttle (see Chapter 4), resemble small bathrooms which accommodate male and female crew members. Provisions are made in the Space Shuttle for personal hygiene activities such as shaving (Figure 8). In longer missions such as Skylab, equipment has been provided for hair grooming (Figure 9) using specially designed vacuum sources to prevent contamination of the cabin environment. Space showers also have been incorporated into the design of space vehicles. Figure 10 shows the shower system used during the Skylab program.

Considerations of human comfort and hygiene will grow in importance as plans for future space stations evolve. These stations will support a much larger work force over longer periods than do current programs. Important as they are now, issues of food service, personal hygiene, waste management, and leisure time scheduling will be of even greater consequence for health maintenance.

Table 10

STS-4 Menu Log

Meal	Day 1 5	Day 2 6	Day 3 7	Day 4
A	Peaches Beef Patty Scrambled Eggs Bran Flakes Cocoa Orange Drink Vitamin Pill Coffee	Applesauce Dried Beef Granola Breakfast Roll Choc. Inst. Brkfst. Orange-Grapefruit Drink Vitamin Pill Coffee Ice Cream	Dried Peaches Sausage Scrambled Eggs Cornflakes Cocoa Orange-Pineapple Drink Vitamin Pill Coffee	Dried Apricots Breakfast Roll Granola w/blueberries Vanilla Inst. Brkfst. Grapefruit Drink Vitamin Pill Coffee
B	Frankfurters Turkey Tetrazzini Bread Bananas Almond Crunch Bar Apple Drink	Corned Beef Asparagus Bread Pears Peanuts Lemonade	Ham Cheese Spread Bread Gr. Beans & Broccoli Crushed Pineapple Shortbread cookies Cashews Tea w/Lemon & Sugar	Ground Beef w/ Pickle Sauce Noodles & Chicken Stewed Tomatoes Pears Almonds Strawberry Drink
C	Shrimp Cocktail Beef Steak Rice Pilaf Broccoli au Gratin Fruit Cocktail Butterscotch Pudding Grape Drink Ice Cream	Beef w/BBQ Sauce Cauliflower w/Cheese Gr. Beans w/Mushrooms Lemon Pudding Pecan Cookies Cocoa	Cream Mushroom Soup Smoked Turkey Mixed Italian Vegetables Vanilla Pudding Strawberries Tropical Punch	Tuna Macaroni & Cheese Peas w/Butter Sauce Peach Ambrosia Chocolate Pudding Lemonade

Table 11
STS-4 Contingency/Pantry List

Beverages

Coffee (R/H)*
Grape Drink (R)
Grapefruit Drink (R)
Instant Breakfast, Chocolate (R)
Instant Breakfast, Vanilla (R)
Lemonade (R)
Orange Drink (R)
Strawberry Drink (R)
Tea w/Lemon & Sugar (R)
Tea w/Sugar (R)

Entrees

Beef, Dried
Beef Patty (R/H)
Beef Steak (H)
Chicken & Rice Soup (R/H)
Eggs, Scrambled (R/H)
Frankfurters (H)
Ham (H)
Meatballs w/BBQ (H)
Turkey, Smoked (H)
Turkey Tetrazzini (R/H)

Fruits

Apricots, Dried
Peaches, Dried
Peach Ambrosia (R)
Strawberries (R)

Snacks/Desserts

Cookies, Butter
Cookies, Pecan
Food Bar, Chocolate Chip
M&M Candy
Nuts, Almonds
Nuts, Cashews
Nuts, Peanuts
Peanut Butter
Soda Crackers
Graham Crackers
Jelly/Jam
Lemon Pudding

Vegetables

Peas w/Butter Sauce (R/H)
Potato Patty (R/H)
Vegetables, Italian (R/H)

---

*R = rehydratable, H = heatable.

315

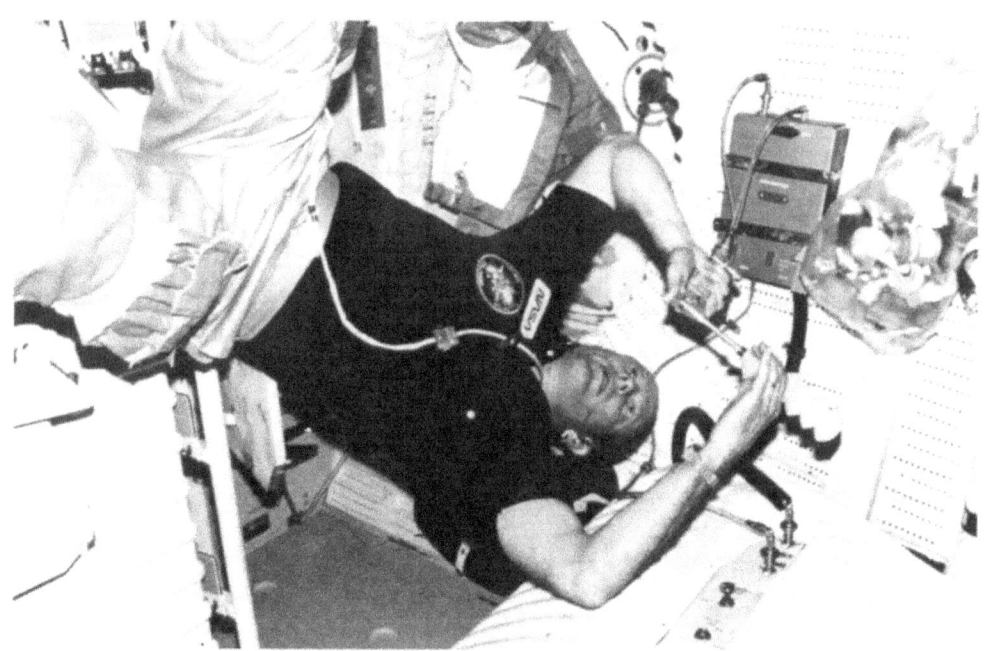

Figure 7. STS-3 astronaut Jack Lousma adds water to prepare a juice drink in weightlessness.

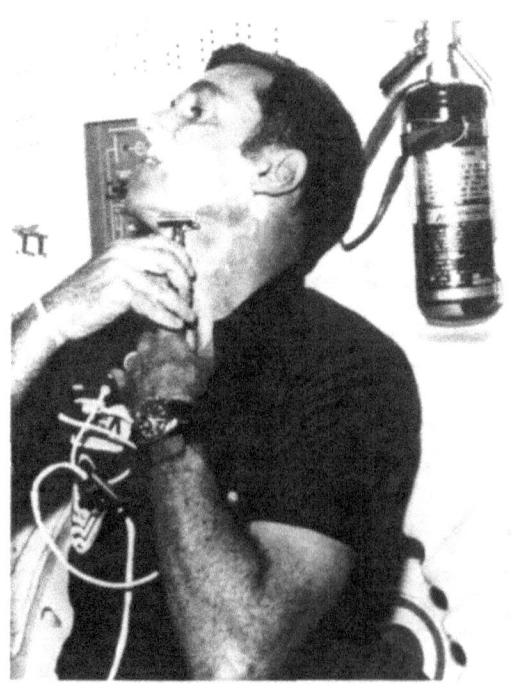

Figure 8. STS-2 astronaut Joe Engle shaving in the mid-deck area of the Space Shuttle.

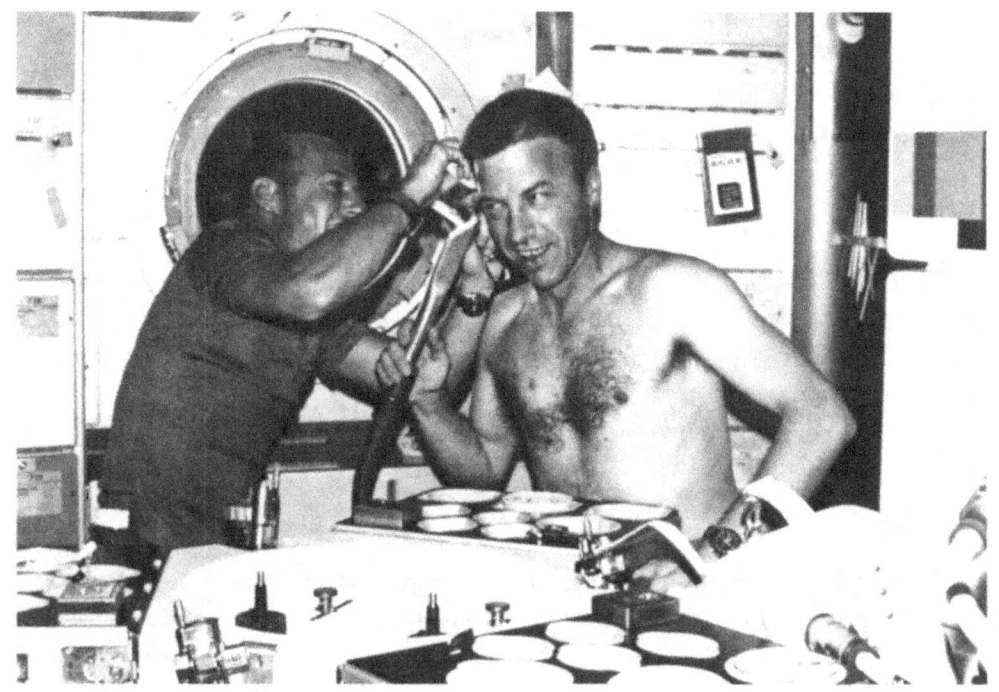

Figure 9. Skylab crewman receives an inflight haircut.

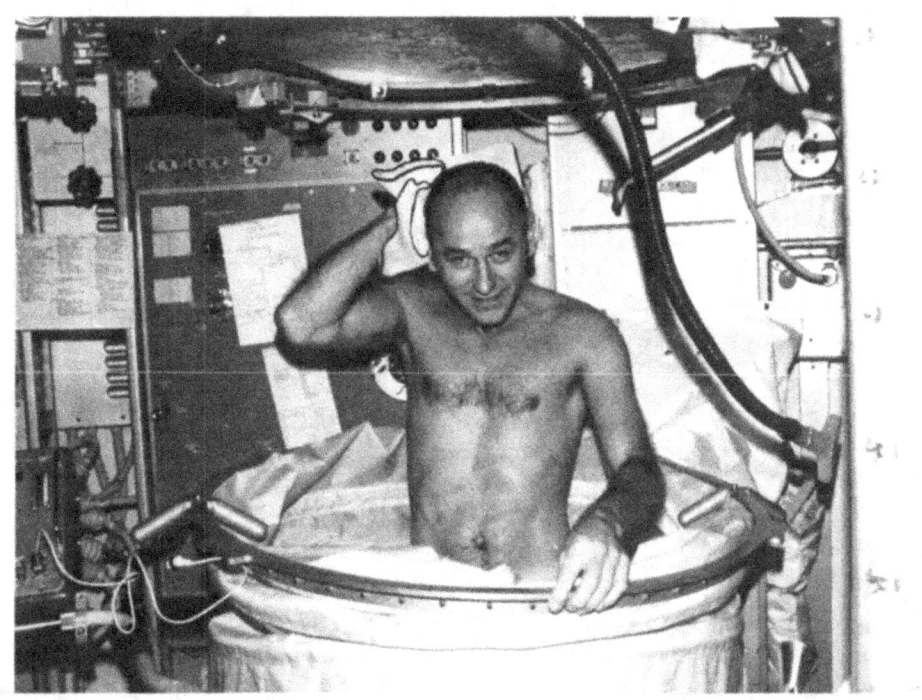

Figure 10. The shower system in use aboard Skylab.

The Future of Space Medicine

The space programs of the next two decades will bring dramatic advances in the utilization of the near-Earth environment. Space stations will be permanently manned, with crews rotating possibly every 90 or 180 days. Indeed, it is conceivable that logistics considerations will dictate that some crew members stay even longer, perhaps for a year or more.

The health and medical care needs of space station crews will place real demands on the growing field of space medicine. For space medicine to fulfill its charter as a specialty within preventive medicine, additional knowledge must be obtained concerning both short- and long-term adaptation to the weightless environment of space. To what extent do problems of space motion sickness, cardiovascular deconditioning, and loss of bone mineral place constraints on the conduct of the space missions planned for the 1990's? What are the critical psychological and comfort needs for persons working routinely in space?

Medical support for space station operations will require new philosophies and new technologies. The epicenter of medical care will shift from ground-based Mission Control Centers to a space-based medical unit. Based on the current estimate of STS turnaround capability, the minimum projected time for arrival of a rescue vehicle following notice by space station personnel is seven to 14 days (Furukawa et al., 1982). In addition to the delay factor, there also is the issue of establishing medical criteria for committing a patient to reentry into a one-gravity environment, following weightlessness, without endangering his or her condition. These considerations mean that a space station must have the personnel, facilities, and technologies to provide adequate medical care and health maintenance services, including provision for such countermeasures as specially tailored exercise programs. Diagnostic systems, laboratory facilities, and medical techniques all must be designed and qualified for use in zero gravity. The challenges for space medicine match those presented by the design of the space station itself.

References

Berry, C.A. Medical care of spacecrews (medical care, equipment, and prophylaxis). In M. Calvin and O.G. Gazenko (General Eds.), *Foundations of space biology and medicine* (Vol. III) (NASA SP-374). Washington, D.C.: U.S. Government Printing Office, 1975.

Furukawa, S., Nicogossian, A., Buchanan, P., & Pool, S. Medical support and technology for long-duration space missions. Paper presented at the 33rd International Congress of the International Astronautical Federation, Paris, France, 27 September–2 October 1982.

Link, M.M. *Space medicine in Project Mercury* (NASA SP-4003). Washington, D.C.: U.S. Government Printing Office, 1965.

National Aeronautics and Space Administration. *MED EQ 2102 medical equipment workbook* (CG3-027). Houston, TX: Lyndon B. Johnson Space Center, Crew Training and Procedures Division, Training Integration Branch, 1979.

Vanderploeg, J.M. Shuttle orbital medical system. In S.L. Pool, P.C. Johnson, Jr., and J.A. Mason (Eds.), *STS-1 medical report* (NASA TM-58240). Washington, D.C.: National Aeronautics and Space Administration, Scientific and Technical Information Branch, December 1981.

www.ingramcontent.com/pod-product-compliance
Lightning Source LLC
Chambersburg PA
CBHW081431170526

45166CB00008B/2168